D0213689

Linear Geometry

Rafael Artzy

Dover Publications, Inc., Mineola, New York

Bibliographical Note

This Dover edition, first published in 2008, is an unabridged republication of the third printing (1974) of the work originally published in 1965 by Addison-Wesley Publishing Company, Inc., Reading, Massachusetts.

Library of Congress Cataloging-in-Publication Data

Artzy, Rafael.
Linear geometry / Rafael Artzy. — ed.
p. cm.
"This Dover edition, first published in 2008, is an unabridged republication of the third printing (1974) of the work originally published in 1965 by Addison-Wesley Publishing Company, Inc., Reading, Massachusetts.
ISBN-13: 978-0-486-46627-9
ISBN-10: 0-486-46627-2
1. Geometry, Modern—Plane. I. Title.

QA474.A76 2008
516'.04—dc22

2007050692

Manufactured in the United States of America
Dover Publications, Inc., 31 East 2nd Street, Mineola, N.Y. 11501

Preface

With the disappearance of the traditional synthetic geometry courses from college curricula, many planners assumed that the mantle of geometry had passed to the various types of Linear Algebra courses. However, most texts in linear algebra were paying only scant attention to geometry, and the special flavor and aesthetics of linear geometry was hardly ever felt in them. Thus arose the need for independent upperclass undergraduate and beginning graduate geometry programs stressing the strong relationship between algebra and linear geometry without, however, making geometry a mere side issue in algebraic discussions. This book is an attempt to fill the need.

The plan for this book took shape when my former teacher, Kurt Reidemeister, the dean of German geometers, kindly placed at my disposal a set of his recent lecture notes. The spirit of these notes found its way into my own Rutgers University notes from which this text developed.

In the first chapter, the complex number plane is used for a propaedeutic discussion of the various transformations and their groups in the euclidean plane. Algebraic concepts (transformation, group, etc.), as needed, are introduced from scratch. This is followed by a short account of Poincaré's model of the hyperbolic plane and its transformation groups.

The second and third chapters contain a systematic treatment of affine, euclidean, and projective spaces over fields. Again algebraic material (vector spaces, matrices, etc.) is explained in full whenever necessary. A few sections deal with projective models of noneuclidean planes and elliptic 3-space. Throughout, the emphasis is on transformations and their groups. The third chapter closes with an outline of results about other geometries.

In the fourth chapter the reader is introduced to the Foundations of Geometry, starting from rudimentary projective incidence planes. We then gradually adjoin axioms, developing various nondesarguesian, desarguesian, and pappian planes, their corresponding algebraic structures (ternary rings, Veblen-Wedderburn systems, alternative division rings, etc.), and their collineation groups.

Then axioms of order, continuity, and congruence make their appearance and lead to euclidean and noneuclidean planes.

Lists of books for suggested further reading follow the third and fourth chapters, and the Appendix provides lists of notations, axioms, and transformation groups.

The material in the first three chapters has been used in a year course for juniors and seniors, and the fourth chapter in a one-semester graduate course. In addition, an Academic Year Institute for high school teachers has studied the first chapter as part of a geometry class. For students with sufficient preparation in algebra the whole book could be covered in a one-year course.

My sincere thanks are due to Roy A. Feinman and Hans W. E. Schwerdtfeger, who read the manuscript and made important comments, to Avraham J. Ornstein, who carefully read the proofs, and to Lynn H. Loomis and Louis F. McAuley, who read parts of the manuscript and contributed valuable remarks. I also gratefully acknowledge the help of my students, in particular those of the 1964–65 Rutgers University Academic Year Institute, who untiringly discovered major and minor mistakes in the lecture notes.

New Brunswick, New Jersey R.A.
April 1965

Contents

Chapter 3. PROJECTIVE GEOMETRY AND NONEUCLIDEAN GEOMETRIES

CHAPTER 1

Transformations in the Euclidean Plane

In this chapter no attempt will be made to construct a geometry from its foundations; this will be left to later chapters. Our aim in the first chapter is the determination of the objectives in our study of geometry. We can best acquire such an orientation by having a close and well-planned look at a kind of geometry which is familiar to us. The euclidean plane serves this purpose because every student has studied it and knows a great deal about its properties. In this chapter we utilize this knowledge of the theorems in plane euclidean geometry as well as facts like the congruence theorems of triangles, Pythagoras' theorem, and others, without inquiring into their sources or their mutual relations. Another tool which will be freely used in this chapter is the analytic geometry of the plane. The student will have discovered that in many instances the methods provided by analytic geometry are easier to apply and less cumbersome than the "synthetic" treatment of geometrical problems which he had to apply in high school. Thus, again without attempting in this chapter to provide the theoretical foundations for analytic geometry, we will use it in a special form, in the plane of complex numbers. To each of the points in the plane a complex number will be assigned. Then the important transformations of the plane will be studied by observing their analytical meaning. By doing so we will be able to introduce and to become acquainted with concepts like isometry, similitude, rotation, and reflection. The algebraic notion of the group will help us in the organization of the details. In the last part of the chapter we will encounter the first example of a noneuclidean geometry.

1-1 THE PLANE OF COMPLEX NUMBERS

In this section we review some well-known properties of the complex numbers without, however, giving proofs for our statements. The proofs are readily available in books on elementary algebra.

An ordered couple of real numbers a and b is called a *complex number*, and the usual notation for it is $a + bi$. Two complex numbers $a + bi$ and $c + di$ are considered equal if and only if $a = c$ and $b = d$. Sometimes it is convenient to distinguish between the *real part* of $z = a + bi$, $\text{Re}(z) = a$, and its *imaginary part*, $\text{Im}(z) = b$. The sum and the product of complex numbers are defined, respectively, by $(a + bi) + (c + di) = (a + c) + (b + d)i$ and $(a + bi)(c + di) = (ac - bd) + (ad + bc)i$. This implies, for instance, $(0 + 1i)(0 + 1i) = -1 + 0i$. If we write $0 + bi = bi$ and identify the complex number $a + 0i$ with the real number a, then the last equation becomes $i^2 = -1$. The arithmetic of complex numbers follows the familiar laws. If Greek letters denote complex numbers, we have $\alpha + \beta = \beta + \alpha$, $\alpha\beta = \beta\alpha$, $(\alpha + \beta) + \delta = \alpha + (\beta + \delta)$, $(\alpha\beta)\delta = \alpha(\beta\delta)$, $\alpha(\beta + \delta) = \alpha\beta + \alpha\delta$, $1 \cdot \alpha = \alpha$, $0 + \alpha = \alpha$, $0 \cdot \alpha = 0$.

We introduce a plane cartesian coordinate system and represent the complex number $a + bi$ by a point with coordinates $x = a$, $y = b$. This representation is a one-to-one correspondence of all the complex numbers and all the points in the plane. In particular, the real numbers are represented by the points on the x-axis, and the imaginary numbers $0 + bi = bi$ by the points on the y-axis. Therefore we call this plane the *plane of complex numbers*, the x-axis the *real axis*, and the y-axis the *imaginary axis*. Because of the correspondence between the complex numbers and the points of the plane we will not have to distinguish between the concepts "point" and "number."

If the cartesian coordinate system is replaced by a polar coordinate system (Fig. 1–1) having the same origin and the real axis as its axis, then the complex number $a + bi$ will be represented by a point whose polar coordinates are ρ and ϕ. The cartesian and the polar coordinates are then connected by the relations $a = \rho \cos \phi$, $b = \rho \sin \phi$, and $\rho = \sqrt{a^2 + b^2}$. Therefore, $a + bi = \rho(\cos \phi + i \sin \phi)$. If we restrict the values of ρ and ϕ by $\rho > 0$ and $-\pi \leqq \phi < \pi$, then every nonzero complex number $a + bi$ determines a unique $\rho = \sqrt{a^2 + b^2}$ and a unique

$$\phi = \arccos \frac{a}{\sqrt{a^2 + b^2}} = \arcsin \frac{b}{\sqrt{a^2 + b^2}} .$$

[The ambiguity of the first equality is removed by considering the second. For instance, for $a = -1$, $b = 1$, $\phi = \arccos(-1/\sqrt{2})$ could be $\frac{3}{4}\pi$ or $-\frac{3}{4}\pi$. However, $\phi = \arcsin(1/\sqrt{2})$ is $\pi/4$ or $\frac{3}{4}\pi$, which leaves only $\phi = \frac{3}{4}\pi$.]

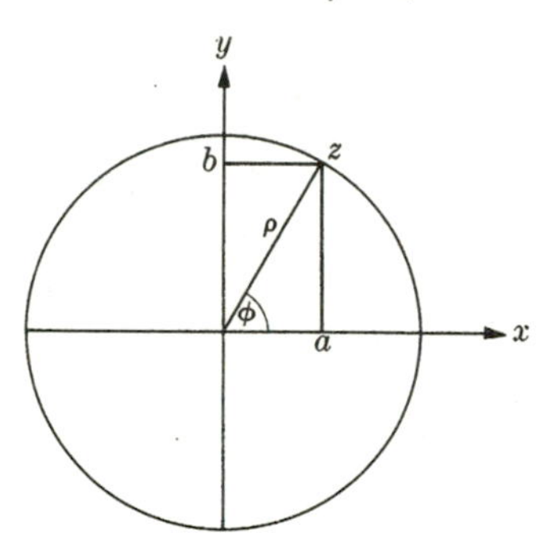

Figure 1–1

The directed segment $(0, z)$ is called the *position vector* of the complex number z.

The value $\rho > 0$ of a nonzero complex number $\rho(\cos \phi + i \sin \phi)$ is called its *absolute value*, and ϕ its *argument*. The absolute value of the number 0 is 0, while its argument is indeterminate. The notation for the absolute value of z is $|z|$, and for its argument is $\arg z$.

If $z = a + bi$ is a complex number, $a - bi$ is called its *conjugate* and is denoted by $\bar{z}$. The points z and $\bar{z}$ are symmetric with respect to the real axis. The product

$$z\bar{z} = (a + bi)(a - bi) = a^2 + b^2 = |z|^2$$

is the square of the absolute value. It is obvious that $|z| = |\bar{z}|$.

The multiplication of complex numbers is considerably facilitated by *De Moivre's theorem* (after the French, later British, mathematician Abraham de Moivre, 1667–1754). Let two complex numbers, z_1 and z_2, be given by their absolute values, ρ_1 and ρ_2, and their arguments, ϕ_1 and ϕ_2. Then

$$z_1 z_2 = \rho_1 \rho_2 [\cos(\phi_1 + \phi_2) + i \sin(\phi_1 + \phi_2)].$$

In words, the absolute value of the product is the product of the absolute values, while the argument of the product is the sum of the arguments. In particular, this implies the relation $|z_1 z_2| = |z_1|\,|z_2|$.

The circle with radius r about the origin has the equation $\rho = r$. In the special case of $r = 1$ this equation represents the so-called *unit circle*.

The length of the segment (z_1, z_2), that is, the distance d between the two points $z_1 = a_1 + b_1 i$ and $z_2 = a_2 + b_2 i$, can be determined by using Pythagoras' theorem,

$$d = \sqrt{(a_1 - a_2)^2 + (b_1 - b_2)^2},$$

and thus we obtain $d = |z_1 - z_2|$. In particular, the length of the position vector of z is $|z|$.

In calculus it is shown that $e^{i\phi} = \cos\phi + i \sin\phi$ for every real ϕ. Hence every complex number with absolute value ρ and argument ϕ can be written as $\rho e^{i\phi}$.

1-2 ISOMETRIES IN THE COMPLEX NUMBER PLANE

Transformations. Let S and T be two sets of points or numbers. We consider ordered pairs (s, t) such that $s \in S$ and $t \in T$. If in a set α of such ordered pairs each element of S appears exactly once as the leading term of a pair, then α is called a *mapping of S into T* or a *function* defined on S with values in T, and the notation is $\alpha\colon S \to T$. If $(s, t) \in \alpha$, we write $(s)\alpha = t$ or, if no ambiguity results, $s\alpha = t$. The element $s\alpha$ is called the *image* of s under α, and s is the *preimage* of $s\alpha$. S is the *domain* of α, and the set of all $s\alpha$ with $s \in S$ is called the *range* of α and is denoted by $S\alpha$. If none of the $s\alpha$ appear more than once in $S\alpha$, that is, if the equation $s_1\alpha = s_2\alpha$ implies that $s_1 = s_2$, then α is *one-to-one* or *injective*. If $S\alpha = T$, that is, if all elements of T are images under α, then α is called *surjective*, or we say that α maps S *onto* T. If a mapping is injective and surjective, it is said to be *bijective*.

One of the most important concepts in geometry is that of a *transformation*. Henceforth by this we will mean a bijective mapping of a set on itself (in contrast to other books in which more general transformations are treated). Therefore, if $\alpha\colon S \to T$ is a transformation, then $S = S\alpha = T$.

Neither of the requirements "injective" or "surjective" is redundant. For instance, the mapping which carries every integer x into the integer $2x$ is injective because for every x there exists one and only one $2x$. However, it is not a transformation in the sense used by us, because its domain is the set Z of all integers, while its range comprises the even integers only. It is "into," not "onto." On the other hand, the mapping of Z that takes the even integers y into $y/2$ while the odd integers are left unchanged is surjective, since the domain and the range are both Z. However, this mapping is not injective, for instance $2 \to 1$ and $1 \to 1$; hence it is not a transformation either.

Transformations can, of course, be very general. The transformations of interest for geometry will always be restricted by the additional requirement that some property be *invariant*, that is, preserved. Thus we might, for instance, deal with transformations that preserve angles, distances, straightness of lines, areas, or volumes.

We shall write $\alpha\beta$ for the transformation which results from performing first α and then β, provided the domains and ranges of α and β all coincide. Thus, in contrast with other usages, we shall always be able to read products from left to right. The notation $(P)\alpha\beta$ makes this possible because $(P)\alpha\beta$ means $[(P)\alpha]\beta$. If α is a transformation, α^{-1} will be the *inverse transformation*, which is defined by $P\alpha^{-1} = Q$ if $Q\alpha = P$.

Isometries. The transformations of the plane which we intend to study now are distance-preserving. They are called congruences, congruent transformations, motions, or, as we will refer to them, *isometries.*

Definition. An *isometry* of the plane is a transformation α with the property that for any two points P and Q, the distance $|PQ|$ is equal to the distance $|P\alpha Q\alpha|$.

In the following, geometric transformations will be denoted principally by capital letters.

Theorem 1–2.1. Let a, b, c, d be complex numbers and $|a| = |c| = 1$. Then the mappings $M\colon z \to az + b$ and $K\colon z \to c\bar{z} + d$ are isometries of the plane.

Proof. It is obvious that K and M are single-valued mappings. In order to show that they are bijective, we start from $zM = az + b$ and obtain $z = (zM - b)/a$, which is always uniquely defined because $a \neq 0$ in view of $|a| = 1$. Thus, for each complex number zM there is exactly one z, namely $(zM - b)/a$, which is mapped on zM. Thus M is bijective. The mapping K also has a unique inverse by which $z \to (\bar{z} - \bar{d})/\bar{c}$. Hence K is also bijective. Now we have to show that distances remain unchanged. If z and w are two points, then

$$
\begin{aligned}
|zM - wM| &= |az + b - aw - b| = |a(z - w)| \\
&= |a|\,|z - w| = |z - w|,
\end{aligned}
$$

in view of $|a| = 1$. Also,

$$
|zK - wK| = |c\bar{z} + d - c\bar{w} - d| = |c|\,|\bar{z} - \bar{w}| = |\overline{z - w}| = |z - w|.
$$

Theorem 1–2.2. Given points z_0, z_1, w_0, and w_1, with $|z_1 - z_0| = |w_1 - w_0| \neq 0$, there exist exactly two transformations of the types $z \to az + b$ and $z \to c\bar{z} + d$, with $|a| = |c| = 1$, which map $z_0 \to w_0$ and $z_1 \to w_1$.

Proof. We have to determine a and b such that

$$az_0 + b = w_0, \qquad az_1 + b = w_1, \qquad |a| = 1.$$

Subtracting the two equations yields

$$w_1 - w_0 = a(z_1 - z_0), \qquad a = \frac{w_1 - w_0}{z_1 - z_0},$$

which is well defined, in view of $|z_1 - z_0| \neq 0$. Also,

$$|a| = \left|\frac{w_1 - w_0}{z_1 - z_0}\right| = \frac{|w_1 - w_0|}{|z_1 - z_0|} = 1.$$

For b we obtain by substitution

$$b = w_0 - az_0 = w_0 - z_0\,\frac{w_1 - w_0}{z_1 - z_0}.$$

Similarly, $w_0 = c\bar{z}_0 + d$ and $w_1 = c\bar{z}_1 + d$ imply that $w_1 - w_0 = c(\bar{z}_1 - \bar{z}_0)$ and $c = (w_1 - w_0)/(\bar{z}_1 - \bar{z}_0)$, where

$$|c| = \left|\frac{w_1 - w_0}{\bar{z}_1 - \bar{z}_0}\right| = \frac{|w_1 - w_0|}{|z_1 - z_0|} = 1.$$

In order to be able to investigate the isometries of the plane we prove first that all isometries are collineations. A *collineation* is a transformation of the points of the plane which carries straight lines into straight lines.

Theorem 1–2.3. Every isometry of the number plane is a collineation.

Proof. If z, w, and t are three distinct points on a straight line, one of them, say w, lies between the two others. Then

$$|w - z| + |t - w| = |z - t|,$$

and if M is an isometry, then the images of z, w, and t behave in the same way, namely

$$|wM - zM| + |tM - wM| = |zM - tM|.$$

But this means that wM, zM, and tM are collinear, because otherwise they would form a triangle in which the sum of two sides is greater than the third side, that is,

$$|wM - zM| + |tM - wM| > |zM - tM|.$$

Corollary. Isometries map parallel lines on parallel lines.

Theorem 1-2.4. An isometry is uniquely determined by a triangle and its congruent image.

Proof (Fig. 1-2). Let z_0, z_1, and z_2 be the vertices of the given triangle, and let w_0, w_1, and w_2 be those of its image, with

$$|z_k - z_l| = |w_k - w_l|, \qquad k, l = 0, 1, 2.$$

Let z be a point distinct from the z_k's. The parallel to the line (z_0, z_1) through z meets the line (z_0, z_2) in a point z'', and the parallel to (z_0, z_2) through z meets (z_0, z_1) in a point z'. If w is the image of z under an isometry which also carries the z_k's into the w_k's, then, according to Theorem 1-2.3 and the Corollary, the lines (z, z') and (z, z'') will be mapped on corresponding lines (w, w') and (w, w''), with w' on (w_0, w_1) and w'' on (w_0, w_2). Then

$$|z_0 - z'| = |w_0 - w'|, \qquad |z_1 - z'| = |w_1 - w'|, \qquad |z_0 - z''| = |w_0 - w''|$$

and $|z_2 - z''| = |w_2 - w''|$, which determines the position of w' and w'' uniquely. But then also the position of w is uniquely defined. Since z was arbitrarily chosen, this proves the theorem.

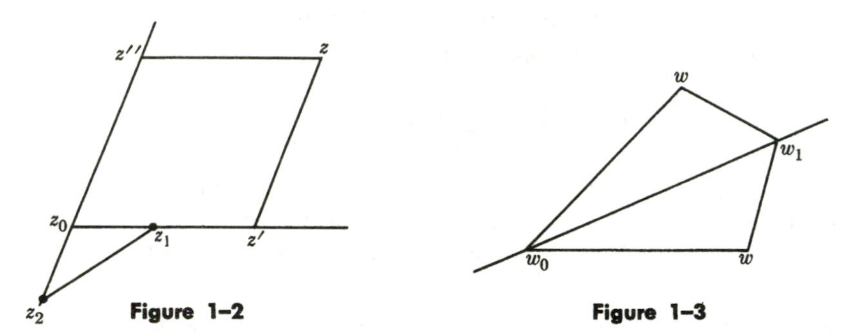

Figure 1-2 Figure 1-3

Theorem 1-2.5. There are exactly two isometries mapping two given points z_0 and z_1 onto two given points w_0 and w_1, provided

$$|z_0 - z_1| = |w_0 - w_1| \neq 0. \tag{1}$$

Proof (Fig. 1-3). In view of Theorems 1-2.1 and 1-2.2, there are at least two such isometries. We have only to show that no more such isometries exist. Now, any isometry mapping z_0 on w_0 and z_1 on w_1 is a collineation, and therefore maps the straight line through z_0 and z_1 onto the straight line through w_0 and w_1. Every point z is mapped on a point w so that

$$|z - z_0| = |w - w_0|, \qquad |z - z_1| = |w - w_1|. \tag{2}$$

If z, z_0, and z_1 are collinear, so are w, w_0, and w_1, in view of Theorem 1-2.3. Then Eqs. (1) and (2) determine uniquely the position of w on the line through

w_0 and w_1. If z is not collinear with z_0 and z_1, then z_0, z_1, and z form a triangle. The triangle with vertices w_0, w_1, and w is congruent to this triangle. The vertices w_0 and w_1 being fixed, there are two possible positions for the triangle, with w lying symmetrically on either side of the line through w_0 and w_1. Thus every point z has one or two images w, according to whether z is collinear or noncollinear with z_0 and z_1. But, by Theorem 1–2.4, each of these triples, w_0, w_1, and w, determines an isometry, and our theorem is proved.

We may summarize the results of this section as follows.

Theorem 1–2.6. The set $\mathfrak{M}$ of all isometries of the number plane is composed of two classes $\mathfrak{M}_+$ and $\mathfrak{M}_-$. The class $\mathfrak{M}_+$ consists of all isometries of the form $z \to az + b$ ($|a| = 1$), and the class $\mathfrak{M}_-$ of all isometries of the form $z \to c\bar{z} + d$ ($|c| = 1$).

The isometries of $\mathfrak{M}_+$ are called *direct*, those of $\mathfrak{M}_-$ *opposite.*

Coordinate transformations. Another concept closely related to the isometries is that of the *coordinate transformations.* By this we mean the introduction of another cartesian coordinate system for assigning values to the points of the plane.

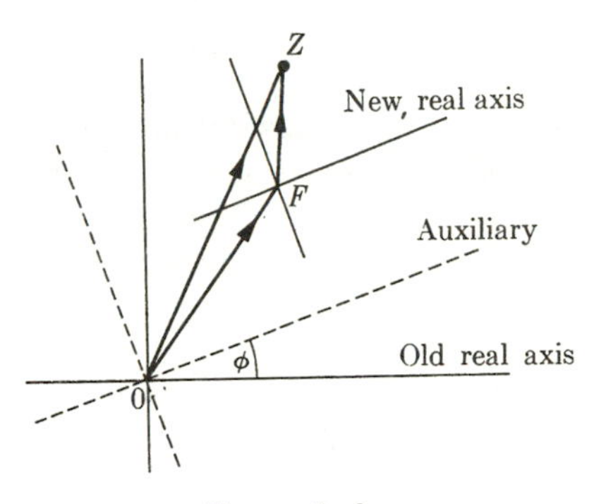

Figure 1–4

Let the origin of the new system be at the point F (Fig. 1–4) whose complex value is f, and let its positive real axis be obtained from the old positive real axis by spinning it through an angle ϕ. Assume also that the positive imaginary axis transforms into the new positive imaginary axis by a rotation through the same angle ϕ. First consider an auxiliary system with origin 0, whose axes are parallel to those of the new system.

If a subscript 1 designates values in the auxiliary system, a subscript 2 those in the new system, and a subscript 0 the old value, let Z be an arbitrary point with $Z_0 = z$. In the auxiliary system all arguments diminish by ϕ, and thus, according to De Moivre's theorem,

$$Z_1 = z[\cos(-\phi) + i \sin(-\phi)] = z(\cos\phi - i \sin\phi),$$

and $F_1 = f(\cos\phi - i\sin\phi)$. The value Z_2 can be obtained by subtraction of F_1 from Z_1 such that

$$Z_2 = Z_1 - F_1 = z(\cos\phi - i\sin\phi) - f(\cos\phi - i\sin\phi) = az + b,$$

say, where a and b are uniquely determined by the choice of f and ϕ, and $|a| = 1$. Conversely, it can be shown (Exercise 9) that for any given a and b with $|a| = 1$

there exists a unique coordinate transformation such that $Z_2 = az + b$ or, for short, $z \to az + b$.

Another case arises if ϕ is again the angle that spins the positive real axis into its new position, but if the positive imaginary axis is turned by this angle into the new *negative* imaginary axis. Then it can be shown (Exercise 10) that the formula becomes $z \to a\bar{z} + b$, $|a| = 1$.

Thus the same formulas $z \to az + b$ and $z \to a\bar{z} + b$ applied to two different operations: to isometries where the points were moved and the coordinate system remained unchanged, and to coordinate transformations where the same points were observed from a new system.

Exercises

1. Are the following mappings transformations of the complex number plane?

   (a) $z \to z^3$ (b) $x + yi \to x^3 + y^3 i$

   (c) $x + yi \to x^2 + y^2 i$

2. Are the following mappings transformations of the number plane? Are they collineations? Find the image and the preimage of i in each of them.

   (a) $z \to |z|$ (b) $z \to \bar{z}$

   (c) $z \to z\bar{z}$ (d) $z \to \mathrm{Re}\,(z)$

3. Show that $z \to z^2$ is not a transformation of the whole number plane. Is this mapping distance-preserving? Are straight lines carried into straight lines?

4. Show that $T\colon z \to 1/z$ is a transformation of the number plane with the point 0 removed. Find the image under T of

   (a) the real axis,
   (b) the imaginary axis,
   (c) the line containing all points with real part 1,
   (d) the circle about 0 with radius r,
   (e) the angle bisector of the first quadrant.

   What are the preimages of all these?

5. Are the transformations $z \to 2iz$ and $z \to 2i\bar{z}$ isometries? Are they angle-preserving (conformal)? Note that an angle α and an angle $-\alpha$ are not to be considered as equal.

6. Find the isometries of the number plane taking

   (a) i into -1, 2 into $2i$, (b) 1 into 2, $1 + i$ into 1.

   Why are the two isometries for (b) related to each other in such a simple way?

7. Are the following isometries direct or opposite?

   (a) $M: 0 \to 0,\ 1 \to i$, and $(i)M > 0$ (b) M^{-1}

   (c) $N: i \to 1 + i,\ 1 \to 2i,\ -1 \to 0$

8. In the proof of Theorem 1–2.5, a geometric argument was used to show that there are two possible positions for the point w, if z_1, z_0, and z are not collinear. Replace this argument by an algebraic reasoning.

9. Prove that for every a and b with $|a| = 1$ there exist unique coordinate transformations $z \to az + b$ and $z \to a\bar{z} + b$.

10. Show that the second type of coordinate transformation described on p. 8 indeed yields a formula $z \to a\bar{z} + b$, $|a| = 1$.

11. Find the transformation if a new system has its origin at $\sqrt{3} - i$, if the new positive real axis passes through 0, and if the new positive imaginary axis intersects the old positive real axis.

12. Describe the coordinate transformations

    (a) $z \to \bar{z}$ (b) $z \to i\bar{z}$ (c) $z \to -iz + 1$.

1-3 GROUPS

Basic concepts. In order to continue our work on isometries we shall need the algebraic concept of a group.

A set S is said to be *closed under a binary operation* $(*)$ if every ordered pair of elements a and b in S determines a unique element of S, which will be called $a * b$. Not every set is closed under every binary operation. For instance, let T be the set of all integers <10 and let the operation $(*)$ be addition. The sum $c = a + b$ of two elements of T might exceed 10 and therefore does not necessarily belong to T. We say then that T is not closed under addition. On the other hand, the set of all integers is closed under addition, and it is also closed under multiplication. Another example is that the set of all points in the plane is closed under the operation of "finding the midpoint."

A set S provided with a binary operation $(*)$ is called a *group* $(S, *)$ if it satisfies the following four postulates.

Gp 1. S is closed under $(*)$.

Gp 2. For all elements a, b, c in S, $(a * b) * c = a * (b * c)$.

This property is called *associativity*.

Gp 3. S contains an element e which satisfies $e * a = a * e = a$, for all a in S.

We call e the *identity* element of the group.

Gp 4. For each element a of S there exists a unique element a' of S such that $a * a' = a' * a = e$.

We call a' the *inverse* of a.

If $a * b$ is replaced by $a + b$ or ab, then the group will be called *additive* or *multiplicative*, respectively, and the identity element e will then be replaced by the zero element 0 or the unity element 1, respectively. The inverse a' will become $-a$ in the additive case and a^{-1} in the multiplicative case. In order to make our notation as short as possible, we will in general write our groups multiplicatively.

If there is no ambiguity as to the operation used, we will write "group S" instead of "the group $(S, *)$."

Examples of groups will be found in the exercises. Here we mention only that the complex numbers form a group under addition, while the nonzero complex numbers form a multiplicative group.

The set S over which the group is defined cannot be the null set because of Gp 3. If S is finite, the number of its elements is called the *order* of the group. It is important to note that ab is not necessarily the same element as ba. If, for all elements of S, $ab = ba$, we call the group *commutative* or *abelian* (in honor of the Norwegian mathematician N. H. Abel, 1802–1829). The property characterizing such a group is referred to as *commutativity*.

The postulate Gp 2 makes the parentheses in $(ab)c$ and $a(bc)$ unnecessary; we may write abc without any ambiguity. It can be proved by induction that in products of more than three elements the parentheses can also be ignored. Consequently, powers of elements can be defined, such as $a^2 = aa$, $a^3 = aaa$, $a^{-2} = a^{-1}a^{-1}$, and so forth.

A few rules for groups follow, with a, b, and c in each case standing for arbitrary group elements.

Theorem 1–3.1. The inverse of a^{-1} is a.

Proof. $a^{-1}a = aa^{-1} = 1$, by Gp 4.

Theorem 1–3.2. $(ab)^{-1} = b^{-1}a^{-1}$.

Proof. $abb^{-1}a^{-1} = 1$.

Corollary. $(a_1a_2 \ldots a_n)^{-1} = a_n^{-1} \ldots a_2^{-1}a_1^{-1}$. Proof by induction.

Theorem 1–3.3. Each of $ac = bc$ and $ca = cb$ implies $a = b$.

Proof. From $ac = bc$ we obtain by right multiplication with c^{-1}, $acc^{-1} = bcc^{-1}$, and hence $a = b$. The other result follows by left multiplication with c^{-1}.

Subgroups and cosets. We define a *subgroup* of a group G as a group whose elements are elements of G and whose operation is the same as that of G. Evidently, a subgroup of a subgroup of G is again a subgroup of G.

Theorem 1–3.4. A nonempty subset H of a group G is a subgroup of G if and only if for every two elements a and b of H, (i) b^{-1} and ab are in H, or (ii) ab^{-1} is in H.

Proof. Suppose that H is a subgroup, containing a and b. Then, by Gp 4, b^{-1} is in H, and, by Gp 1, ab and ab^{-1} are as well. For the converse, in the case (ii), suppose that $ab^{-1} \in H$ whenever a and $b \in H$. From $a \in H$ we obtain $aa^{-1} = 1 \in H$. Now, 1 and $b \in H$ imply $1b^{-1} = b^{-1} \in H$, and a and $b^{-1} \in H$ imply $a(b^{-1})^{-1} = ab \in H$. Thus Gp 3, Gp 4, and Gp 1 are satisfied for H. Since the operation is the same as in G, Gp 2 is fulfilled. Hence H is a group. When the conditions (i) hold, then $ab^{-1} \in H$, and H is a subgroup.

Definition. In a group G with a subgroup H the set C of elements ha (h arbitrary in H, a fixed in G) is denoted Ha and called a *right coset* with respect to H. Correspondingly, aH is a *left coset.*

Every element of G belongs to a coset with respect to H because obviously $b \in Hb$ and $b \in bH$. A further consequence of this fact is the following theorem.

Theorem 1–3.5. If the cosets Ha and Hb have an element in common, they coincide.

Proof. Let c be in Ha and Hb. Then there are elements h and k in H such that $c = ha$ and $c = kb$. This implies $ha = kb$, $a = h^{-1}kb$. Now $h^{-1}k \in H$, and therefore $a \in Hb$ and, with l an arbitrary element of H, $la \in Hb$. This implies $Ha \subseteq Hb$. On the other hand, $b = k^{-1}ha$, which results in $Hb \subseteq Ha$. Hence $Ha = Hb$.

Theorem 1–3.6. If the number of left or right cosets with respect to a subgroup is finite, then there are as many left as right cosets.

Proof. There is a one-to-one correspondence between left and right cosets because $(aH)^{-1} = H^{-1}a^{-1} = Ha^{-1}$; that is, the inverses of a left coset form a right coset.

Definition. The number of cosets with respect to H is the *index* of H in G.

It should be remarked that Theorem 1–3.6 in its present formulation does not hold for infinitely many cosets. However, if "cardinality" is substituted for "number," the theorem becomes valid also for this case, and the proof remains correct without any change.

Corollary. The cosets of a group G with respect to a subgroup H are disjoint and exhaust G.

Theorem 1–3.7. Two elements a and b of G belong to the same coset with respect to H if and only if $ab^{-1} \in H$.

Proof. $ab^{-1} \in H$ implies $a \in Hb$ and therefore $Ha = Hb$. Conversely, if $Ha = Hb$, then $a \in Hb$ and $ab^{-1} \in H$.

As an example let us mention that $\mathfrak{M}_-$, the set of all opposite isometries, is a coset of $\mathfrak{M}$ with respect to $\mathfrak{M}_+$. A detailed proof of this statement will be given later.

Equivalence. At this point, a very useful abstract concept will be introduced, that of an *equivalence relation.*

In a set S, let a relation $(\sim)$ between elements be given such that for all x, y, and z in S,

Ev 1. $x \sim y$ implies $y \sim x$ (*symmetry*).
Ev 2. $x \sim x$ (*reflexivity*).
Ev 3. If $x \sim y$ and $y \sim z$, then $x \sim z$ (*transitivity*).

Then this relation is an equivalence relation.

It can be shown that the existence of an equivalence relation in a set brings about the partition of the set into *equivalence classes* such that each set element belongs to one and only one class to each of whose elements it is related, whereas it is related to none of the elements of other classes.

If, in a group G with subgroup H, $x \sim y$ means $xy^{-1} \in H$, then the relation $(\sim)$ can be shown to satisfy the postulates Ev 1 through 3. The equivalence classes of G are then exactly the cosets with respect to H. The proof will be left for the exercises.

Conjugate and normal subgroups. If again H is a subgroup of a group G, the elements $g^{-1}hg$, with g in G and h running through H, form a subset of G that may be written $g^{-1}Hg$. Moreover the following result holds.

Theorem 1–3.8. For a given subgroup H of G and a given element g in G, $K = g^{-1}Hg$ is a subgroup.

Proof. Let h and k be in H. Then $g^{-1}kg$ and $g^{-1}hg$ are in K, and

$$g^{-1}kg(g^{-1}hg)^{-1} = g^{-1}kgg^{-1}h^{-1}g = g^{-1}kh^{-1}g \in K.$$

By Theorem 1–3.4, K is a subgroup.

We call $g^{-1}ag$ a *conjugate* of a and $g^{-1}Hg$ a *conjugate subgroup* of H. Important in particular are those subgroups which coincide with all their conjugates. Such subgroups are called self-conjugate, invariant, or, as we shall say, *normal subgroups.* If H is a normal subgroup of G, we write $H \lhd G$.

For a normal subgroup H we have, therefore, $H = g^{-1}Hg$ for all elements g, or $gH = Hg$. This means that the right and left cosets with respect to a normal subgroup H coincide. It does not mean, however, that for every element h of H, $gh = hg$ for every element g. On the contrary, in general we will have $gh = h'g$, where h' is an element of H other than h.

Every group G has two trivial normal subgroups, namely G itself and the group consisting of 1 alone. Indeed, $gG = G = Gg$ and $g \cdot 1 = g = 1 \cdot g$. In an abelian group all multiplication is commutative, and thus every subgroup is normal.

Another definition of a normal subgroup is equivalent to our definition although it seems weaker. By this definition H is a normal subgroup of G if, whenever $h \in H$, $g^{-1}hg$ is also in H for all $g \in G$. Now it follows that $g^{-1}Hg \subseteq H$ for all g in G, that is, also for g^{-1}. Hence $(g^{-1})^{-1}Hg^{-1} = gHg^{-1} \subseteq H$, and therefore $H \subseteq g^{-1}Hg$. But $H \subseteq g^{-1}Hg$ and $g^{-1}Hg \subseteq H$ imply the equality of H and $g^{-1}Hg$, and thus we have come back to our original definition. Conversely, it is obvious that our first definition had the weaker definition as a consequence.

Theorem 1–3.9. A subgroup H of index 2 in G is always normal.

Proof. There is only one left coset, gH, distinct from H, and only one such right coset. Since g is not in H, g is in the right coset which is, therefore, Hg. Thus $Hg = gH$, and H is normal in G.

Isomorphism. One further concept that is of great importance is that of *isomorphism.* Two groups G and $G\alpha$ are *isomorphic,* in symbols $G \cong G\alpha$, if there exists a bijective mapping α of $(G, \circ)$ on $(G, *)$ taking the elements 1, $a, b, \ldots$ of G into the elements $1\alpha, a\alpha, b\alpha, \ldots$ of $G\alpha$ such that

$$(a \circ b)\alpha = a\alpha * b\alpha,$$

for all elements a and b of G. The significance of this requirement is that multiplication is preserved under the isomorphism. It is easily verified that isomorphism is an equivalence relation, namely that (a) $G \cong G$, (b) $G \cong H$ implies $H \cong G$, (c) $G \cong H$ and $H \cong K$ imply $G \cong K$. At times it will seem that establishing an isomorphism between two groups is just another way of saying that they are "essentially the same group," or that they are "abstractly identical." Indeed, the isomorphism notion only provides a well-defined term to describe the situation. Examples of isomorphic groups will be mentioned in the exercises.

A special kind of isomorphism is the *automorphism.* An automorphism is an isomorphism of a group with itself. In the next theorem we will meet special automorphisms.

Theorem 1–3.10. The mappings α_g: $G \to g^{-1}Gg$, where g is a fixed element of the group G, are automorphisms of G. If H is a subgroup of G, then the mappings α_g, for all elements g in H, form a group.

The mappings α_g are called *inner automorphisms.*

Proof. The mapping α_g has as inverse α_g^{-1}: $G \to gGg^{-1}$ which is single-valued. Thus α_g is a bijective mapping of G on itself. For establishing an automorphism we have to prove, for two arbitrary elements a and b of G, that $ab \to (g^{-1}ag)(g^{-1}bg)$. But indeed $ab \to g^{-1}(ab)g = g^{-1}agg^{-1}bg$. In order to prove that the inner automorphisms form a group, we observe first that the identity mapping α_1 is an automorphism, and that $\alpha_{g^{-1}}$ is the inverse of α_g.

The closure is obvious. Namely, let g and h be in H. Then

$$\alpha_g \alpha_h: G \to h^{-1}(g^{-1}Gg)h = (gh)^{-1}G(gh),$$

and $\alpha_g \alpha_h = \alpha_{gh}$. Since associativity of one-to-one mappings of a set onto itself can be taken for granted, the theorem is proved.

If $b = g^{-1}cg$, we say that c has been *transformed by the element* g to yield b. The inner automorphism α_g, therefore, transforms the group G by the element g.

Transformation groups. We are now ready to introduce *transformation groups.* We mentioned already that the result of performing a transformation α on a set S, followed by a second transformation β, is a third transformation which we call $\alpha\beta$. We always assume that S is the domain and range of α, β, and $\alpha\beta$. We may consider this composition of transformations as an operation and inquire whether the transformations form a group under this operation. Since $\alpha\beta$ is uniquely determined, Gp 1 is fulfilled. In order to verify Gp 2 ,we observe that

$$(S\alpha\beta)\delta = [(S\alpha)\beta]\delta = (S\alpha)\beta\delta.$$

The transformation which leaves all elements of S unchanged is called the *identity transformation* I and has the property required in Gp 3. To each transformation α, as a bijective mapping of S on itself, there exists a unique α^{-1} such that $\alpha\alpha^{-1} = \alpha^{-1}\alpha = I$. This *inverse transformation* of α maps each element of S on its preimage under α.

We can summarize these results as:

Theorem 1-3.11. The transformations of a set S form a group $\mathcal{G}(S)$ under composition.

$\mathcal{G}(S)$ and its subgroups are called *transformation groups.*

In the special case in which S is finite, every transformation becomes a permutation of the elements of S. The group $\mathcal{G}(S)$ is then a so-called *symmetric group* of permutations. If S has n elements, $\mathcal{G}(S)$ has order $n!$.

Exercises

1. Which of the following sets are groups?
   (a) All rationals under addition
   (b) All rationals under multiplication
   (c) The complex numbers z with $|z| = 1$, under multiplication
   (d) All integers under subtraction
   (e) All even integers under addition
2. Let $n > 1$ be an integer. Do the n roots of the equation $x^n = 1$ form a group under addition? under multiplication?

3. Over the set of all rationals distinct from -1 let an operation be defined by $a * b = a + b + ab$ (usual addition and multiplication).
   (a) Show that the result is a group.
   (b) Is the group abelian?
   (c) Do the integers $\neq -1$ form a subgroup?
4. Is the multiplicative group of nonzero reals isomorphic with the additive group of all reals?
5. Prove:
   (a) Under an isomorphism between two groups the identity elements correspond, and all the inverses of corresponding elements correspond themselves.
   (b) A group isomorphic to an abelian group is itself abelian.
6. (a) Let n be an integer >0. Two integers are to be considered as identical when their difference is divisible by n. Then prove that the n integers $0, 1, 2, \ldots, n - 1$ form an additive group G.
   Prove:
   (b) The integers $1, 2, \ldots, n - 1$ form a multiplicative group H, provided n is prime.
   (c) G, with $n = 4$, and H, with $n = 5$, are isomorphic.
7. Prove that the set of all even integers forms a group under addition. What kind of a set is formed by the odd integers?
8. Prove: The group of all even integers is isomorphic to the group of all integers under addition.
9. Prove: If a subgroup H is finite, each coset Ha of H has exactly as many elements as H has.
10. Prove: A nonempty subset C of a group is a coset if and only if the following holds true. If a, b, and c are elements of C, then $ab^{-1}c$ is also in C.
11. If a and b are elements of a group, are ab and ba conjugate?
12. G is the group of all transformations of the set of reals that can be expressed by $x \rightarrow ax + b$, where $a \neq 0$ and b are real.
    (a) Prove: All the elements of G with $a = 1$ form a normal subgroup H of G.
    (b) Describe the right and left cosets of G with respect to H.
    (c) Does the subset of all elements of G with $b = 0$ form a normal subgroup of G?
    (d) Is G abelian? is H abelian?
13. Prove: If H and K are normal subgroups of a group G, so are $H \cap K$ and HK (the set of all products hk with $h \in H$ and $k \in K$).
14. (a) Do $K \lhd H$ and $H \lhd G$ imply $K \lhd G$?
    (b) Do $H \lhd G$ and $K \lhd G$ and $K \subset H$ imply $K \lhd H$?
15. Let Z (the "center") be the set of all elements z of a group G with the property that $zg = gz$ for all elements g in G. Is Z a normal subgroup of G?

16. Prove: The transformation of all group elements by an element g brings about the identity automorphism if and only if g is in the center (Exercise 15).
17. Are " is father of" and "is brother of" equivalence relations for the set of all human males?
18. If parallel lines in the plane are defined as lines with equal slope, show that parallelism is an equivalence relation for the set of all lines in the plane. What are the equivalence classes?
19. Prove the existence of the equivalence classes for an equivalence relation, as described in the text.
20. (a) Prove that $(\sim)$ is an equivalence relation if $x \sim y$ in a group G with subgroup H means $xy^{-1} \in H$.
    (b) Prove that the equivalence classes are the cosets of G with respect to H.

1–4 COLLINEATION GROUPS OF THE COMPLEX NUMBER PLANE

Translations. The transformations which we met in Section 1–2 were of the types $z \rightarrow az + b$ and $z \rightarrow a\bar{z} + b$, with the restriction $|a| = 1$. In this section we will, temporarily, remove the restriction $|a| = 1$ and study the resulting transformations, with special attention to the various groups that are formed by them. The requirement $|a| = 1$ was necessary and sufficient for the transformations to be distance-preserving. Thus the transformations $z \rightarrow az + b$ and $z \rightarrow a\bar{z} + b$ will be of a more general type, and the properties preserved by them, namely, similarity of triangles and of figures in general will, therefore, turn out to be weaker than the equality of distance. It is well known from elementary geometry that congruent figures are also similar. Thus isometries are special similarity-preserving transformations, and in this section we will see how the isometries fit into the more general picture.

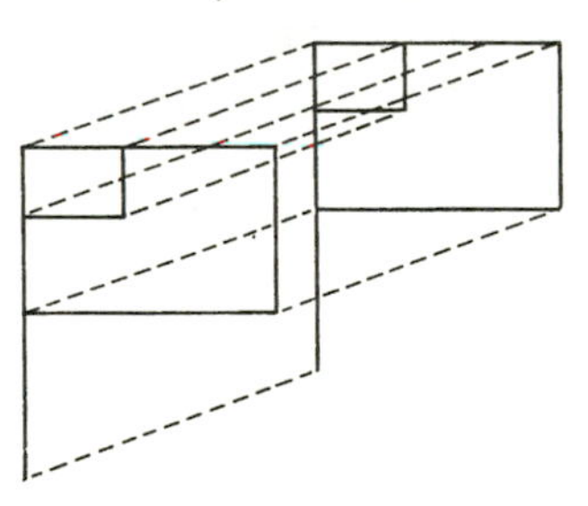

Figure 1–5

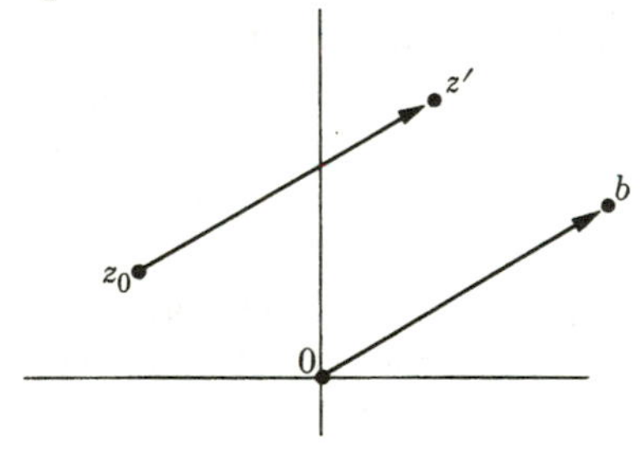

Figure 1–6

We start from one of the simplest types of transformation, the so-called *translations*, represented by $z \rightarrow z + b$. By Theorem 1–2.1, translations are isometries. Geometrically it is easy to see (Fig. 1–5) that a translation takes every point z into a point z' such that the directed segment from z to z' is parallel and equal in length to the position vector of b and equally directed (Fig. 1–6).

Theorem 1–4.1. The translations of the number plane form an abelian group $\mathcal{T}$ which is isomorphic to the additive group of all complex numbers.

Proof. Let T_b: $z \to z + b$ and T_c: $z \to z + c$ be two arbitrary translations. In view of Theorem 1–3.4, we have only to prove that $T_bT_c^{-1}$ is a translation. Now, T_c^{-1} is obviously the translation $z \to z - c$, and thus $T_bT_c^{-1}$: $z \to z + b - c$, which again is a translation. The isomorphism uses the one-to-one mapping $T_b \leftrightarrow b$. The translation T_bT_c corresponds to $z \to (z + b) + c = z + (b + c)$, and hence $T_bT_c = T_{b+c}$, which establishes the desired isomorphism. The additive group of complex numbers is abelian, and therefore so is $\mathcal{T}$.

Dilatations and rotations. An obvious analog to the translations, with multiplication as operation, are the transformations $z \to az$, where a is a nonzero complex number. Of course, $|a|$ is not necessarily 1, and hence these transformations are not always isometries. They are called *dilative rotations about* 0 (dilate = expand, stretch) (Fig. 1–7), and the motivation for this term will soon become clear.

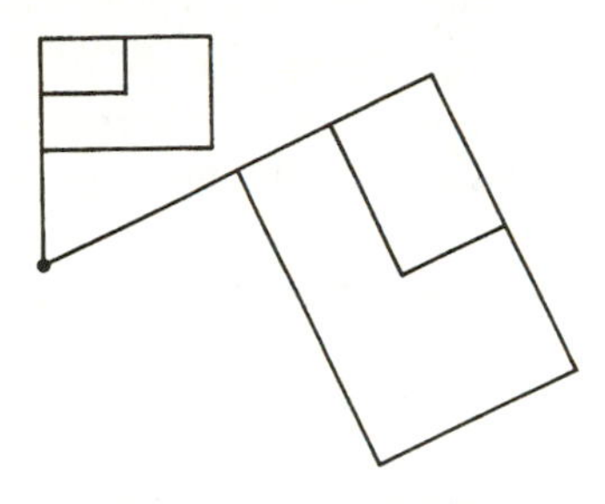

Figure 1–7

Theorem 1–4.2. The dilative rotations of the number plane about 0 form an abelian group $\mathfrak{D}$ which is isomorphic to the multiplicative group of nonzero complex numbers.

Proof. Composition of $z \to az$ and $z \to cz$ yields $z \to caz$, and $z \to a^{-1}z$ is the inverse of $z \to az$. Now the proof follows the line of that of Theorem 1–4.1.

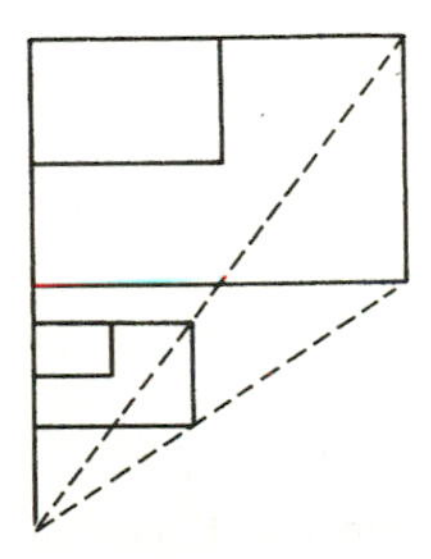

Figure 1–8

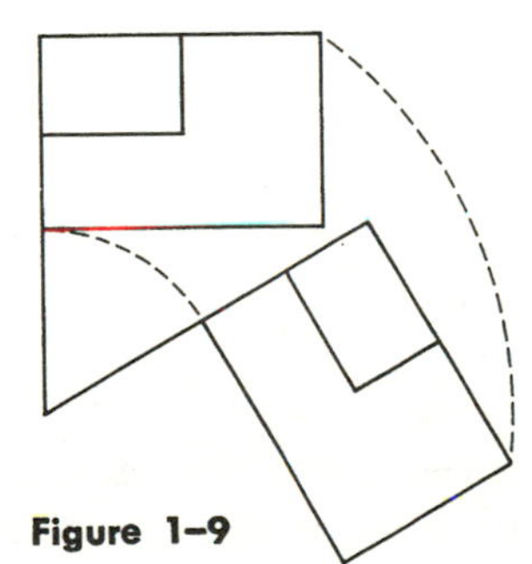

Figure 1–9

We will attempt to learn more about the structure of $\mathfrak{D}$ by considering subgroups of $\mathfrak{D}$. The dilative rotations $z \to az$ with real nonzero a will be referred to as *dilatations* from 0 (Fig. 1–8), and those with $|a| = 1$ as *rotations about* 0 (Fig. 1–9). Indeed, if $|a| = 1$, a can be written as $\cos \alpha + i \sin \alpha$, where $\alpha = \arg a$. If $z = \rho(\cos \phi + i \sin \phi)$, then

$$az = \rho(\cos \phi + i \sin \phi)(\cos \alpha + i \sin \alpha) = \rho[\cos (\phi + \alpha) + i \sin (\phi + \alpha)],$$

and hence ρ stays unchanged, while the argument increases by α. In short, this is a rotation through α about 0. Geometrically a dilatation from 0 carries every point z into a point z' lying on the straight line which joins 0 and z. The meaning of a rotation about 0 is the usual one. A dilative rotation is a dilatation from 0 followed by a rotation about 0, because

$$z' = \frac{a}{|a|}\,(|a|z), \qquad \text{and} \qquad \left|\frac{a}{|a|}\right| = 1.$$

Theorem 1–4.3. The dilatations form a subgroup $\mathfrak{D}^*$ of $\mathfrak{D}$. The rotations about 0 form a subgroup $\mathfrak{R}_0$ of $\mathfrak{D}$. Both are abelian.

Proof. If $a \neq 0$ is real, so is a^{-1}, and the product of two real numbers is real. This proves the first statement. The second part of the theorem follows from the fact that $|a| = |c| = 1$ implies $|a^{-1}| = 1$, and $|ac| = 1$. As subgroups of the abelian group $\mathfrak{D}$, the groups $\mathfrak{D}^*$ and $\mathfrak{R}_0$ are also abelian.

In a similar manner we now consider groups of isometries.

Theorem 1–4.4. The set $\mathfrak{M}$ of isometries of the number plane forms a group with the set $\mathfrak{M}_+$ of direct isometries as a normal subgroup. The set $\mathfrak{M}_-$ of opposite isometries forms a coset with respect to $\mathfrak{M}_+$. Both $\mathfrak{M}$ and $\mathfrak{M}_+$ are nonabelian.

Proof. The product of two distance-preserving transformations is distance-preserving, and so is the inverse of an isometry. Hence $\mathfrak{M}$ is a group. Let M_1: $z \to az + b$ and M_2: $z \to cz + d$, with $|a| = |c| = 1$, be two transformations of $\mathfrak{M}_+$. Then we obtain

$$M_1M_2\colon z \to c(az + b) + d = acz + bc + d.$$

Since $|ac| = 1$, again $M_1M_2 \in \mathfrak{M}_+$. In order to find M_1^{-1}, suppose $M_1M_2 = I$; that is, $acz + bc + d = z$. Then $c = a^{-1}$, $d = -bc = -ba^{-1}$, and $M_1^{-1} = M_2$: $z \to a^{-1}z - a^{-1}b$. In view of $|a^{-1}| = 1$, $M_1^{-1} \in \mathfrak{M}_+$, and $\mathfrak{M}_+$ is a subgroup.

Now N: $z \to \bar{z}$ is in $\mathfrak{M}$, by Theorem 1–2.1. We have NM_1: $z \to a\bar{z} + b$, and since all transformations of $\mathfrak{M}_-$ are of this form by Theorem 1–2.5, they can all be represented in the form NM, where $M \in \mathfrak{M}_+$. Hence $\mathfrak{M}_- = N\mathfrak{M}_+$, a coset. However, $\mathfrak{M}$ is completely made up of $\mathfrak{M}_+$ and $\mathfrak{M}_-$. Thus $\mathfrak{M}_-$ is the only coset, and by Theorem 1–3.9, $\mathfrak{M}_+ \lhd \mathfrak{M}$.

To show that $\mathfrak{M}_+$ is not abelian, combine M: $z \to az$ ($|a| = 1$) and M': $z \to z + b$. Then MM': $z \to az + b$ and

$$M'M\colon z \to a(z + b) = az + ab.$$

For $a \neq 1$ and $b \neq 0$, M and M' are different, and thus $\mathfrak{M}_+$ cannot be abelian. Having a nonabelian subgroup, $\mathfrak{M}$ cannot be abelian either.

Similitudes. Let us now consider the more general class of transformations $z \to az + b$ and $z \to a\bar{z} + b$, where $a \neq 0$, but not necessarily of absolute value 1. These are the so-called *similitudes*. The transformations $z \to az + b$ alone are called *direct similitudes*, the others *opposite similitudes*.

The following theorem can be proved by the same method as Theorem 1-4.4.

Theorem 1-4.5. The similitudes form a group $\mathcal{S}$ with the direct similitudes forming a normal subgroup $\mathcal{S}_+$. The opposite similitudes form a coset $\mathcal{S}_-$ with respect to $\mathcal{S}_+$. Neither $\mathcal{S}$ nor $\mathcal{S}_+$ is abelian. $\mathfrak{M}$ is a subgroup of $\mathcal{S}$, and $\mathfrak{M}_+$ a subgroup of $\mathcal{S}_+$.

The diagram (Fig. 1-10) summarizes the relations between the transformation groups which have been mentioned. A group appearing below another group and connected to it by a line is one of its subgroups.

What are the similitudes from a geometrical point of view? The following theorems will give an answer to this question and also motivate the name "similitude."

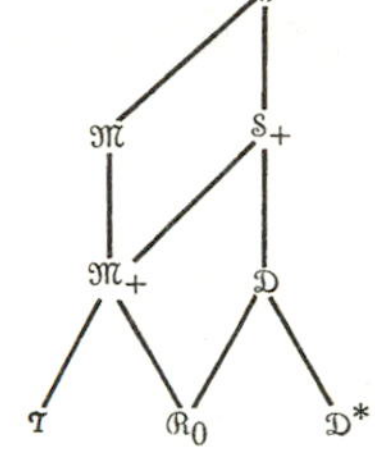

Figure 1-10

Theorem 1-4.6. The vertices z_1, z_2, z_3 of a triangle are taken by a direct similitude into the corresponding vertices w_1, w_2, w_3 of another triangle if and only if

$$\frac{z_2 - z_1}{z_3 - z_1} = \frac{w_2 - w_1}{w_3 - w_1}. \tag{1}$$

Proof. Suppose there is a direct similitude S: $z \to az + b$, with $z_kS = w_k$ where $k = 1, 2, 3$, and $a \neq 0$. Then

$$\frac{w_2 - w_1}{w_3 - w_1} = \frac{az_2 - az_1}{az_3 - az_1} = \frac{z_2 - z_1}{z_3 - z_1}.$$

Conversely, assume Eq. (1). This implies that $w_3 \neq w_1$. Consider the similitude V: $z \to (z - z_1)/(z_3 - z_1)$. V maps z_1 on 0, z_2 on $(z_2 - z_1)/(z_3 - z_1)$, and z_3 on 1. Similarly let W be $z \to (z - w_1)/(w_3 - w_1)$. Then, under W, w_1 goes into 0, w_2 into $(w_2 - w_1)/(w_3 - w_1)$, and w_3 into 1. The product VW^{-1}, itself a similitude, acts as follows:

$$z_1VW^{-1} = 0W^{-1} = w_1,$$

$$z_2VW^{-1} = \left(\frac{z_2 - z_1}{z_3 - z_1}\right) W^{-1} = \left(\frac{w_2 - w_1}{w_3 - w_1}\right) W^{-1} = w_2,$$

$$z_3VW^{-1} = 1W^{-1} = w_3.$$

Thus VW^{-1} is the required similitude. Both V and W are direct, hence also VW^{-1} is direct. Explicitly we can compute VW^{-1} as

$$z \to \frac{w_3 - w_1}{z_3 - z_1}(z - z_1) + w_1.$$

An immediate consequence of this theorem is:

Theorem 1-4.7. A similitude carries any triangle into a similar triangle.

Proof. Let the similitude be either $S: z \to az + b$ or $S': z \to a\bar{z} + b$, with $a \neq 0$. By Theorem 1-4.6, we have Eq. (1) if $z_k S = w_k$, with $k = 1, 2, 3$. This implies

$$\frac{|z_2 - z_1|}{|z_3 - z_1|} = \frac{|w_2 - w_1|}{|w_3 - w_1|}. \tag{2}$$

Moreover, by subtraction of 1 from both sides of Eq. (1) and consideration of absolute values, we obtain

$$\frac{|z_2 - z_3|}{|z_3 - z_1|} = \frac{|w_2 - w_3|}{|w_3 - w_1|},$$

and hence

$$\frac{|z_1 - z_2|}{|w_1 - w_2|} = \frac{|z_2 - z_3|}{|w_2 - w_3|} = \frac{|z_3 - z_1|}{|w_3 - w_1|},$$

and the triangles are similar.

In the second case we now let $w_k = z_k S' = a\bar{z}_k + b$. Then we cannot use Theorem 1-4.6 which deals with direct similitudes only. However, here also

$$\frac{w_2 - w_1}{w_3 - w_1} = \frac{\bar{z}_2 - \bar{z}_1}{\bar{z}_3 - \bar{z}_1} = \overline{\left(\frac{z_2 - z_1}{z_3 - z_1}\right)},$$

which implies at once Eq. (2). The rest of the proof is like that of the direct case.

Orientation. The meaning of the distinction between "direct" and "opposite" has not been clarified geometrically. We will do this for isometries by discussing *orientation*. By Theorem 1-2.4 there exists exactly one isometry mapping the vertices of a given triangle onto the corresponding vertices of a given congruent triangle. If this isometry is direct, we say that the triangles have the *same orientation;* if it is opposite, their orientations are different. The property of being equally oriented is invariant under the isometries, as the following theorem will show.

Theorem 1-4.8. Let z_k, with $k = 1, 2, 3$, be the vertices of a triangle, w_k be those of a congruent triangle, and let both triangles have the same orientation. If M is any isometry, then the triangles with the vertices $z_k M$ and $w_k M$ are equally oriented. If, on the other hand, the original triangles are differently oriented, so also are the images under M.

Proof. If the orientation of the triangles is the same, then there is a direct isometry N such that $z_k N = w_k$. Then $w_k M = z_k NM = z_k M(M^{-1}NM)$. Now, according to Theorem 1-4.4, $\mathfrak{M}_+$ is a normal subgroup of $\mathfrak{M}$. As a consequence, $N \in \mathfrak{M}_+$ implies that $M^{-1}NM \in \mathfrak{M}_+$. Thus the $z_k M$ are carried into the $w_k M$ by a direct isometry, and hence the triangles are equally oriented.

In the case of different orientation, suppose that z_kM and w_kM are equally oriented. Then there is a direct isometry N such that $z_kM = w_kMN$ and $z_k = w_kMNM^{-1}$. Again MNM^{-1} is in $\mathfrak{M}_+$, a contradiction.

It should be noted that we have by no means defined the concept of orientation. All that we have accomplished is the definition of equal or opposite orientation of congruent triangles.

EXERCISES

1. Describe the dilatations from 0 which are isometries.
2. Prove: The square of an opposite isometry is a translation.
3. What is the center of $\mathfrak{M}_+$? (Cf. Section 1–3, Exercise 15.)
4. Find an infinite subgroup of $\mathcal{T}$ and its cosets. Also find a finite subgroup of $\mathcal{R}_0$ and its cosets.
5. Prove: Similitudes are collineations, and direct similitudes are angle-preserving. What about opposite similitudes?
6. Prove that there are exactly two similitudes which carry two given distinct points A and B into their given distinct images A' and B'. Find these similitudes when $A = 1$, $B = i$, $A' = -1$, and $B' = 1$, and find the image of 0 under them. Draw a diagram and check the similarity of triangles.
7. Prove: A dilatation from 0 transforms each straight line into a parallel line. Is the same true for dilative rotations about 0? for similitudes?
8. Prove: The square of an opposite similitude is a dilatation from 0 followed by a translation.
9. Prove: If two parallel line segments AB and $A'B'$ are given, there exists a unique direct similitude taking A into A' and B into B'. The similitude is the product of a dilatation from 0 and a translation.
10. In Fig. 1–10, which of the 9 relations "is subgroup of" can be strengthened to read "is normal in"?
11. (a) Is $\mathcal{T}$ a normal subgroup of $\mathcal{S}$?
    (b) Is $\mathfrak{D}^*$ normal in $\mathcal{S}$?
    (c) Is $\mathcal{R}_0$ normal in $\mathcal{S}$?
12. Are the following pairs of congruent triangles ABC and $A'B'C'$ equally or differently oriented?
    (a) $A = A' = 0, \quad B = B' = 2, \quad C = i, \quad C' = -i$
    (b) $A = A' = 1 + i, \quad B = C' = 2 - i, \quad C = B' = -1$
    (c) $A = 1, \quad B = 3 + i, \quad C = C' = 1 + i, \quad A' = 2 + i, \quad B' = 1 + 3i$
13. Give a definition of equal and different orientation which applies also to similitudes, not only to isometries. Show that Theorem 1–4.8 can be generalized accordingly.

14. Prove that having the same orientation is an equivalence relation for congruent triangles.
15. How would you go about defining a dilatation from a point $w \neq 0$? Express it in terms of a dilatation from 0 and a translation.

1-5 TRANSLATIONS AND ROTATIONS

Geometries and their transformations. In the last section a general overall picture of transformations in the complex number plane was attempted. The complex number plane was used as a convenient description of the plane of *euclidean geometry* (named after the Greek mathematician Euclid, about 300 B.C.), by which we mean the geometry of everyday physical experience as it is studied in high school.

It should be clear by now that the group $\mathfrak{M}$ of isometries is of special interest for us because isometries preserve distances, and therefore every figure will be carried by an isometry into a congruent figure. The role of congruence in euclidean geometry can be understood if we observe that we sometimes identify congruent figures which are in fact different. We say (perhaps loosely) that AB and CD are "equal", and that ABC and DEF are "the same triangle in different places", when actually we mean that they are congruent or, in other words, that they may be carried into each other by an isometry. Thus preservation of distances is strongly connected to the study of euclidean geometry, and the isometries can be considered as the transformation group that "belongs to euclidean geometry". The similitudes carry every figure into a similar figure whose dimensions are $|a|$ times those of the preimage if $z \to az + b$ is the transformation used. If, for instance, $|a| > 1$, we would not consider a figure and its image figure as "the same." We would rather call the image "an enlarged copy." We could, of course, deal with a geometry where only shape, not size, counts. Such a geometry, which is different from euclidean geometry, would have $\mathcal{S}$ as its "own" group of transformations.* This characterization of geometries by groups of transformations is the essence of the "Erlangen program" proposed in 1872 by the German mathematician Felix Klein (1849–1925) in his inaugural address at the University of Erlangen.

Now transformations can be of a very general nature. In topology, for instance, the only restriction imposed on transformations is that they be continuous. In topology the interiors of all circles and all triangles are considered copies of the same object, whereas the annular region between two concentric circles is a different object.

In this course, we will deal only with *linear geometry*, in which the transformations have to map straight lines on straight lines and planes on planes. Thus the transformations of linear geometry in the plane are collineations. Since

*Some authors call this geometry euclidean.

straight lines may be mapped on curves by topological and differential-geometrical transformations, topology and differential geometry do not belong to linear geometry and will not be treated in this course.

It appears imperative to study the group $\mathfrak{M}$ and its subgroups if we wish to gain more knowledge about the euclidean plane. Other linear geometries and their groups will be treated later.

Groups of translations and rotations. We start with the study of the simplest isometries, the translations. Until now a transformation and a preimage were always given, and we had to find the image. But, conversely, it might be interesting to know whether there is an element of the transformation group which carries a given point into a given image point. If for every point z_0 there exists a transformation in the group which takes 0 into z_0, then this transformation group will be called *transitive.*

An equivalent, though seemingly stronger, requirement for a transitive transformation group is that it contain an element which takes an arbitrary point w into another arbitrary point w'.

We are now able to state the following about translations.

Theorem 1–5.1. The group $\mathcal{T}$ of translations is transitive and is a normal subgroup of $\mathfrak{M}_+$.

Proof. The translation $z \rightarrow z + z_0$ takes 0 into z_0. This proves the transitivity. Now let M: $z \rightarrow az + b$, with $|a| = 1$, and let T_d be an arbitrary translation. Then

$$M^{-1}\colon z \rightarrow a^{-1}(z - b),$$

$$M^{-1}T_dM\colon z \rightarrow a[a^{-1}(z - b) + d] + b = z + ad,$$

and hence $M^{-1}T_dM = T_{ad} \in \mathcal{T}$ for an arbitrarily chosen T_d. In other words,

$$M^{-1}\mathcal{T}M \subseteq \mathcal{T}, \qquad \mathcal{T} \lhd \mathfrak{M}_+.$$

A more general definition of rotations will now be given, and we will, of course, have to show that our previously introduced "rotations about 0" indeed fit into the new definition.

Intuitively, a rotation may be best described as an isometry that leaves unchanged one, and only one, point of the plane. If we want to study rotations, it will be worth while to consider such unchanged points, which we will call "invariant points."

Definition. A point z is called an *invariant point* of transformation β if $z\beta = z$. A line l is an *invariant line* if β maps every point of l on a point of l.

The definition of an invariant line might cause misunderstandings. We do not require every point of l to be an invariant point, in other words, β does not necessarily leave l fixed pointwise. Not the points of l, but rather l as a whole, is preserved under β.

It is obvious that nonidentical translations do not have invariant points. Algebraically this amounts to the fact that $z = z + b$, with $b \neq 0$, cannot be solved. Therefore it is reasonable to start with the investigation of invariant points for the direct isometries which are not translations.

Theorem 1–5.2. Let M be the isometry $z \to az + b$ with $|a| = 1$, but $a \neq 1$. Then M has exactly one invariant point

$$w = b(1 - a)^{-1}, \tag{1}$$

and M can be written

$$z \to a(z - w) + w. \tag{2}$$

If T_w is the translation $z \to z + w$, and R_a is the rotation $z \to az$, then $M = T_w^{-1} R_a T_w$.

Proof. If w is an invariant point, then we get $w = aw + b$ and Eq. (1). Since the operation is reversible, this is the only invariant point. Now substitute $b = w(1 - a)$ into $az + b$. The result is Eq. (2). But this could be expressed as the result of performing consecutively T_w^{-1}: $z \to z - w$, R_a and T_w. Hence $M = T_w^{-1} R_a T_w$.

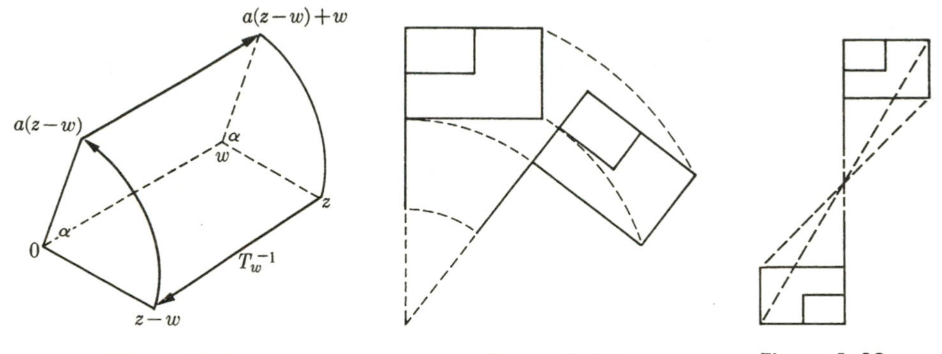

Figure 1–11 Figure 1–12 Figure 1–13

What then is the geometrical meaning of a direct isometry which is not a translation? Figure 1–11 shows the three stages used for constructing M. The geometrical significance of the construction is a rotation about w through an angle $\alpha = \arg a$. It is, therefore, reasonable to use the term *rotation about w through* $\arg a$ for every direct isometry (Eq. 2) which is not a translation (Fig. 1–12). In particular, a rotation about w through π is a *half turn* about w (Fig. 1–13). In this case, $a = -1$ and $z \to -z + 2w$. Incidentally, the definition of rotations fits the name "rotation about 0" which we gave previously to $z \to az$ with $|a| = 1$. The special case of the identity transformation $z \to z$ may be considered as a rotation about 0 through an angle 0.

Another way of looking at the direct isometries with $a \neq 1$ involves a coordinate transformation. If we move the origin to the invariant point w and let the new axes be parallel and equally oriented to the old axes, then in the new system the isometry becomes $z \to az$. This is geometrically obvious in view of Fig. 1–11. Algebraically, Eq. (2) implies $z + w \to az + w$, and after the coordinate transformation this indeed turns into $z \to az$. We may consider this as a standard or *canonical* form for all direct isometries which are not translations. This form makes it even more obvious that every direct isometry which is not a translation is actually a rotation about some point.

A general definition of a canonical form is the following. Let S be a set and $\mathcal{G}(S)$ a transformation group over S. Suppose there exists a subset C of S with the property that for every $s \in S$ there is exactly one element $c \in C$, and some transformation $T \in \mathcal{G}(S)$ such that $sT = c$. Then c is called the *canonical form* of s under $\mathcal{G}(S)$. In the special case mentioned above, s is the direct isometry $z \to az + b$, and c is the isometry $z \to az$.

We shall now investigate the set of all rotations about a point w.

Theorem 1–5.3. The set $\mathcal{R}_w$ of rotations about the point w is a group, and the groups $\mathcal{R}_w$ for all values of w are isomorphic to each other.

Proof. We know from Theorem 1–5.2 that every $M \in \mathcal{R}_w$ can be written as $M = T_w^{-1}RT_w$, where $R \in \mathcal{R}_0$. Let $M = T_w^{-1}RT_w$ and $M' = T_w^{-1}R'T_w$ be two rotations about w. In order to show that $\mathcal{R}_w$ forms a subgroup of $\mathfrak{M}$, we have to prove that $M'M^{-1}$ is in $\mathcal{R}_w$. Now

$$M'M^{-1} = T_w^{-1}R'T_w(T_w^{-1}RT_w)^{-1} = T_w^{-1}R'T_wT_w^{-1}R^{-1}T_w = T_w^{-1}R'R^{-1}T_w.$$

By Theorem 1–4.3, $\mathcal{R}_0$ is a group, and thus $R'R^{-1}$ is another element of $\mathcal{R}_0$, and $M'M^{-1}$ is in $\mathcal{R}_w$. If β is the bijective mapping $R \to T_w^{-1}RT_w = M$ of $\mathcal{R}_0$ onto $\mathcal{R}_w$, then MM' can be expressed as $R\beta R'\beta$ as well as $(RR')\beta$, which proves that β is an isomorphism. Thus every $\mathcal{R}_w \cong \mathcal{R}_0$, and the $\mathcal{R}_w$'s are isomorphic to each other.

Corollary. All the rotation groups $\mathcal{R}_w$ form a complete set of conjugate subgroups of $\mathcal{R}_0$ in the group of all isometries.

Proof. Let M: $z \to az + b, N$: $z \to a\bar{z} + b$, and R: $z \to cz$, with $|a| = |c| = 1$. Then it is easy to verify that

$$M^{-1}RM = T_b^{-1}RT_b, \qquad N^{-1}RN = T_b^{-1}R'T_b,$$

where R': $z \to \bar{c}z$. Both transformations are elements of $\mathcal{R}_b$. This implies the assertion.

Finally it may be remarked that the concept of the invariant point provides also a geometrical characterization of the translations, which at the start of Section 1–4 were defined in a purely algebraic way.

Theorem 1–5.4. An isometry of the euclidean plane is a translation if and only if it has no invariant points and preserves orientation.

Proof. If the isometry M preserves orientation, it is direct; $M: z \to az + b$. A direct isometry has an invariant point w satisfying Eq. (1) unless $a = 1$. Thus in the absence of invariant points we have necessarily $a = 1$, and M carries z into $z + b$, a translation. Conversely, a translation is a direct isometry and cannot have invariant points.

Exercises

1. Show that the two definitions of a transitive group given in the text are in fact equivalent.
2. Check the transitivity of the groups $\mathcal{R}_w$ (for a given w), $\mathfrak{D}^*$, and $\mathfrak{D}$. Could any of the intransitive groups among them be regarded as transitive if the origin were disregarded?
3. Show that $\mathcal{R}_0$ has infinitely many finite subgroups while $\mathcal{T}$ has none.
4. Let R be the rotation about 0 through $\pi/3$, and T: $z \to z + 1$. Let $\mathcal{G}$ be the group *generated* by R and T, that is, the group consisting of all possible products using any of R, R^{-1}, T, T^{-1} as factors, any number of times. Is $\mathcal{G}$ finite or infinite? If z is a fixed point, what would its images be under the elements of $\mathcal{G}$?
5. If M is the product of a half turn about w and a half turn about w', what is the product in reversed order, when $w \neq w'$?
6. Let A, B, and C be half turns. Prove that $ABC = CBA$.
7. Prove: Every product of three half turns is a half turn.
8. Let R be the rotation about w through α and R' the rotation about w' through $-\alpha$. What is RR'?
9. (a) Prove: If w is an arbitrary point, then any direct isometry can be expressed as the product RT of a rotation R about w and a translation T. Is this representation unique? (b) Is the same true for TR instead of RT?
10. A mirror at $Re(z) = 1$ and another mirror at $Re(z) = -1$, are directed toward each other. A point z is mapped by consecutive alternating reflections in the two mirrors, starting with the first mirror. What is the image after $2n$ reflections? after $2n + 1$ reflections? (n is a positive integer.)
11. Develop a canonical form for all direct similitudes that are not translations. What does this form show for the geometrical nature of these similitudes?

1–6 REFLECTIONS

Line reflections. After having discussed the direct isometries with respect to their invariant points, it is natural to ask the analogous questions for opposite isometries. We will do so in this section.

We start with the special opposite isometries of the type $z \to a\bar{z}$, $|a| = 1$. An invariant point w of this transformation satisfies $w = a\bar{w}$, that is, $w = 0$ or $w\bar{w}^{-1} = a$. Let us compare the arguments of the two sides of this equation. If α is the argument of w then the argument of $\bar{w}$ is $-\alpha$, and α is the argument of $\bar{w}^{-1}$. Hence arg $(w\bar{w}^{-1}) = \alpha + \alpha = 2\alpha$, and $2\alpha = \arg a$ and $\alpha = (\arg a)/2$. All the points whose argument is $(\arg a)/2$, as well as the point 0, are invariant. Since the procedure which yielded this result can be reversed, these are the only invariant points. All of them lie on the straight line through 0, forming an angle $(\arg a)/2$ with the real axis, and all points on this line are invariant. Every other point z will be taken by the isometry $z \to a\bar{z}$ into a point z' symmetrical to z with respect to the invariant line.

A *line reflection* will be defined as an isometry leaving invariant every point of a fixed line l and no other points. We refer to this line reflection also as the *reflection in the line* l. An optical reflection along l in a mirror having both sides silvered, would yield the same result (Fig. 1–14).

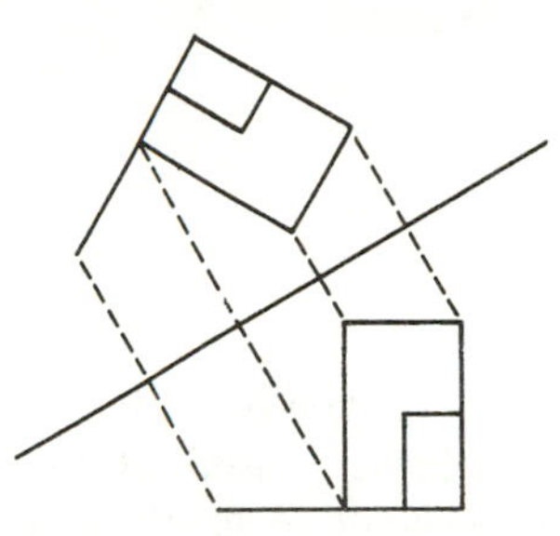

Figure 1–14

We have, therefore, proved

Theorem 1–6.1. The isometry U_a: $z \to a\bar{z}$, with $|a| = 1$, is a reflection in the straight line through 0, enclosing with the positive real axis an angle of $(\arg a)/2$.

We are now able to attack the more general transformation $z \to a\bar{z} + b$, with $|a| = 1$. Does it have invariant points? For an invariant point w, $w = a\bar{w} + b$. Using conjugates, we obtain $\bar{w} = \bar{a}w + \bar{b}$. We combine these two equations and obtain

$$w = a(\bar{a}w + \bar{b}) + b = a\bar{a}w + a\bar{b} + b = |a|^2w + a\bar{b} + b = w + a\bar{b} + b,$$

and hence,

$$a\bar{b} + b = 0, \tag{1}$$

as a necessary condition for the existence of an invariant point of $z \to a\bar{z} + b$. The condition of Eq. (1) is also sufficient, because if it holds, $w = b/2$ is always an invariant point, namely,

$$a\bar{w} + b = a\bar{b}/2 + b = (a\bar{b} + b)/2 + b/2 = b/2 = w.$$

Now suppose M: $z \to a\bar{z} + b$ has an invariant point w. Then $w = a\bar{w} + b$. If z' is the image of z, then $z' = a\bar{z} + b$. Subtraction of the two last equations yields

$$z' - w = a(\bar{z} - \bar{w}) = a(\overline{z - w}), \qquad z \to a(\overline{z - w}) + w.$$

Thus

$$M = T_w^{-1}U_aT_w. \tag{2}$$

What are the invariant points of M? If $zT_w^{-1}U_aT_w = z$, then $zT_w^{-1}U_a = zT_w^{-1}$, and $zT_w^{-1} = z - w$ is an invariant point of U_a. But, according to Theorem 1–6.1, $z - w$ is invariant under U_a if and only if it lies on the invariant line l of U_a. If we write $v = z - w$, then we see that all invariant points of Eq. (2) are the points $v + w$, where v assumes the values of all the points on l, and w is a fixed invariant point of M. By Theorem 1–2.3, these points $v + w$ lie also on a straight line, and M has to be a line reflection.

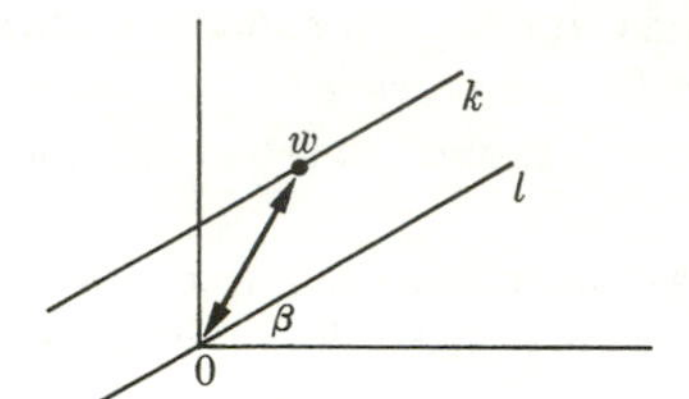

Figure 1–15

If, on the other hand (Fig. 1–15), a reflection U in a line k is given, let w be a point on k and let U_a be a reflection in the line l which passes through 0 and is parallel to k. Such a U_a exists because, if β is the angle between l and the positive real axis, there exists a complex number $a = \cos 2\beta + i \sin 2\beta$, satisfying the conditions $\beta = (\arg a)/2$ and $|a| = 1$. Then M of Eq. (2) leaves invariant all points of k. By Theorem 1–2.5, there are two isometries mapping all points of k on themselves. One of these isometries is the identity. Now Eq. (2) is certainly not the identity, and is uniquely determined as the isometry which leaves the points on k, and only these points, invariant. So U is Eq. (2). We have therefore

Theorem 1–6.2. An opposite isometry $z \to a\bar{z} + b$ has invariant points if and only if Eq. (1) holds. If an opposite isometry has an invariant point, it is a line reflection, and every line reflection is an opposite isometry.

Here also a canonical form may be developed by using a suitable coordinate transformation. Let $z \to a\bar{z} + b$, with $|a| = 1$ and Eq. (1). Then

$$z\sqrt{a} + b/2 \to a\overline{(z\sqrt{a} + b/2)} + b = \bar{z}\sqrt{a} + a\bar{b}/2 + b/2 + b/2 = \bar{z}\sqrt{a} + b/2.$$

Under the coordinate transformation $z \to z\sqrt{a} + b/2$, this becomes $z \to \bar{z}$, which may be considered as the canonical form of every opposite isometry which has invariant points. Geometrically it is a line reflection whose mirror is the real axis.

Involutory isometries. The line reflections have the property of being *involutory*, which means that their period is two; that is, for every involutory transformation V, $V^2 = I$. Points are interchanged by V; in other words, $zV = z'$ if and only if $z'V = z$. The question before us is whether line reflections are the only involutory isometries. As it turns out, there is one other kind of involutory isometry which somehow also deserves the name "reflection." This is the *point reflection*, which we have already met under the name half turn. A half turn about w is a reflection in the point w. The next theorem will justify this alternative name for half turns. A line reflection leaves invariant one line and all the points lying on it. Thus it would be a fitting analog if the point reflection were to leave invariant a point and all the lines passing through it.

Theorem 1–6.3. A half turn about a point w leaves invariant all lines passing through w.

Proof. It is intuitively evident that a rotation about w through π maps points of a line through w on points of the same line. However, an analytic proof will be given. The distance between a point z and w is $|z - w|$, and since a half turn is an isometry, the image point $z' = -z + 2w$ satisfies $|z' - w| = |z - w|$. Now

$$|z - z'| = |z + z - 2w| = 2|z - w| = |z - w| + |z' - w|.$$

Thus for the triangle whose vertices are z, z', and w, one side equals the sum of the two remaining sides. This can hold only when the three vertices are collinear.

Theorem 1–6.4. Every involutory isometry is a line reflection, or a point reflection, and conversely, these reflections are involutory isometries.

Proof. For an involutory direct isometry $z \to az + b$, we have

$$z = a(az + b) + b = a^2z + ab + b$$

identically. This implies $a^2 = 1$ and $ab + b = 0$. Then $a = 1$ and $b = 0$, or $a = -1$ and b arbitrary. Consequently the isometry is the identity or a point reflection. For opposite isometries $z \to a\bar{z} + b$, we get

$$z = a\overline{(a\bar{z} + b)} + b = a(\bar{a}z + \bar{b}) + b = a\bar{a}z + a\bar{b} + b,$$

and hence Eq. (1). But then Theorem 1–6.2 implies that the transformation is a line reflection. The converse statement is trivial.

Isometries generated by reflections. We shall now deal with transformations resulting from consecutively performed line reflections. First we consider two such reflections.

Theorem 1–6.5. Two reflections in lines k and l result in (i) a translation if and only if k and l are parallel, (ii) a rotation about the common point of k and l if and only if they intersect.

Proof. Let the reflection in k be $z \to a\bar{z} + b$, with $|a| = 1$ and Eq. (1), and let the reflection in l be $z \to c\bar{z} + d$, with $|c| = 1$ and

$$c\bar{d} + d = 0. \tag{3}$$

Then, after performing both successively, we obtain

$$z \to c\overline{(a\bar{z} + b)} + d = c(\bar{a}z + \bar{b}) + d = \bar{a}cz + \bar{b}c + d.$$

We have to distinguish between two cases.

(i) $\bar{a}c = 1$. Then $z \to z + \bar{b}c + d$ is a translation. From $\bar{a}c = 1$ we obtain $a\bar{a}c = a$ and $c = a$. This implies $(\arg c)/2 = (\arg a)/2$, and so k and l have the same slope with respect to the positive x-axis. Hence in this case $k \parallel l$. Conversely, $k \parallel l$ implies $a = c$ and $a\bar{c} = 1$.

(ii) $\bar{a}c \neq 1$. Then $z \to \bar{a}cz + (\bar{b}c + d)$ is a rotation about the point $w = (\bar{b}c + d)/(1 - \bar{a}c)$, by Theorem 1–5.2. For the converse, intersecting k and l imply $a \neq c$ and $a\bar{c} \neq 1$. In order to prove that w is the point of intersection of k and l, we show that both reflections leave w invariant. For reflection in k,

$$w \to a\bar{w} + b = a(b\bar{c} + \bar{d})(1 - a\bar{c})^{-1} + b = \frac{a\bar{d} + b}{1 - a\bar{c}}.$$

Multiplying both the numerator and the denominator by $\bar{a}c$ and recalling that $a\bar{a} = c\bar{c} = 1$, we get

$$w \to \frac{c\bar{d} + \bar{a}bc}{\bar{a}c - 1}.$$

We now apply Eqs. (1) and (3) and obtain $(\bar{b}c + d)/(1 - \bar{a}c)$, which is w. Thus w lies on k.

A reflection in l maps w on $c\bar{w} + d$. By substitution we get

$$\frac{c(b\bar{c} + \bar{d})}{1 - a\bar{c}} + d = \frac{c(\bar{c}b + \bar{d}) + d - a\bar{c}d}{1 - a\bar{c}} = \frac{b + c\bar{d} + d - a\bar{c}d}{1 - a\bar{c}}.$$

By Eq. (3), this becomes $(b - a\bar{c}d)/(1 - a\bar{c})$, by multiplication with $\bar{a}c$ in both the numerator and the denominator, $(\bar{a}bc - d)/(\bar{a}c - 1)$, and again by Eq. (1), $(\bar{b}c + d)/(1 - \bar{a}c)$, which is w. Hence w is invariant under the reflections in k and l, and it must be their point of intersection.

In our classification of isometries we encountered the translations ($a = 1$) and rotations ($a \neq 1$) in the direct case, and the line reflections with invariant points in the opposite case. However, the case of an opposite isometry without invariant points has not been dealt with. In this case, M: $z \to a\bar{z} + b$ and $M = U_aT_b$, a line reflection followed by a translation. Such an opposite isometry without invariant points is called a *glide reflection*. Geometrically we obtain the following characterization.

Theorem 1-6.6 (Fig. 1-16). A reflection in a line, followed by a translation T_b, results in a glide reflection, provided the reflecting line l is not perpendicular to the position vector of b. Conversely, every glide reflection is the product of a reflection in a line l followed by a translation parallel to l.

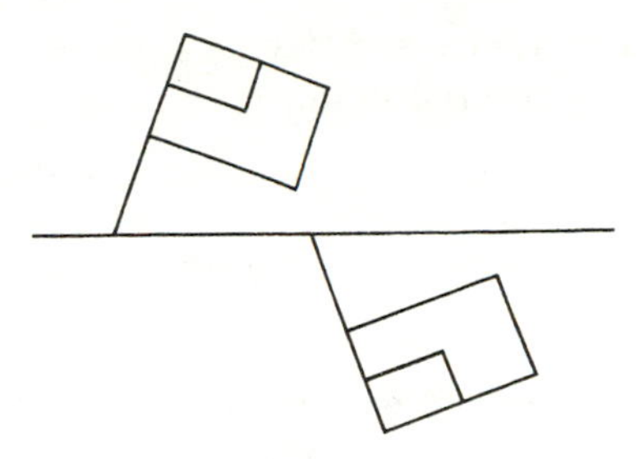

Figure 1-16

Proof. If $M = U_a T_b$ is not a glide reflection, M must have an invariant point, and Eq. (1) holds. Then $a\bar{b} = -b$ and $\arg a\bar{b} = \arg(-b)$, which becomes $\arg a - \arg b = \arg b \pm \pi$, and finally $(\arg a)/2 = \arg b \pm \pi/2$. Now, by Theorem 1-6.1, $(\arg a)/2$ is the angle between the positive real axis and l, while $\arg b$ is the angle between the positive real axis and the position vector of b. Hence l is then perpendicular to this position vector.

Conversely, let the glide reflection be

$$z \to a\bar{z} + b = a\left(\bar{z} - \frac{\bar{b}}{2}\right) + \frac{b}{2} + \frac{a\bar{b} + b}{2}$$

$$= zT_{\bar{b}/2}^{-1} U_a T_{b/2} T_d, \qquad \text{with} \qquad d = \frac{a\bar{b} + b}{2} \neq 0.$$

It remains to show that $2 \arg d = \arg a$. Indeed,

$$\left(\frac{d}{|d|}\right)^2 = \frac{d^2}{d\bar{d}} = \frac{d}{\bar{d}} = \frac{a\bar{b} + b}{\bar{a}b + \bar{b}} = a,$$

which implies

$$\arg\left(\frac{d}{|d|}\right)^2 = 2 \arg \frac{d}{|d|} = 2 \arg d = \arg a.$$

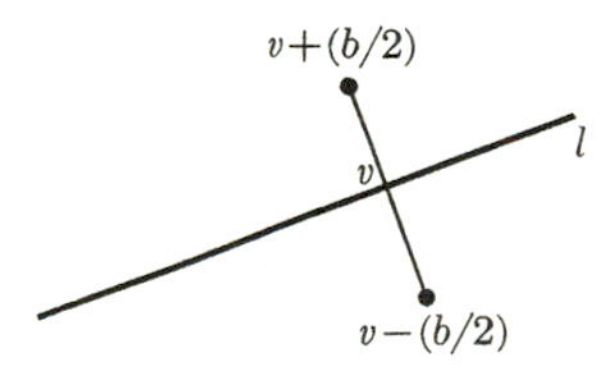

Figure 1-17

This completes the proof.

A simple geometrical consideration (Fig. 1-17) shows what happens in the case when l is perpendicular to the position vector of b. If v is any point on l, $v + b/2$ is invariant because it is carried into $v - b/2$ by the reflection in l and back into $v + b/2$ by T_b. Hence this transformation has a line of invariant points and is a line reflection rather than a glide reflection.

Theorem 1-6.7. Every product of three line reflections is either a line reflection or a glide reflection.

Proof. Let the three reflections be $z \to a_j\bar{z} + b_j$, with $j = 1, 2, 3$. Then their product is

$$z \to a_3(\overline{a_2(\overline{a_1\bar{z} + b_1}) + b_2}) + b_3 = a_1\bar{a}_2a_3\bar{z} + \bar{a}_2a_3b_1 + a_3\bar{b}_2 + b_3,$$

an opposite isometry. Now every opposite isometry is either a line reflection or a glide reflection.

The next theorem sheds additional light on invariant points of isometries.

Theorem 1–6.8. All isometries with a common invariant point form a group.

Proof. The inverse of an isometry M has the same invariant point as M. If M and N are two isometries leaving w invariant, then $wMN = wN = w$.

There exists an interesting relation between perpendicularity and reflections.

Theorem 1–6.9. Two distinct lines k and l are perpendicular to each other if and only if the product of the reflections in k and l is involutory (Figs. 1–18 and 1–19).

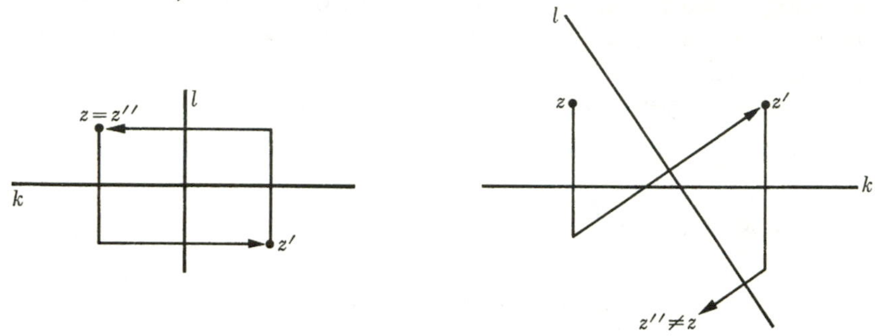

Figure 1–18

Figure 1–19

Proof. Suppose that $k \perp l$, and w is their point of intersection. Then the product that is asserted to be involutory is $M = T_w^{-1}U_aT_wT_w^{-1}U_cT_w$, where $T_w^{-1}U_aT_w$ is the reflection in k and $T_w^{-1}U_cT_w$ that in l. We have

$$M = T_w^{-1}U_aU_cT_w\colon z \to c\overline{(a\overline{(z - w)})} + w = \bar{a}cz - \bar{a}cw + w.$$

Now $k \perp l$ implies

$$\frac{\arg a}{2} - \frac{\arg c}{2} = \pm\frac{\pi}{2}, \qquad \arg a - \arg c = \arg\frac{a}{c} = \pm\pi,$$

and consequently $a/c = -1$ and $a = -c$. After substitution, this is

$$M\colon z \to \bar{a}(-a)z - \bar{a}(-a)w + w = -z + 2w.$$

This is a reflection in w, and hence is involutory.

Now suppose the product M of the reflections in k and l to be involutory. By Theorem 1–6.5, k and l are not parallel; otherwise the product would be a translation which is not involutory. Let k and l intersect in w. Then

$$M\colon z \to \bar{a}cz - \bar{a}cw + w.$$

As an involutory direct isometry, $M \neq 1$ can only be a reflection in a point, and $\bar{a}c = -1$, $z \to -z + 2w$. From $\bar{a}c = -1$ we get $\arg c - \arg a = I \pm \pi$, $(\arg c)/2 - (\arg a)/2 = \pm\pi/2$, and this proves that $k \perp l$.

Another theorem connects with reflections the *incidence* of a point and a line, that is, the relation between a point and a line, which is expressed by saying "the point lies on the line" or "the line passes through the point."

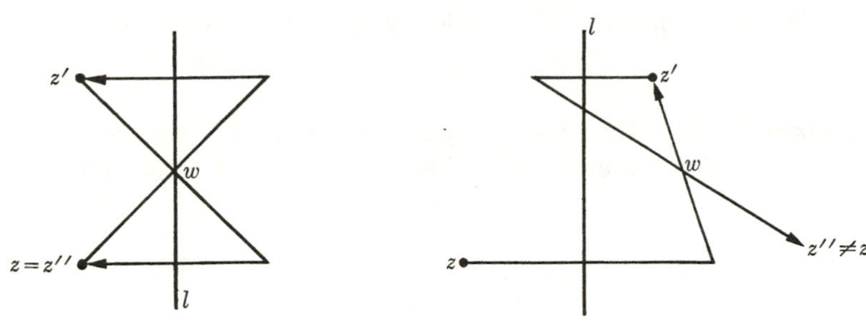

Figure 1–20 **Figure 1–21**

Theorem 1–6.10. (Figs. 1–20 and 1–21). The product of the reflection in a point w and the reflection in a line l is involutory if and only if w lies on l ("w is incident with l").

Proof. Let the reflection in l be $U: z \to a\bar{z} + b$, with Eq. (1). The reflection in w is $V: z \to -z + 2w$. Their product is $VU = M: z \to a(\overline{-z + 2w}) + b = -a\bar{z} + 2a\bar{w} + b$. Now assume w on l. Then w is invariant under U; that is, $w = a\bar{w} + b$. When we substitute, we have $M: z \to -a\bar{z} + 2w - b$, and

$$\begin{aligned} M^2: z \to -a(\overline{-a\bar{z} + 2w - b}) + 2w - b \\ = z - 2a\bar{w} + a\bar{b} + 2w - b \\ = z - (2a\bar{w} + 2b) + (b + a\bar{b}) + 2w \\ = z - 2w + 0 + 2w = z, \end{aligned}$$

and M is involutory.

To prove the converse, assume $M^2 = I$. Then $M: z \to -a\bar{z} + 2a\bar{w} + b$, as an involutory opposite isometry, has to be a line reflection and has an invariant point. The condition for this is, by Eq. (1),

$$\begin{aligned} -a(\overline{2a\bar{w} + b}) + 2a\bar{w} + b = 0, \\ -2w - a\bar{b} + 2a\bar{w} + b = 0, \\ -2w - (a\bar{b} + b) + 2a\bar{w} + 2b = 0, \\ -2w + 2a\bar{w} + 2b = 0, \end{aligned}$$

and finally $w = a\bar{w} + b$, which makes w invariant under U. Hence w lies on l.

The last two theorems show that very important relations can be expressed in terms of reflections. It is, therefore, reasonable to attempt a construction of the whole geometry based on the concept of the reflection. This has indeed been achieved by a school of geometers, the foremost of whom was the Danish mathematician J. T. Hjelmslev (1873–1950). For more information on this subject, the reader is referred to F. Bachmann, *Aufbau der Geometrie aus dem Spiegelungsbegriff*, Berlin: Springer Verlag, 1959.

The following theorem shows clearly the fundamental significance of the reflections.

Theorem 1–6.11. Every isometry is the product of at most three line reflections. If the isometry has an invariant point, at most two line reflections suffice.

Remark. The condition is not necessary: every translation is the product of two reflections, but has no invariant points.

Proof. If the isometry is direct, it is either a translation T_b or a rotation $T_w^{-1} R_a T_w$. In T_b, with $b \neq 0$ (the identity as a trivial case can be ignored), there is a unique a such that Eq. (1) holds. Consider then the two reflections U_a and $T_{b/2}^{-1} U_a T_{b/2}$. Their product is

$$U_a T_{b/2}^{-1} U_a T_{b/2}\colon z \to a(\overline{a\bar{z} - b/2}) + b/2 = z - (a\bar{b})/2 + b/2$$
$$= z - (a\bar{b} + b)/2 + b = z + b,$$

which is T_b. Now turn to the rotational case,

$$T_w^{-1} R_a T_w = (T_w^{-1} U_1 T_w)(T_w^{-1} U_a T_w),$$

which is the product of two line reflections, since $U_1\colon z \to \bar{z}$, $U_a\colon z \to a\bar{z}$, and therefore $U_1 U_a\colon z \to a\bar{\bar{z}} = az$, which is R_a.

If, on the other hand, the isometry is opposite, it is either a line reflection or a glide reflection. A glide reflection is the product of a line reflection and a translation, but the translation is the product of two reflections. Hence the glide reflection may be written as the product of at most three reflections.

If there is an invariant point, the isometry is either a rotation or a line reflection, that is, the product of two reflections at most.

Exercises

1. Prove Theorem 1–6.10 with the modification that the point reflection be performed after the line reflection instead of before it.
2. If A and B are two distinct points and C their midpoint, determine the isometries that take
   (a) A into C and C into B,

(b) A into B and B into A,
(c) A into B and C into C.
Describe each isometry geometrically.

3. In each of the following isometries, determine whether it has an invariant point. If it is a reflection, find its reflecting line or point.

$$\text{(a) } z \to i\bar{z}, \qquad \text{(b) } z \to \bar{z} + i - 1, \qquad \text{(c) } z \to -z + 4 - 2i.$$

4. (a) Use the geometrical meaning of Eq. (1) in order to prove the following for every line reflection $z \to a\bar{z} + b$, with $b \neq 0$: The join of any point, not on the mirror line, and its image is perpendicular to the mirror line. (b) Prove the same for $b = 0$.
5. Prove: The product of reflections in three lines that pass through one point w is a reflection in a line through w.
6. Prove: The product of reflections in two parallel lines at a distance d from each other is a translation T_b with $|b| = 2d$.
7. Let M be a glide reflection which can be written as UT, U the reflection in a line l, T a translation along l. Then prove that the midpoint of the segment joining z and zM lies on l.
8. Prove: Every direct isometry is the product of two line reflections. Every opposite isometry is the product of a line reflection and a point reflection.
9. What is the product of three glide reflections?
10. If U, V, W are three line reflections, what is $(UVW)^2$?
11. Let U and V be reflections in lines k and l, respectively, and let k and l be distinct and not perpendicular to each other. If there is an invariant line for UV, prove that it is perpendicular to k and l.
12. Prove: The product of the reflections in three distinct lines is never the identity.
13. Prove: Every glide reflection is the product of a line reflection and a point reflection.
14. Compute the transformation $M = UT$ if U is the reflection in a line l of slope angle $\pi/4$ through i, and T the translation through $\sqrt{2}$ units along l in the direction of increasing real part. Find also TU. What kind of transformation is M?
15. Let U_1, U_2, U_3, and U_4 be, respectively, the reflections in the real axis, in the angle bisector of the first and third quadrants, in the imaginary axis, and in the line through i enclosing an angle of $\pi/4$ with the positive real axis. Compute $U_1U_2U_3$, $U_1U_2U_4$, and $U_1U_3U_4$. What kind of isometry is each of them?
16. Find one way of decomposition of the translation T_i into a product of line reflections. Do the same with $z \to i\bar{z}$ and $z \to i\bar{z} + i$.

17. Use the same coordinate transformation as in the case of the opposite isometries with invariant points for establishing a canonical form for those without invariant points. What is the geometrical significance of the result?
18. Find canonical forms for opposite similitudes and discuss them geometrically.

1-7 VECTORS

Vectors and translations. Vectors are known to the reader from calculus, physics, and vector analysis. Here we shall introduce them in a slightly different way.

If w and z are two points, then, by Theorem 1-5.1, there exists exactly one translation, T_{z-w}, which carries w into z. We know that translations T_b and T_c are *multiplied* by *adding* b and c, that is, the *multiplicative* group of translations is isomorphic to the *additive* group of complex numbers. It would simplify the handling of translations if the multiplicative notation were replaced by an additive one. For this reason a translation T_b will alternatively be designated as the *vector* $\mathbf{b}$. The operation between vectors is addition, and thus $\mathbf{b} + \mathbf{c}$ corresponds to the translation T_{b+c}. Thus the vectors form an additive group $\mathcal{V}$ isomorphic to $\mathcal{T}$, and the *null vector* $\mathbf{0}$ corresponds to the identity transformation. How is this new definition of a vector related to the familiar description of a vector as a set of parallel segments, equal in length and direction? How does the addition of these vectors fit into our picture? The answer is contained in the next theorem.

Theorem 1-7.1. The multiplicative group of translations, $\mathcal{T}$, the additive group of vectors, $\mathcal{V}$, and the group of all classes of equal, parallel, and equally directed segments under vector addition are isomorphic.

Proof. Every directed segment is equal, parallel, and equally directed to one and only one position vector of a complex number, and no two distinct complex numbers have the same position vector. Therefore there exists a one-to-one correspondence between vectors and complex numbers. The familiar vector addition according to the "parallelogram law" is performed componentwise. That is, if vectors $\mathbf{v}_1$ and $\mathbf{v}_2$ have components (x_1, y_1) and (x_2, y_2) in a cartesian coordinate system, then $\mathbf{v}_1 + \mathbf{v}_2$ has the components $(x_1 + x_2, y_1 + y_2)$. This is the same addition as that of the complex numbers, which proves the isomorphism between $\mathcal{V}$, $\mathcal{T}$, and the vectors, in view of Theorem 1-4.1. We may also define the length $|\mathbf{v}|$ of $\mathbf{v}$ as the absolute value of the corresponding complex number.

How do vectors behave under isometries? If we denote the vector leading from a point P to a point Q by $\overrightarrow{PQ}$, and if M is an isometry, what is the vector $\overrightarrow{(P)M(Q)M}$? Let p and q be the complex numbers corresponding to P and Q, respectively. Then the translation belonging to $\overrightarrow{PQ}$ is T_{q-p}, that belonging to $\overrightarrow{(P)M(Q)M}$ is T_{qM-pM}, which after a short computation can be shown to be equal to $M^{-1}T_{q-p}M$. By Theorem 1–5.1, $M^{-1}T_{q-p}M$ is in $\mathcal{T}$, and $|qM - pM| = |q - p|$. So we have proved:

Theorem 1–7.2. An isometry carries a vector into a vector of the same length.

The *scalar product* of a vector $\mathbf{v}$ and a real number ρ is defined in the well-known way. If $\rho > 0$, then $\rho\mathbf{v} = \mathbf{v}\rho$ is the vector parallel to $\mathbf{v}$ and equally directed, such that $|\rho\mathbf{v}| = \rho|\mathbf{v}|$. For $\rho = 0$, $\rho\mathbf{v}$ is the null vector. If $\rho < 0$, we introduce $\rho\mathbf{v} = \mathbf{v}\rho = (-\rho)(-\mathbf{v})$, where $-\mathbf{v}$ is the additive inverse of $\mathbf{v}$. The rules $\rho(\sigma\mathbf{v}) = (\rho\sigma)\mathbf{v}$, $\rho(\mathbf{v} + \mathbf{w}) = \rho\mathbf{v} + \rho\mathbf{w}$, and $(\rho + \sigma)\mathbf{v} = \rho\mathbf{v} + \sigma\mathbf{v}$ and their proofs are well known from calculus texts. They imply the following theorem.

Theorem 1–7.3. The mapping $\mathbf{v} \to \alpha\mathbf{v}$ for any nonzero real α is an automorphism of $\mathcal{V}$:

$$(\mathbf{v}_1 + \mathbf{v}_2)\alpha = \mathbf{v}_1\alpha + \mathbf{v}_2\alpha, \qquad \text{and} \qquad (\rho\mathbf{v})\alpha = \rho(\mathbf{v}\alpha).$$

The set of vectors $\alpha\mathbf{v}$ with fixed nonzero $\mathbf{v}$ forms an abelian group under addition. This group is isomorphic to the additive group of real numbers.

The *inner product* of two vectors, $\mathbf{vw}$, is defined as $|\mathbf{v}|\,|\mathbf{w}| \cos \measuredangle(\mathbf{v}, \mathbf{w})$. The commutativity of this product follows from $\cos(-\alpha) = \cos\alpha$. The following rules are well known:

$$(\rho\mathbf{v})\mathbf{w} = \rho(\mathbf{vw}), \qquad \mathbf{u}(\mathbf{v} + \mathbf{w}) = \mathbf{uv} + \mathbf{uw}, \qquad \mathbf{v}^2 = \mathbf{vv} = |\mathbf{v}|^2.$$

Two vectors, both nonzero, have a vanishing inner product if and only if they are perpendicular to each other. A vector of length 1 is called a *unit vector.*

Warning. With the introduction of the inner product, the correspondence between vectors and complex numbers breaks down. If $\mathbf{v}$ and $\mathbf{w}$ correspond, respectively, to complex numbers v and w, then the complex product does *not* correspond to the inner product $\mathbf{vw}$, which is not even a vector.

Theorem 1–7.4. If $\mathbf{e}_1$ and $\mathbf{e}_2$ are two perpendicular unit vectors, then every vector $\mathbf{v}$ has a unique representation

$$\mathbf{v} = x_1\mathbf{e}_1 + x_2\mathbf{e}_2, \tag{1}$$

with real x_1 and x_2. If $\mathbf{w} = y_1\mathbf{e}_1 + y_2\mathbf{e}_2$, then the inner product has the value $\mathbf{vw} = x_1y_1 + x_2y_2$.

Proof. Let $\mathbf{v}$ be given, $\mathbf{ve}_1 = x_1$, and $\mathbf{ve}_2 = x_2$. Then, in view of $\mathbf{e}_1\mathbf{e}_2 = 0$ and $\mathbf{e}_j^2 = 1$, with $j = 1, 2$, we have

$$(\mathbf{v} - x_1\mathbf{e}_1 - x_2\mathbf{e}_2)\mathbf{e}_j = x_j - x_j = 0,$$

substituting for j successively 1 and 2. The term in the parentheses cannot be perpendicular to both $\mathbf{e}_1$ and $\mathbf{e}_2$, and it must, therefore, be zero. The result is Eq. (1). Then it follows that

$$\mathbf{vw} = x_1y_1\mathbf{e}_1^2 + x_2y_2\mathbf{e}_2^2 + x_1y_2\mathbf{e}_1\mathbf{e}_2 + x_2y_1\mathbf{e}_1\mathbf{e}_2 = x_1y_1 + x_2y_2.$$

In Eq. (1), x_1 and x_2 are called the *components* of $\mathbf{v}$, and $\mathbf{e}_1$ and $\mathbf{e}_2$ are called *basis vectors.*

Traces. We shall now be interested in the application of vectors to set up the so-called parametric representation of a straight line l (Fig. 1-22). Let $\mathbf{a}$ be the position vector of a point of l, and $\mathbf{b}$ be a vector parallel to l; then for any point of l the position vector $\mathbf{r}$ can be written in the form

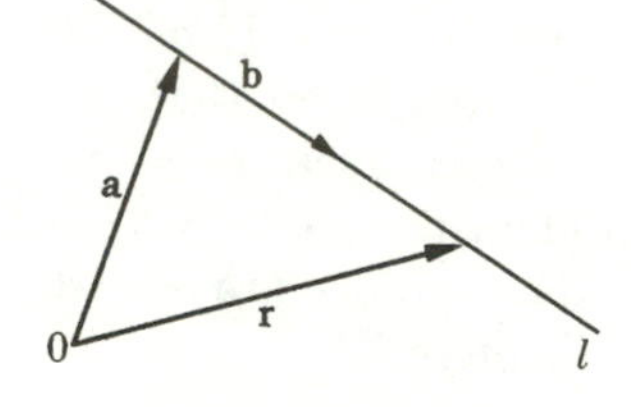

Figure 1-22

$$\mathbf{r} = \mathbf{a} + \rho\mathbf{b}, \qquad \rho \text{ real.} \tag{2}$$

This provides the equation of a straight line in the plane of complex numbers, and we have:

Theorem 1-7.5. If a is any point on l, then every point r on l can be represented as $r = a + \rho b$, where b is a complex number whose position vector is parallel to l, and where ρ is a *real* parameter.

We have encountered here a situation which is worthy of closer examination. Let us write Eq. (2) in translation notation, $r = aT_{\rho b}$. A reasonable definition of the ρth *power of the translation* T_b would be $(T_b)^\rho = T_{\rho b}$ because this coincides with the meaning of $(T_b)^\rho$ with integral ρ, the only case already defined. Besides, this new definition which replaces the additive operation in the subscript by the multiplicative operation between translations fits the fact of the isomorphism between the multiplicative group $\mathcal{T}$ and the additive group of complex numbers. By allowing ρ to assume all real values, we recognize now that the group $\mathcal{T}'$ of all powers of T_b depends on the continuous parameter ρ. By operating on a with the ρth power of T_b, we obtain exactly all points r on the straight line l, when ρ ranges through all real numbers. We say then that l is the *trace* of a under the group $\mathcal{T}'$. We know that the group $\mathcal{T}'$ is a *one-parameter group.*

Another example of a one-parameter group is $\mathcal{R}_0$. We write again $r = aR^\rho$, with the restriction $a \neq 0$, and assume

$$R\colon\ z \to z(\cos\alpha + i\sin\alpha).$$

We then obtain

$$r = a(\cos\alpha + i\sin\alpha)^\rho = a(\cos\alpha\rho + i\sin\alpha\rho).$$

If ρ ranges through all reals, r runs through all points on a circle about 0 through a. Thus this circle is the trace of a under the group consisting of all powers of R which is identical with $\mathcal{R}_0$. The group $\mathcal{R}_0$ also is a one-parameter group.

EXERCISES

1. Are vectors invariant under
   (a) translations? (b) direct isometries? (c) opposite isometries?
2. Are inner products invariant under
   (a) direct isometries? (b) opposite isometries?
3. Use the equation of the straight line in the form of Theorem 1–7.5 to show that similitudes are collineations.
4. Find the parametric representation of the straight line passing through the two points with position vectors **v** and **w**.
5. Prove: If z_k, with $k = 1, 2, 3$, are distinct collinear points and z_k' their images under an isometry, then the midpoints of the segments (z_k, z_k') are either collinear or all three coincide.
6. What is the trace of a point c under $\mathcal{R}_w$, when $c \neq w$?
7. What is the trace of a nonreal point z_0, with $|z_0| > 1$, in the complex number plane under the transformation group $z \to z^\rho$, where ρ is a real parameter?

1–8 BILINEAR TRANSFORMATIONS

Basic concepts. We return now to the plane of complex numbers. We adjoin to this plane one additional point, $z = \infty$, called the *point at infinity*, and we stipulate that $1/z = \infty$ when $z = 0$. This procedure is customary in the theory of complex functions and will be most useful in our exposition. We have, therefore, to deal with the euclidean plane, augmented by the new point, ∞. Practical computation with the symbol ∞ will be performed according to the following rules: For any finite complex number c,

$$c \pm \infty = \infty, \qquad c/\infty = 0,$$

and for nonzero c,

$$c \cdot \infty = \infty, \qquad c/0 = \infty.$$

As usual, the expressions $\infty - \infty$, ∞/∞, $0 \cdot \infty$, and $0/0$ are to be considered as undefined.

The transformations that will interest us here are the so-called *bilinear transformations* or Möbius transformations (after the German mathematician August F. Möbius, 1790–1868),

$$B\colon z \to \frac{az + b}{cz + d}, \tag{1}$$

where a, b, c, and d are complex numbers, with the restriction that

$$\delta = ad - bc \neq 0.$$

The number δ is called the *determinant* of B.

First of all, let us prove that Eq. (1) is a transformation in the augmented complex number plane. Every finite z yields an image, in particular

$$(-d/c)B = \infty,$$

unless $a(-d/c) + b = 0$, which would contradict the condition $\delta \neq 0$. For $z = \infty$, note that

$$\frac{az + b}{cz + d} = \frac{a + b/z}{c + d/z} = \frac{a}{c}.$$

On the other hand, $w = (az + b)/(cz + d)$ implies $z = (-dw + b)/(cw - a)$, which makes a unique z correspond to every given w because this again is a bilinear transformation, in view of $(-d)(-a) - bc = ad - bc \neq 0$. Hence the mapping is bijective, as asserted.

A special case of a bilinear transformation is the mapping $N\colon z \to 1/z$. With the help of N we are going to decompose B in a most convenient way.

Theorem 1–8.1. The transformation Eq. (1), with $ad - bc \neq 0$, can be written as

$$B = T_{d/c}ND_fT_{a/c},$$

where

$$f = \frac{bc - ad}{c^2} = \frac{-\delta}{c^2},$$

and D_f designates the dilative rotation $z \to fz$.

Proof.

$$zT_{d/c}ND_fT_{a/c} = \left(z + \frac{d}{c}\right)^{-1}D_fT_{a/c}$$

$$= \frac{bc - ad}{c^2(z + d/c)} + \frac{a}{c} = \frac{bc - ad}{c^2z + cd} + \frac{a}{c},$$

which results in Eq. (1). D_f is a dilative rotation only if $f \neq 0$; this is assured by the requirement $\delta \neq 0$.

It should be noted that in general B is no isometry and not even a similitude, because D_f is not necessarily an isometry, and N is certainly no similitude, while the rest of the factors are similitudes.

Theorem 1–8.2. All bilinear transformations form a group $\mathcal{B}$. If B and C are two bilinear transformations, and δ_B and δ_C their determinants, then $\delta_{BC} = \delta_B\delta_C$ is the determinant of BC.

Proof. We have already shown that B^{-1} is a bilinear transformation if B is. The identity transformation is bilinear. Let

$$C\colon z \to \frac{a'z + b'}{c'z + d'}, \qquad \text{with} \qquad \delta_C \neq 0.$$

Then

$$BC\colon z \to \frac{a'(az+b)/(cz+d)+b'}{c'(az+b)/(cz+d)+d'} = \frac{(aa'+b'c)z+a'b+b'd}{(ac'+cd')z+bc'+dd'} = \frac{\alpha z+\beta}{\gamma z+\epsilon},$$

say, and $\alpha\epsilon - \beta\gamma = (ad-bc)(a'd'-b'c') \neq 0$. This completes the proof.

Theorem 1-8.3. A bilinear transformation is uniquely determined by three distinct points z_j and their three distinct image points w_j, with $j = 1, 2, 3$.

Proof. Consider

$$B\colon z \to \frac{(z-z_2)(z_1-z_3)}{(z-z_1)(z_2-z_3)}.$$

It is a bilinear transformation, since its determinant is

$$\delta_B = (z_1-z_2)(z_2-z_3)(z_3-z_1) \neq 0.$$

Under B, $z_1 \to \infty$, $z_2 \to 0$, and $z_3 \to 1$. Now define another transformation

$$C\colon z \to \frac{(z-w_2)(w_1-w_3)}{(z-w_1)(w_2-w_3)}.$$

Its determinant $\delta_C = (w_1-w_2)(w_2-w_3)(w_3-w_1) \neq 0$ so that C is a bilinear transformation, which carries $w_1 \to \infty$, $w_2 \to 0$, and $w_3 \to 1$. The product BC^{-1} is again a bilinear transformation that takes z_j into w_j. Is BC^{-1} the only bilinear transformation fulfilling this requirement? Suppose that A is another such bilinear transformation. Then $BC^{-1}A^{-1}$ is not the identity, but it leaves the z_j invariant. Let $BC^{-1}A^{-1}$ be $z \to (\alpha z+\beta)/(\gamma z+\epsilon)$. Then the equation

$$z = \frac{\alpha z+\beta}{\gamma z+\epsilon}$$

has three distinct solutions. However, $(\gamma z+\epsilon)z = \alpha z+\beta$ is a quadratic equation which has at most two solutions, unless $\gamma = 0$. In this case we would have $\epsilon z = \alpha z+\beta$, which can have three solutions only if it is an identity, namely $\epsilon = \alpha$ and $\beta = 0$. Then the transformation is $z \to \alpha z/\alpha = z$, the identity transformation.

Cross ratio. We shall now define a concept of great importance for further work.

Definition. The *cross ratio* of four points z_j, with $j = 1, 2, 3, 4$, is

$$(z_1, z_2|z_3, z_4) = \frac{(z_1-z_3)(z_2-z_4)}{(z_1-z_4)(z_2-z_3)}.$$

A few properties of the cross ratio follow.

Theorem 1–8.4. The cross ratio of four points is invariant under bilinear transformations.

Proof. Let the four points z_j, with $j = 1, 2, 3, 4$, be taken into w_j. Then, using the transformations B and C mentioned in the proof of Theorem 1–8.3, $z_jBC^{-1} = w_j$, and consequently $z_jB = w_jC$. That is,

$$\frac{(z_j - z_2)(z_1 - z_3)}{(z_j - z_1)(z_2 - z_3)} = \frac{(w_j - w_2)(w_1 - w_3)}{(w_j - w_1)(w_2 - w_3)},$$

and substituting $j = 4$ we obtain the required result

$$(z_1, z_2|z_3, z_4) = (w_1, w_2|w_3, w_4).$$

An alternative proof is based on the fact that the bilinear transformation is the product of certain translations, a dilative rotation and a mapping $N\colon z \to 1/z$. It can easily be confirmed that the cross ratio is preserved under each of these types of transformation.

Theorem 1–8.5. The cross ratio of four distinct points is real if and only if the four points lie on a straight line or a circle.

Proof. Of the four points z_j, with $j = 1, 2, 3, 4$, let z_1, z_2, and z_3 lie on a straight line. Then $\arg(z_1 - z_3)$ and $\arg(z_2 - z_3)$ differ only by a multiple of π, that is, $\arg(z_1 - z_3)/(z_2 - z_3)$ is a multiple of π, and $(z_1 - z_3)/(z_2 - z_3)$ is real. The cross ratio c is therefore real in this case if and only if the remaining factor $(z_2 - z_4)/(z_1 - z_4)$ is real. This in turn happens exactly when $\arg(z_2 - z_4)$ and $\arg(z_1 - z_4)$ differ by a multiple of π, that is, when z_1, z_2, and z_4 are collinear. Thus if z_1, z_2, and z_3 lie on one straight line, c is real if and only if z_4 also lies on this line. All these considerations will hold also if one of the four points is ∞.

Now suppose that z_1, z_2, and z_3 are not collinear. Then there exists a circle passing through them. In Fig. 1–23,

$$\alpha = \arg(z_2 - z_3) - \arg(z_1 - z_3)$$

$$= \arg(z_2 - z_3)/(z_1 - z_3),$$

up to a multiple of 2π. Also

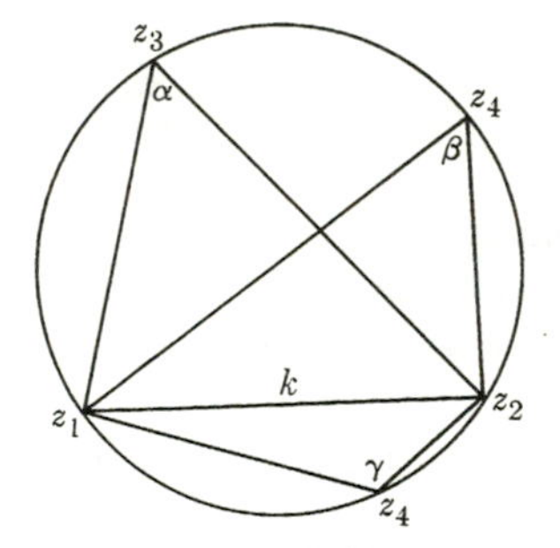

Figure 1–23

$$\beta = \arg(z_2 - z_4) - \arg(z_1 - z_4),$$

and $\beta - \alpha = 0$; that is, $\arg c = 0$ if z_4 lies on the circle, on the same side as z_3 of the chord k. If z_4 lies on the circle, but on the other side of k, the angle is

$$\gamma = \arg(z_1 - z_4) - \arg(z_2 - z_4), \qquad \text{and} \qquad \alpha + \gamma = \pi,$$

which implies

$$\arg 1/c = \pi, \qquad \text{and} \qquad \arg c = -\pi.$$

To these values of $\arg c$, any multiple of 2π might have to be added. If z_4 does not lie on the circle, $\arg c$ cannot have any multiple of π as its value. Hence c is real if and only if z_4 lies on the circle.

Now the main theorem of this section follows easily.

Theorem 1-8.6. A bilinear transformation takes straight lines and circles into straight lines and circles.

This does not mean that lines are mapped on lines and circles on circles. In a stricter wording the theorem could be expressed as "takes into itself the set of all straight lines and circles of the plane." Consider, for instance, the circle about the center 1 with radius 1, transformed by $z \to 1/z$. The points $2, 1 + i$, and $1 - i$ of the circle are carried into $1/2, 1/2 - i/2$, and $1/2 + i/2$, respectively, and the three image points are collinear. Every straight line contains the point ∞, and therefore the image (or preimage) of a straight line under $z \to 1/z$ is a circle or a straight line passing through 0. In our example the circle indeed passed through the point 0.

Proof. Bilinear transformations preserve the cross ratio of four points. If $z_j \to w_j$, with $j = 1, 2, 3, 4$, and if z_1, z_2, and z_3 lie on a circle or a straight line which contains also z_4, then $(z_1, z_2|z_3, z_4)$ is real and so is $(w_1, w_2|w_3, w_4)$. The points w_1, w_2, and w_3 lie on a circle or a straight line which contains also w_4 because $(w_1, w_2|w_3, w_4)$ is real. Thus the theorem is proved.

Real bilinear transformations. What are the bilinear transformations that leave the real axis invariant? From Theorem 1-8.3 we know that three points and their distinct image points uniquely determine just such a transformation. Let these three points be 0, 1, and ∞, all real, and let their images be p, q, and r, respectively, all three real and distinct. Then, using the construction in the proof of Theorem 1-8.3, the transformation is $z \to w$, with

$$\lim_{t\to\infty} \frac{(z-1)(0-t)}{(z-0)(1-t)} = \frac{(w-q)(p-r)}{(w-p)(q-r)},$$

or

$$\frac{z-1}{z} = \frac{(w-q)(p-r)}{(w-p)(q-r)}$$

which yields eventually

$$w = \frac{(p-q)rz + p(q-r)}{(p-q)z + (q-r)}.$$

Here the determinant $\delta = (p - q)(q - r)(r - p) \neq 0$. The coefficients of the resulting transformation are all real. Conversely, every bilinear transformation (Eq. 1) with real a, b, c, and d takes every real z into a real image. We have proved:

Theorem 1–8.7. A bilinear transformation (Eq. 1) with real a has the real axis as an invariant line if and only if b, c, and d are also real.

Such a bilinear transformation will be called a *real bilinear transformation*. The provision that a be real avoids the difficulty which arises when in a real bilinear transformation all four coefficients are multiplied by a nonreal number, so that the real transformation seems to be nonreal at the same time. All bilinear transformations may be considered as having $\delta = \pm 1$ because

$$\frac{az + b}{cz + d} = \frac{az/\sqrt{|\delta|} + b/\sqrt{|\delta|}}{cz/\sqrt{|\delta|} + d/\sqrt{|\delta|}},$$

with determinant $\delta/|\delta| = \pm 1$. A *real* bilinear transformation has either $\delta > 0$ or $\delta < 0$, never both. It can be reduced to a transformation with $\delta = 1$ or -1. Let, for instance, $\delta = 1$ and $(az + b)/(cz + d) = (akz + bk)/(ckz + dk)$, with real k. The determinant now is k^2, which cannot assume the value -1.

The statement of Theorem 1–8.7 can be further extended to deal with the images of the half-planes created by the real axis.

Theorem 1–8.8. A real bilinear transformation with positive determinant preserves the upper and the lower half-planes, while a real bilinear transformation with negative determinant interchanges them.

Proof. Let $z = x + yi$, with x and y real, be transformed by Eq. (1), with $\delta \neq 0$. Then

$$z \to \frac{a(x + yi) + b}{c(x + yi) + d} = u + vi,$$

with real u and v. A short computation shows that

$$v = \frac{y\delta}{(cx + d)^2 + c^2y^2}.$$

For a preimage in the upper half-plane, $y > 0$. Then the sign of v is equal to that of δ. For a preimage in the lower half-plane, $y < 0$. The image $u + vi$ lies then in the upper half-plane if $\delta = -1$ and in the lower half-plane if $\delta = 1$.

Theorem 1–8.9. The real bilinear transformations form a subgroup of $\mathcal{B}$.

Proof. The inverse of a real transformation and the product of two real transformations are again real transformations.

Theorem 1–8.10. The real bilinear transformations with positive determinant form a group, $\mathcal{H}_+$.

Proof. By Theorems 1–8.2 and 1–8.9, the product of two such transformations is again real, with $\delta > 0$. The identity has a positive determinant, and hence the inverse of a real bilinear transformation with $\delta > 0$ has a positive determinant and is real.

Bilinear transformations and their groups play an important role in the theory of complex analytic functions. It is not difficult to show that the bilinear transformations preserve angles. Angle-preserving mappings are called *conformal.*

In analogy with what we did in the first sections, it is reasonable to investigate transformations of the type G: $z \to (a\bar{z} + b)/(c\bar{z} + d)$, with real a, b, c, and d, and with determinant $ad - bc = -1$. Obviously, if Eq. (1) holds, then $G = UB$, where U is the reflection in the real axis, $z \to \bar{z}$ (U is the U_1 of Theorem 1–6.1).

Theorem 1–8.11. The set $\mathcal{H}_+$ of all real bilinear transformations with positive determinant and the set $\mathcal{H}_-$ of all transformations $z \to (a\bar{z} + b)/(c\bar{z} + d)$ with real a, b, c, and d and with $ad - bc = -1$, form a group $\mathcal{H}$ in which $\mathcal{H}_+$ is normal of index 2.

Proof. First we intend to show that $\mathcal{H}_-$ is a coset with respect to $\mathcal{H}_+$. Every element H of $\mathcal{H}_-$ can be represented as a product UBF, where B is $z \mapsto -z$ and F is a suitable element of $\mathcal{H}_+$. Namely, from $H = UBF$ we can determine $F = B^{-1}UH$, with F real bilinear and $\delta = (-1)(-1) = 1$. (Note that U commutes with every real bilinear transformation, and $U = U^{-1}$.) On the other hand, every UBF, for all $F \in \mathcal{H}_+$, is in $\mathcal{H}_-$. In other words, all elements of $\mathcal{H}_-$, and only these, are products of one of them, UB, with all elements of $\mathcal{H}_+$. Hence they form a coset. If we can prove that $\mathcal{H}_+$ and $\mathcal{H}_-$ form a group $\mathcal{H}$, we shall have proved that $\mathcal{H}_-$ is the only coset of $\mathcal{H}$ with respect to the subgroup $\mathcal{H}_+$, and thus $\mathcal{H}_+ \triangleleft \mathcal{H}$. Now if C is a real bilinear transformation with $\delta = -1$, the inverse of an element UC is UC^{-1} which is in $\mathcal{H}_-$. The product of two elements of $\mathcal{H}_+$ is in $\mathcal{H}_+$. An element F of $\mathcal{H}_+$ and an element UC of $\mathcal{H}_-$ yield a product $FUC = UFC \in \mathcal{H}_-$, or $UCF \in \mathcal{H}_-$. Two elements of $\mathcal{H}_-$, UC and UC', yield $UCUC' = U^2CC' = CC' \in \mathcal{H}_+$. This completes the proof.

Theorem 1–8.12. The transformations of $\mathcal{H}$ preserve the real cross ratio of point quadruples and take into itself the set of circles and straight lines of the plane.

Proof. In view of Theorem 1–8.6 this has to be proved only for transformations of $\mathcal{H}_-$. Every such transformation is of the form UC with $C \in \mathcal{B}$. Nonreal cross ratios are complex numbers which are taken by U into conjugate complex numbers; real cross ratios are left unchanged by U and hence also by UC. Thus, by Theorem 1–8.5, UC takes circles and lines into circles and lines.

Theorem 1–8.13. The transformations of $\mathcal{H}$ leave the real axis invariant and map the lower and upper half-planes into themselves.

Proof. Again only the transformations of $\mathcal{H}_-$ offer any difficulty. U interchanges the half-planes, and so by Theorem 1-8.8, does a real bilinear transformation B with $\delta = -1$. Therefore UB leaves axis and half-planes invariant.

Theorem 1-8.14. The transformations of $\mathcal{H}$ map onto itself the set $\mathcal{C}$ of circles with real center and of straight lines parallel to the imaginary axis.

Proof. The theorem follows immediately from the fact that the transformations of $\mathcal{H}$ preserve right angles. However, we will give a direct proof. The set $\mathcal{C}$ contains the circles and straight lines, and only those, which are symmetric to the real axis. In other terms, they are invariant under the reflection in the real axis; that is, invariant under the transformation U. For every H in $\mathcal{H}$ we have $HU = UH$. Let z be a point belonging to a circle or a straight line from $\mathcal{C}$. The point zH lies then on the image of this line under H. In order to show that this image belongs to $\mathcal{C}$ we have to prove that it is invariant under U, that is, that zHU is a point of this image. But $zHU = zUH$, and zUH is a point of the image line because zU is a point of the original line.

The group $\mathcal{H}$ will play a major role in the considerations of the next section where significant analogies between $\mathcal{H}$ and the group $\mathfrak{M}$ of isometries will be discussed and applied. An analogy between $\mathcal{H}_+$ and $\mathfrak{M}_+$ will be shown, and this will help the reader to understand why $\mathcal{H}_+$ had to be introduced in a somewhat artificial way; $\mathcal{H}_-$ will be the analog of $\mathfrak{M}_-$.

For further and more advanced discussion of these subjects the reader is referred to H. Schwerdtfeger's, *Geometry of Complex Numbers*, Toronto: University of Toronto Press, 1962.

Exercises

1. Find the bilinear transformations which take
   (a) the three points 0, 1, and i into the points -1, $-i$, and i, respectively
   (b) $\infty \to 0$, $i \to i$, $0 \to \infty$
   (c) $-i \to -1$, $0 \to i$, $i \to 1$.
   (d) Into what curve does the transformation (c) take the imaginary axis?
2. Find the general form of a bilinear transformation with two distinct invariant points w_1 and w_2.
3. Prove: If under a bilinear transformation a circle is carried into a circle, the center of the first circle will in general not be mapped on the center of the image circle.
4. Find the invariant points of

$$\text{(a) } z \to \frac{z - 1 - i}{z + 2} \qquad \text{(b) } z \to \frac{6z - 9}{z}$$

5. The *stereographic* mapping of the points on a sphere onto a plane π is defined as follows. Let N (North Pole) be a fixed point on the sphere and

S (South Pole) be its antipode. The tangent plane to the sphere at S is called π. If P is a point on the sphere, its image P' is the point where the line NP intersects π.

Show that the stereographic mapping yields a one-to-one mapping of the points on the sphere onto the plane π augmented by a point ∞. What is the preimage of ∞? Show also that the great circles through N on the sphere are mapped onto the straight lines of π that pass through S.

6. What characterizes the circles that are taken into straight lines by the transformation Eq. (1)?
7. Find necessary and sufficient conditions for a bilinear transformation to map the right half-plane $\text{Re}(z) > 0$, the left half-plane $\text{Re}(z) < 0$ and the imaginary axis, respectively, onto themselves. Suppose that a in Eq. (1) is real.
8. (a) Show that the cross ratio $(z_1, z_2|z_3, z_4)$ becomes the ratio of two segments when one of the z_j is ∞. (b) What are the possible values of the cross ratio when two of the z_j coincide?
9. A circle circumference is divided by four of its points z_j, with $j = 1, 2, 3, 4$, into four equal parts. Find $(z_1, z_2|z_3, z_4)$ if the four points are cyclically arranged on the circle. Why is the cross ratio independent of the choice of the circle and of the position of the points?
10. If $(z, w|v, u) = c$, find

    (a) $(z, w|u, v)$ (b) $(z, v|w, u)$ (c) $(z, v|u, w)$
    (d) $(w, z|u, v)$ (e) $(v, u|z, w)$

11. Prove: $(z, w|v, u) = -1$ if and only if $(z, w|v, u) = (z, w|u, v)$.
12. Prove: If z, w, v, and u are distinct points, then

$$(w, v|z, u)(v, z|w, u)(z, w|v, u) = -1.$$

13. Check the conformity of the transformation $z \to 1/z$ by finding the angle of intersection of the circles $|z - i| = 1$ and $|z - \frac{1}{2}| = \frac{1}{2}$, and by comparing it with the angle of intersection of their images.

1–9 POINCARÉ'S EUCLIDEAN MODEL OF HYPERBOLIC GEOMETRY

Description of the model. The subject of this section is a new "geometry," involving new "lines" and a new "distance," and we propose to show that our concepts of isometries and groups of isometries can be applied here just as we did in the euclidean plane. The geometry will be called *hyperbolic geometry*, its lines *hyperbolic lines*, and its distance *hyperbolic distance*. Here the term "hyperbolic" does not reflect the existence of any direct connection with hyperbolas. The reason for the name is mainly historical.

The points of the new geometry will be restricted to the upper half-plane alone. The points with positive imaginary part will be called *proper points* (that is, points properly belonging to the geometry), while the real points and the point ∞ will be referred to as *improper*. The remaining points of the euclidean plane will not be mentioned at all. The hyperbolic lines will be all the half-circles and half-lines in the upper half-plane which are perpendicular to the real axis. In the case of the circles this means that we consider only half-circles whose centers are improper points, lying on the real axis. This implies that through every two proper points there exists exactly one hyperbolic line, namely a straight line when the points have equal real parts, and otherwise one of those half-circles. Lines which intersect on the real axis (including the point ∞) will be called *parallel hyperbolic lines* (Fig. 1–24).

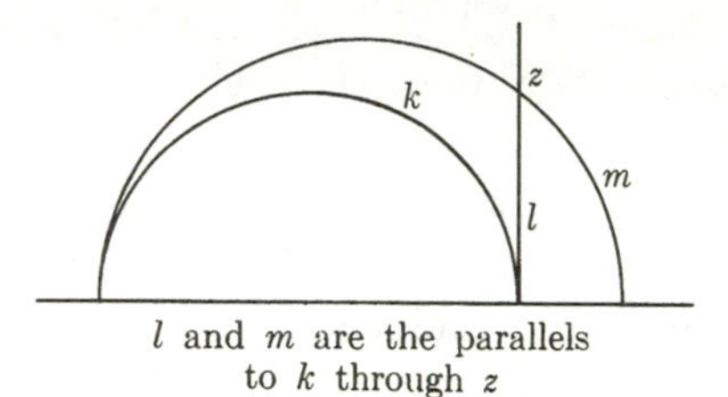

l and m are the parallels to k through z

Figure 1–24

We come now to the definition of a hyperbolic distance between two points. Let z and w be two proper points. There is a unique hyperbolic line passing through them which intersects the real axis in two real points p and q. If the line is straight, either p or q is ∞. The *hyperbolic distance* between z and w is now defined as $d(z, w) = |\log (p, q|z, w)|$, where the basis of the logarithm is arbitrary, but > 1. This implies that $d(z, w)$ is defined up to a constant factor. For working purposes we may assume temporarily that the basis is e.

This definition of hyperbolic distance will strike the reader as farfetched, to say the least. In the following observations and theorems we will try to make it more palatable by showing that it behaves in a reasonable way in many important respects.

Observation 1–9.1. The order of the points p and q was arbitrarily chosen. Does this affect the value of $d(z, w)$? We have

$$|\log (p, q|z, w)| = \left|\log \frac{(p - z)(q - w)}{(p - w)(q - z)}\right| = \left|- \log \frac{(p - w)(q - z)}{(p - z)(q - w)}\right|$$
$$= |\log (q, p|z, w)|.$$

Thus, indeed, it does not matter in which order p and q are used.

Observation 1–9.2. The same argument shows that $d(z, w) = d(w, z)$.

Observation 1–9.3. The hyperbolic distance between a point and itself is 0 because $d(z, z) = |\log 1| = 0$.

Theorem 1–9.4. The logarithm in the distance expression is real.

Proof. The logarithm of a positive real number is real. Thus we want to show that the cross ratio always has a real and positive value. Its reality is assured by Theorem 1-8.5. When the line through z and w is straight we have $p = \infty$, and hence

$$\lim_{p\to\infty} \frac{(p - z)(q - w)}{(p - w)(q - z)} = \lim_{p\to\infty} \frac{(1 - z/p)(q - w)}{(1 - w/p)(q - z)} = \frac{q - w}{q - z},$$

where q, w, and z have equal real parts. Let $q = \alpha$, $w = \alpha + \beta i$, and $z = \alpha + \epsilon i$. Then $(q - w)/(q - z) = \beta/\epsilon$. Both β and ϵ are real and positive, thus β/ϵ is positive. In the second case where p, q, z, and w lie on a half-circle, express the cross ratio in the form $\rho(\cos\phi + i\sin\phi)$, with positive ρ. The argument of $(z - p)/(z - q)$ is (Fig. 1-25)

$$\arg\,(z - p) - \arg\,(z - q) = \alpha - \beta = -\pi/2,$$

and

$$\arg\frac{w - q}{w - p} = \arg\,(w - q) - \arg\,(w - p) = \gamma - \epsilon = \frac{\pi}{2}\cdot$$

Hence $\phi = -\pi/2 + \pi/2 = 0$, and the cross ratio is $\rho(\cos 0 + i\sin 0) = \rho > 0$.

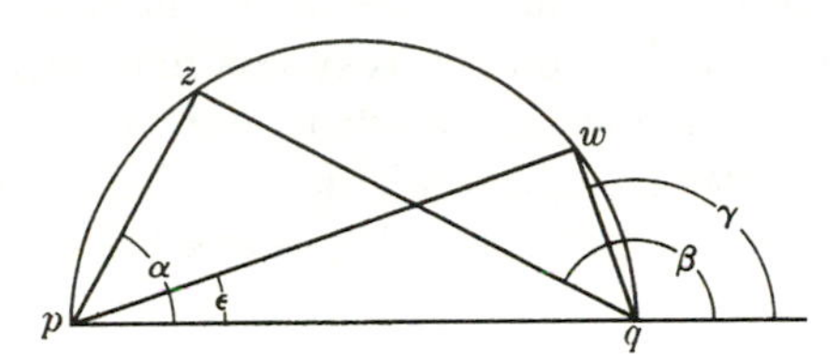

Figure 1-25

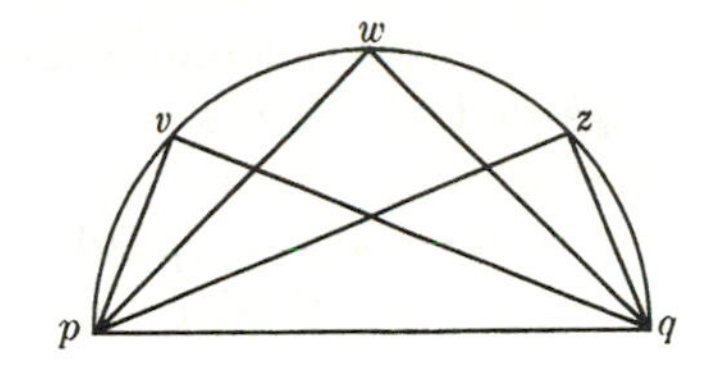

Figure 1-26

Theorem 1-9.5. If z, w, and v are points on a hyperbolic line, with w between v and z, then $d(z, w) + d(w, v) = d(z, v)$.

Proof. Consider first the case of a straight line passing through v, w, and z. Then $v = \alpha + \beta i$, $w = \alpha + \gamma i$, $z = \alpha + \epsilon i$, with $0 < \beta < \gamma < \epsilon$ or $\beta > \gamma > \epsilon > 0$, and

$$d(z, w) + d(w, v) = \left|\log\frac{\epsilon}{\gamma}\right| + \left|\log\frac{\gamma}{\beta}\right|\cdot$$

Now ϵ/γ and γ/β are either both >1 or both <1, and thus

$$d(z, w) + d(w, v) = \left|\log\frac{\epsilon}{\gamma} + \log\frac{\gamma}{\beta}\right| = \left|\log\frac{\epsilon}{\beta}\right| = d(z, v).$$

In the case of v, w, and z lying on a circle (Fig. 1-26), we may assume, without loss of generality, that $\mathrm{Re}(v) < \mathrm{Re}\,(w) < \mathrm{Re}\,(z)$. This implies

$$0 < |v - p| < |w - p| < |z - p|, \qquad \text{and} \qquad |v - q| > |w - q| > |z - q| > 0.$$

Then the real number

$$\frac{(z-p)(w-q)}{(w-p)(z-q)} = \left|\frac{z-p}{w-p}\right| \left|\frac{w-q}{z-q}\right| > 1,$$

because each of the two factors is greater than 1. Similarly we obtain

$$\frac{(w-p)(v-q)}{(v-p)(w-q)} > 1.$$

Hence $\log (p, q|\, z, w)$ and $\log (p, q|\, w, v)$ are both positive, and

$$\begin{aligned} d(z, w) + d(w, v) &= \log \frac{(z-p)(w-q)}{(w-p)(z-q)} + \log \frac{(w-p)(v-q)}{(v-p)(w-q)} \\ &= \log \frac{(z-p)(v-q)}{(z-q)(v-p)} = d(z, v). \end{aligned}$$

Observation 1-9.6. The hyperbolic distance between a fixed point z and another point moving perpendicularly toward the real axis tends toward infinity. This might at first sound unreasonable, but we have to recall that the real axis is the boundary of the plane, and beyond it no more proper points exist. Therefore the improper points on the real axis may well be regarded as "points at infinity." Let us look at this behavior of points z and w in the special case when w moves along a straight line which passes through z and is perpendicular to the real axis, and therefore $z = \alpha + \beta i$, $w = \alpha + \epsilon i$. We have then

$$\begin{aligned} \lim_{\epsilon \to 0} d(z, w) &= \lim_{\epsilon \to 0} \left|\log \frac{z-\alpha}{w-\alpha}\right| = \lim_{\epsilon \to 0} \left|\log \frac{\beta}{\epsilon}\right| \\ &= \log \beta - \lim_{\epsilon \to 0} \log \epsilon = \infty, \end{aligned}$$

because $\lim_{\epsilon \to 0} \log \epsilon = -\infty$.

How do the hyperbolic lines behave with respect to intersection with each other? Since they are either half-circles or straight lines parallel to the imaginary axis, two lines cannot intersect more than once. On the other hand, it is quite possible that two lines do not intersect and yet are not parallel ("parallel" meaning intersecting in a point of the real axis). Thus, in hyperbolic geometry, we encounter a new situation: a pair of lines is intersecting (one common proper point), or parallel (one common improper point) or not intersecting (no common point). The following theorem also indicates a basic difference between euclidean and hyperbolic geometry.

Theorem 1-9.7. For every proper point Z and every hyperbolic line l not passing through Z there exist exactly two hyperbolic lines passing through Z, both hyperbolically parallel to l.

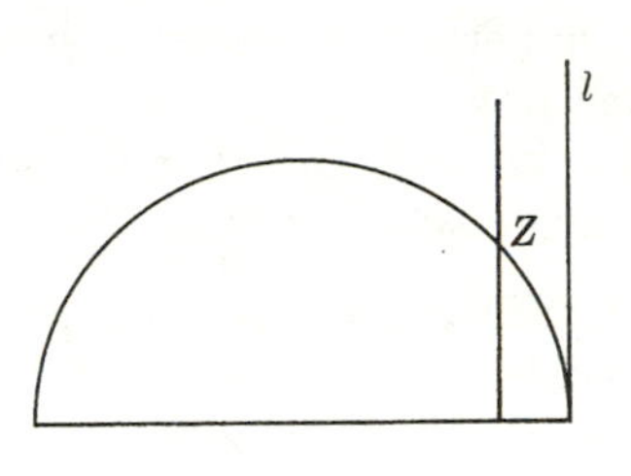

Figure 1–27

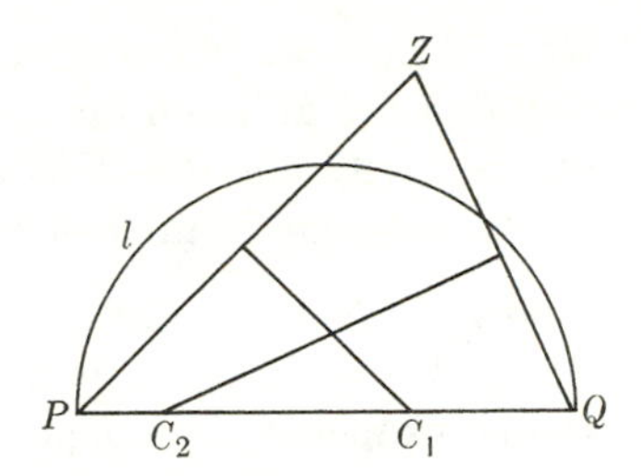

Figure 1–28

Proof (Fig. 1–27). When l is a straight line, the parallel straight line through Z is also a hyperbolic parallel. The second hyperbolic parallel is the half-circle passing through Z and the finite point of intersection of l and the real axis. We now let l be a half-circle intersecting the real axis in P and Q (Fig. 1–28). The perpendicular bisectors of the segments PZ and QZ meet the real axis in C_1 and C_2, respectively. C_1 and C_2 do not coincide because, if they did, then $C_1 = C_2$ would be the center of the half-circle PQ and Z would be on it, a contradiction. The half-circle with C_1 as center, which passes through P, and the half-circle about C_2, which passes through Q, are the hyperbolic parallels of l through Z. If C_1 or C_2 is the point ∞, one of these half-circles becomes a straight line.

Evidently hyperbolic parallelism is not transitive; that is, if $j \| k$ and $k \| l$, then not necessarily $j \| l$.

Hyperbolic isometries. Our next concern will be *hyperbolic isometries*, that is, transformations of the hyperbolic plane which preserve hyperbolic distance. The developments of Section 1–8 will provide us with a group of hyperbolic isometries. However, in a different manner from our procedure in the euclidean plane, here we shall take it for granted and not attempt to prove that there are no hyperbolic isometries other than these.

Theorem 1–9.8. The group $\mathcal{H}$ is a group of hyperbolic isometries.

Proof. By Theorem 1–8.12, the transformations of $\mathcal{H}$ preserve cross ratios of point quadruples and, therefore, hyperbolic distance.

Theorem 1–9.9. The transformations of $\mathcal{H}$ are hyperbolic collineations.

Proof. By Theorem 1–8.14, hyperbolic lines are carried into hyperbolic lines. At first, there might be some doubt whether or not the wrong parts of circles (only half-circles are needed!) appear as images. However, Theorem 1–8.13 assures us that the upper half-plane is preserved.

After having established the analogy between the isometry groups $\mathcal{H}$ and $\mathcal{M}$ it is reasonable to investigate subgroups of $\mathcal{H}$. By Theorem 1–8.11 the transformations from $\mathcal{H}_+$ play the role of the direct hyperbolic isometries, in analogy with $\mathcal{M}_+$, and the opposite hyperbolic isometries $\mathcal{H}_-$ correspond to

$\mathfrak{M}_-$. We will now study invariant points for hyperbolic isometries, just as we did in Sections 1–5 and 1–6 for euclidean isometries.

Let us start with the direct isometries, $\mathfrak{H}_+$. We will find that the situation here is quite different from that encountered in the study of $\mathfrak{M}_+$.

Theorem 1–9.10 (Fig. 1–29). A direct hyperbolic isometry [Eq. (1) of Section 1–8], with $\delta = 1$, which is not the identity, has
(i) one proper invariant point if and only if $|a + d| < 2$,
(ii) one improper invariant point if and only if $|a + d| = 2$,
(iii) two improper invariant points if and only if $|a + d| > 2$.

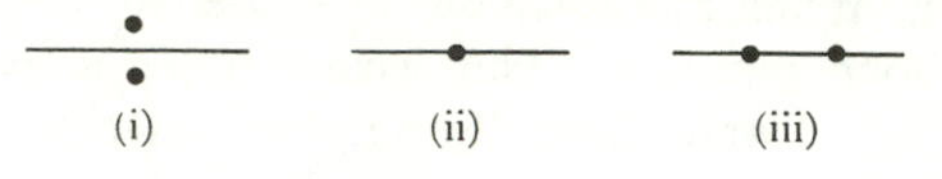

Figure 1–29

Proof. Let us exclude at first the case $c = 0$. For an invariant point w, $w = (aw + b)/(cw + d)$, with $c \neq 0$, we obtain $cw^2 + w(d - a) - b = 0$ and

$$w = \frac{1}{2c}\left[a - d \pm \sqrt{(a - d)^2 + 4bc}\right]. \tag{1}$$

(i) When $(a - d)^2 + 4bc < 0$, there are two solutions, and only one of them lies in the upper half-plane. Then also

$$a^2 + d^2 - 2ad + 4bc = a^2 + d^2 + 2ad - 4,$$

in view of $\delta = 1$. We have, therefore, $(a + d)^2 - 4 < 0$ and $|a + d| < 2$. Conversely, $|a + d| < 2$ implies $(a - d)^2 + 4bc < 0$.

(ii) Now let $|a + d| = 2$. Then $(a + d)^2 = 4$,

$$a^2 + 2ad + d^2 - 4ad + 4bc = 4 - 4\delta = 0,$$

and the quadratic equation has one single solution $w = (a - d)/2c$, which is real. Conversely, $|a + d| = 2$ when this is the only solution.

(iii) The case $|a + d| > 2$ is equivalent to a positive discriminant of Eq. (1). Hence there are two real solutions.

Finally consider $c = 0$. Then $\delta = ad = 1$, and the transformation is $z \to a^2 z + ab$. If $a = \pm 1$, $z \to z \pm b$ is a translation which has ∞ as its only improper invariant point. This belongs to case (ii), and indeed $|a + d|$ has the value 2. If, on the other hand, $a^2 \neq 1$, an invariant point w has $w = a^2 w + ab$, $w = ab/(1 - a^2)$. Another invariant point is ∞, so that we have here two improper invariant points if and only if $a^2 \neq 1$, which is equivalent to

$$|a + 1/a| = |a + d| > 2.$$

The direct hyperbolic isometry with one proper invariant point, as in case (i), is called a *hyperbolic rotation*. This is in accordance with the analogous isometries of the group $\mathfrak{M}_+$. However, in contrast to the situation in the euclidean plane, the cases (ii) and (iii) constitute two different categories of direct hyperbolic isometries without proper invariant points, instead of the one subgroup of translations there.

Let us find canonical forms for the three cases. We treated euclidean rotations by representing them in the form $T_w^{-1}RT_w$, where R was a rotation about 0. In the hyperbolic case 0 has, of course, to be replaced by a suitable proper point. We choose i. If $w = u + vi$ with $v > 0$ is to be the proper invariant point, the hyperbolic isometry $A: z \rightarrow vz + u$ maps i on w, and every hyperbolic rotation can be written as $A^{-1}RA$, where R is a hyperbolic rotation about i. Now the invariant point i of R is given by Eq. (1), if we disregard the minus sign. Then we obtain $a = d$ and $\sqrt{b/c} = i$, which implies $b = -c$. Hence R is of the form $z \rightarrow (az + b)/(-bz + a)$, with the provision $\delta = a^2 + b^2 = 1$.

The case (ii) of Theorem 1–9.10 is called a *parallel displacement*. Here again we introduce a standard parallel displacement, this time with ∞ as invariant point. When $w = \infty$ and $(a - d)^2 + 4bc = 0$, then $c = 0$ and $a = d$, and we obtain this standard form as $T: z \rightarrow z + b$. The parallel displacement, whose invariant point has the real value w, is CTC^{-1}, where C is some hyperbolic isometry mapping w on ∞. One obvious possible choice of C would be $z \rightarrow (-1)/(z - w)$.

The case (iii) is called a *hyperbolic translation*. Here we transfer the two invariant points to 0 and ∞. The result is EDE^{-1}, where E is a hyperbolic isometry taking the real invariant values w_1 and w_2 into 0 and ∞, respectively, and $D: z \rightarrow az$. One possible choice for E would be $z \rightarrow (z - w_1)/(z - w_2)$.

We summarize these results.

Theorem 1–9.11. Each of the types (i), (ii), and (iii) of direct isometries of Theorem 1–9.10 can be transformed by direct hyperbolic isometries into a standard form which is

(i) $z \rightarrow (az + b)/(-bz + a)$, a hyperbolic rotation about i,
(ii) $z \rightarrow z + b$, a parallel displacement with ∞ as invariant point,
(iii) $z \rightarrow az$, a hyperbolic translation with 0 and ∞ as invariant points.

Next we propose to study the invariant points of opposite hyperbolic isometries. As they did in the euclidean plane, they will lead us to reflections. However, hyperbolic lines being euclidean half-circles, it appears obvious that the notion of reflection has to be generalized. Thus we arrive at the *inversion* with respect to a circle.

An invariant point w under an opposite hyperbolic isometry

$$z \rightarrow (a\bar{z} + b)/(c\bar{z} + d)$$

has to satisfy $w = (a\bar{w} + b)/(c\bar{w} + d)$, which becomes $cw\bar{w} + dw - a\bar{w} - b = 0$. Taking conjugates and subtracting, we get $(d + a)(w - \bar{w}) = 0$ as a necessary condition for the existence of an invariant point w. The condition $w - \bar{w} = 0$ yields only real points, which we may disregard as not being proper points of the hyperbolic plane. When, however, $a + d = 0$ is fulfilled, the transformation becomes $H: z \to (a\bar{z} + b)/(c\bar{z} - a)$. An easy computation shows this to be involutory, $H^2 = I$. What are the invariant points w of H now? We have $cw\bar{w} - aw - a\bar{w} - b = 0$, and with $w = x + yi$, $c(x^2 + y^2) - 2ax - b = 0$. If $c = 0$, this is the equation of a straight line $x = -b/2a$, parallel to the y-axis. Thus in the case $c = 0$ we obtained a pointwise invariant hyperbolic line.

Now assume $c \neq 0$. Then the equation is that of a circle whose center is the point a/c. The square of the radius of the circle is

$$\frac{b}{c} + \frac{a^2}{c^2} = \frac{bc + a^2}{c^2} = \frac{1}{c^2} > 0.$$

The upper half of the circle is again a pointwise invariant hyperbolic line.

Again we define a *line reflection* as an isometry that leaves a hyperbolic line pointwise invariant and has no other proper invariant points. Thus we have:

Theorem 1–9.12. An opposite hyperbolic isometry has proper invariant points if and only if it has the form

$$z \to \frac{a\bar{z} + b}{c\bar{z} - a}.$$

In this case it is a reflection in a hyperbolic line.

What are these reflections in hyperbolic lines? In the first case, where $c = 0$, we had a transformation

$$z \to \frac{a\bar{z} + b}{-a} = -\bar{z} - \frac{b}{a},$$

with $a^2 = 1$. Now, $z \to -\bar{z} \pm b$ is a euclidean reflection as well as a hyperbolic reflection in the line $x = \pm b/2$.

The second case is more complicated because there the reflecting line is a circle, and reflections in circles have yet to be introduced geometrically.

Using Theorem 1–8.1, we see that $H: z \to (a\bar{z} + b)/(c\bar{z} - a)$ can be broken up into

$$H = UT_{-a/c}ND_{1/c^2}T_{a/c} = T_{a/c}^{-1}(UND_{1/c^2})T_{a/c},$$

where $N: z \to 1/z$. We will have to investigate the transformation UND_{1/c^2}, which appears here transformed by $T_{a/c}$. The meaning of UND_{1/c^2} is $z \to 1/(c^2\bar{z})$ which by Theorem 1–9.12 is a reflection in the circle about 0 with the radius $1/c$. What is the geometrical significance of this reflection? A point z is mapped into $\bar{z}$ by U. Now $\bar{z}$ is taken by N into

$$\frac{1}{\bar{z}} = \frac{z}{|z|^2} = \frac{z}{|z|} \cdot \frac{1}{|z|}.$$

But this is a point on the same straight line through 0 as z, with absolute value $1/|z|$. The dilatation D_{1/c^2} moves this point along the same straight line through 0 to a point at a distance of $1/(c^2|z|)$ from 0. This whole operation is called an inversion in the circle of radius $1/c$ about 0.

Definition. An *inversion in a circle* of radius r and center C carries a point P into a point Q such that C, P, and Q are collinear and that $|CP|\,|CQ| = r^2$ (Fig. 1-30).

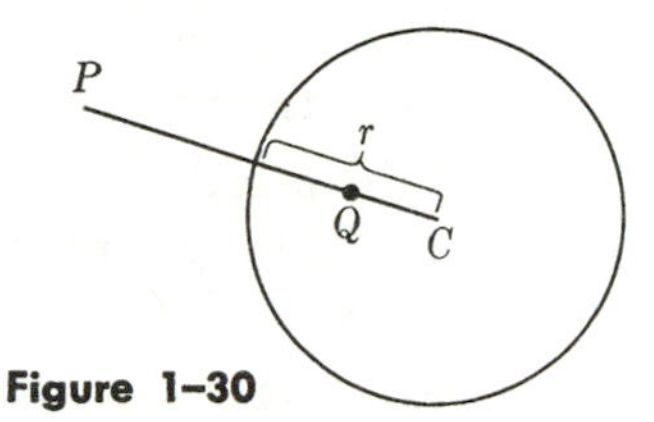

Figure 1-30

Theorem 1-9.13. Every hyperbolic line reflection is either a euclidean reflection in a straight line or an inversion in a circle whose center lies on the real axis.

Proof. We need to show only that $T_{a/c}^{-1}UND_{1/c^2}T_{a/c}$ is an inversion. First $T_{a/c}^{-1}$ takes the reflecting circle into the circle about 0 with the same radius. Then the inversion is performed and finally the circle is translated back. But the translations are euclidean isometries which cannot change the purely geometrical operations of the inversion (purely geometrical in the sense of euclidean geometry, namely preserving euclidean distance). Hence our hyperbolic reflection is an inversion in the circle about a/c with radius $1/c$.

We have thus had a first glimpse of hyperbolic geometry. Actually we have used the euclidean plane in order to construct in it the new geometry. We could have proceeded differently, and later we will meet other realizations of the hyperbolic plane, where the same relations between points, lines, distances, parallels, etc., hold and where in particular the group of isometries is isomorphic to $\mathcal{H}$ and therefore has the same abstract structure of subgroups. However, the lines will not necessarily be euclidean half-circles again, and the hyperbolic plane might well look very different from a euclidean half-plane. Thus what we have had here was only a *model* of a hyperbolic plane. This model was euclidean, and it is called the *conformal model* or *Poincaré's model* of the hyperbolic plane (after the French mathematician Henri Poincaré, 1854–1912). The concept of hyperbolic geometry goes back to the independent simultaneous work of two mathematicians, the Hungarian János Bolyai (1802–1860) and the Russian Nicolai I. Lobachevsky (1793–1856). Hyperbolic geometry is therefore frequently called *Bolyai-Lobachevsky geometry.*

EXERCISES

1. Verify the validity of the following postulates for hyperbolic lines and proper points.
   (a) There is exactly one line passing through two arbitrarily given points.
   (b) There is at most one point of intersection for two arbitrarily given lines.

(c) To a given line there are exactly two parallel lines through a given point not on the line.

2. Find the hyperbolic distances between the vertices of the square, $2i$, i, $i + 1$, $2i + 1$. Which is the "longest," which the "shortest" side?
3. Find $d(z, w)$, where $z = 5i$, $w = 3 + 4i$, and find a point v on the hyperbolic line through z and w such that w is the hyperbolic midpoint of z and v.
4. Find the parallels
   (a) to the imaginary axis through the point $1 + i$,
   (b) to the unit circle through the point $2i$.
5. Assign "hyperbolic coordinates" to all points on the positive imaginary axis in the following way. Let i be the origin and assign to every point yi a coordinate $h(y)$ that is the hyperbolic distance of yi from i. If $y > 1$, let the coordinate be positive, and if $y < 1$, let it be negative. Determine the function $h(y)$. What is the coordinate of $200i$? of i? of $10^{-3}i$?
6. Show that the hyperbolic rotations about i form a group. Is this group isomorphic to the multiplicative group of complex numbers of absolute value 1, like $\mathcal{R}_0$? Show that the transformations depend on one parameter only.
7. Show that the trace of $2i$ under the hyperbolic rotations about i is a (euclidean) circle. Is i its center?
8. Prove: The hyperbolic rotations about a proper point w form a group isomorphic to the group of hyperbolic rotations about i.
9. In the euclidean plane the translations have straight lines as traces. Is there one type of hyperbolic isometry that has hyperbolic lines as traces?
10. What are the involutory hyperbolic isometries?
11. Prove: Every real bilinear transformation is either a similitude or the product of an inversion and a euclidean isometry.
12. Prove: A straight line not through the center C of the reflecting circle is transformed by an inversion into a circle through C.
13. Prove: A straight line through the center C of a reflecting circle is mapped on itself.
14. Prove: The reflecting circle of an inversion is uniquely determined by two given points and their images under the inversion.
15. Prove: An inversion carries parallel straight lines not through the center C of the reflecting circle into two circles which touch each other at C.
16. Prove: An inversion in a circle of increasing radius tends toward a reflection in a straight line.

CHAPTER 2

Affine Geometry and Euclidean Geometry

2–1 FINITE-DIMENSIONAL VECTOR SPACES OVER A FIELD

Skew fields and fields. In contrast to the procedure in Chapter 1, we will now build up our geometries from first principles. For this purpose we need several basic algebraic concepts, which will be given in this section.

A set S, together with two binary operations, $+$ (addition) and $\cdot$ (multiplication), is called a *skew field* if the following postulates hold.

Fd 1. S has at least two elements, 0 and 1.

Fd 2. $(S, +)$ is an abelian group with 0 as identity element.

Fd 3. If S' is the set S with the element 0 excluded, then $(S', \cdot)$ is a group with 1 as identity element.

Fd 4. For all a, b, c in S, $a(b + c) = ab + ac$ and $(b + c)a = ba + ca$ (*distributivity*).

The skew field is called a *field* if, in addition to Fd 1–4, the following postulate holds.

Fd 5. $(S', \cdot)$ is abelian.

All the real numbers obviously form a field, which will be denoted by R. The complex numbers also form a field. Other fields will be discussed in Exercise 14.

Linear equations. Much of the following will be connected with systems of linear equations. A *linear equation* in n unknowns $x_1, \ldots, x_n$ over a field K is of the form

$$c_1x_1 + \cdots + c_nx_n = b$$

with b and the c's elements of K. If $b = 0$, the equation is called *homogeneous*. If several equations are to be considered at the same time, they are called *simultaneous*. An ordered n-tuple $x_1, \ldots, x_n$ of elements of K satisfying an equation or several simultaneous equations is a *solution*. The solution of homogeneous equations with all $x_k = 0$ is the *trivial solution;* any other solution is *nontrivial.*

Theorem 2–1.1. A system of r simultaneous homogeneous linear equations in n unknowns over K always has a nontrivial solution if $r < n$.

Proof. Let the system be

$$c_{11}x_1 + c_{12}x_2 + \cdots + c_{1n}x_n = 0,$$
$$c_{21}x_1 + c_{22}x_2 + \cdots + c_{2n}x_n = 0,$$
$$\vdots$$
$$c_{r1}x_1 + c_{r2}x_2 + \cdots + c_{rn}x_n = 0.$$

We employ induction on r. For $r = 0$ no conditions are imposed on the unknowns, so the theorem is trivially fulfilled. For instance, all x may be $=1$. Assume now that the theorem has been proved for all numbers of equations up to and including $r - 1$. If all coefficients c are zero, the theorem is again trivially true. Let us, therefore, suppose that one of the c's is nonzero. Without loss of generality take $c_{11} \neq 0$, which can always be obtained by renumbering equations or unknowns, a process that does not affect the existence or nonexistence of solutions. For short we denote the left-hand side of the kth equation by L_k, and consider the new system

$$L_k - c_{k1}c_{11}^{-1}L_1 = 0, \qquad \text{with} \qquad k = 2, \ldots, r,$$

consisting of $r - 1$ equations. Since x_1 has been eliminated from all equations, the system now has $n - 1$ unknowns. We have $r < n$ and therefore $r - 1 < n - 1$, and the induction hypothesis yields the existence of a nontrivial solution for $x_2, \ldots, x_n$. By substitution of this solution into $L_1 = 0$ we obtain a value for x_1, as well. Obviously this solution is also a solution of the original system $L_k = 0$, with $k = 1, \ldots, n$, and the theorem is proved.

Basic facts about vector spaces. Next we define a *vector space* V over a field K. Its elements are called *vectors* and are denoted by small boldface letters. They are connected with each other and with the elements of K (called *scalars*) by two binary operations, $+$ (addition) and $\cdot$ (scalar multiplication). The following postulates hold in V.

VS 1. $(V, +)$ is an abelian group with the null vector $\mathbf{0}$ as identity element.

VS 2. Every $r \in K$ and $\mathbf{v} \in V$ determine a unique scalar product $r\mathbf{v} = \mathbf{v}r$ such that for all r and s in K and all $\mathbf{v}$ and $\mathbf{w}$ in V,

VS 2.1. $r(\mathbf{v} + \mathbf{w}) = r\mathbf{v} + r\mathbf{w}$, $(r + s)\mathbf{v} = r\mathbf{v} + s\mathbf{v}$ (*distributivity*),

VS 2.2. $(rs)\mathbf{v} = r(s\mathbf{v})$ (*associativity*),

VS 2.3. $1\mathbf{v} = \mathbf{v}$.

The vector space is called *n-dimensional* ($n > 0$) if, in addition to VS 1 and 2, VS 3 also holds:

VS 3. There exist n vectors $\mathbf{e}_1, \ldots, \mathbf{e}_n$ in V such that every vector $\mathbf{v}$ of V can be represented in a unique way as $\mathbf{v} = x_1\mathbf{e}_1 + \cdots + x_n\mathbf{e}_n$, with all x_i elements of K.

K itself, considered as a vector space, is one-dimensional, and $\mathbf{0}$ may be defined as a zero-dimensional vector space.

The vectors of Section 1-7 formed a two-dimensional vector space over the real field $K = R$. Thus the set $\mathcal{V}$ of vectors in Section 1-7 constitutes a special case of a vector space satisfying the postulates VS 1 through 3.

The n vectors $\mathbf{e}_1, \ldots, \mathbf{e}_n$ mentioned in Axiom VS 3 will be called a *basis* of V.

A *linear combination* of vectors $\mathbf{v}_1, \ldots, \mathbf{v}_r$ is any vector of the form $x_1\mathbf{v}_1 + \cdots + x_r\mathbf{v}_r$ with scalar x's.

Another important definition is the following. A set of nonzero vectors $\mathbf{v}_1, \ldots, \mathbf{v}_r$ is *linearly dependent* if there exist scalars $x_1, \ldots, x_r$, not all zero, such that $x_1\mathbf{v}_1 + \cdots + x_r\mathbf{v}_r = \mathbf{0}$. If all the x are necessarily zero, then the set is said to be *linearly independent.*

If one or more of the vectors are null vectors, the set may be said to be linearly dependent in a trivial way.

Theorem 2-1.2. In a linearly dependent set of nonzero vectors, there is at least one vector which is a linear combination of the rest.

Proof. We have $x_1\mathbf{v}_1 + \cdots + x_r\mathbf{v}_r = \mathbf{0}$. Not all x are zero. Suppose that $x_1 \neq 0$; otherwise renumber. Then

$$\mathbf{v}_1 = -x_1^{-1}x_2\mathbf{v}_2 - \cdots - x_1^{-1}x_r\mathbf{v}_r.$$

The following example should serve as a warning that in a linearly dependent set of vectors not every vector has to be a linear combination of the rest. Let $\mathbf{v}_1 = \mathbf{e}_1 + \mathbf{e}_3$, $\mathbf{v}_2 = 2\mathbf{v}_1$, and $\mathbf{v}_3 = \mathbf{e}_2$. Then $2\mathbf{v}_1 - \mathbf{v}_2 + 0\mathbf{v}_3 = \mathbf{0}$, and the three vectors are linearly dependent. However, $\mathbf{v}_3$ cannot be represented as a linear combination of $\mathbf{v}_1$ and $\mathbf{v}_2$.

Corollary. If two nonzero vectors are linearly dependent, each is a scalar multiple of the other.

Subspaces and bases.

Definition. A subset of a vector space which is itself a vector space is called a *subspace* of the vector space.

If a vector space consists entirely of linear combinations of a set of vectors $\mathbf{v}_1, \ldots, \mathbf{v}_r$, then $\mathbf{v}_1, \ldots, \mathbf{v}_r$ are said to *span* the space.

Theorem 2-1.3. All the linear combinations of a subset $\mathbf{v}_1, \ldots, \mathbf{v}_s$ of an n-dimensional vector space V form a subspace of V.

Proof. To show, according to VS 1, that the linear combinations form an additive abelian group, let us first check closure. Indeed, the sum of two linear combinations is itself a linear combination. Associativity and commutativity hold in V and hence also in every subset of V. The null vector $0\mathbf{v}_1$ is a linear combination and so is the additive inverse of every linear combination. The requirements VS 2 carry over from V.

Theorem 2-1.4. The intersection $W \cap U$ of two subspaces W and U of a vector space V is itself a vector space.

Proof. Every linear combination of elements of $W \cap U$ is in W as well as in U, and therefore in $W \cap U$.

Theorem 2-1.5. In an n-dimensional vector space V, any $n + 1$ or more nonzero vectors are linearly dependent.

Proof. Let $\mathbf{b}_k$, with $k = 1, \ldots, n + 1$, be $n + 1$ nonzero vectors whose representations in terms of the basis $\mathbf{e}_1, \ldots, \mathbf{e}_n$ are

$$\mathbf{b}_k = b_{k1}\mathbf{e}_1 + \cdots + b_{kn}\mathbf{e}_n. \tag{1}$$

In order to check the linear dependence of the vectors $\mathbf{b}_k$, we consider the equation

$$x_1\mathbf{b}_1 + \cdots + x_n\mathbf{b}_n + x_{n+1}\mathbf{b}_{n+1} = \mathbf{0}.$$

Substituting Eq. (1) for the $\mathbf{b}_k$, we obtain

$$(x_1b_{11} + \cdots + x_{n+1}b_{n+1,1})\mathbf{e}_1 + \cdots + (x_1b_{1n} + \cdots + x_{n+1}b_{n+1,n})\mathbf{e}_n = \mathbf{0}.$$

Since the representation of $\mathbf{0}$ is unique, each of the expressions in parentheses has to be 0. This yields a simultaneous system of n homogeneous linear equations in the $n + 1$ unknowns x_k. By Theorem 2-1.1, there is a solution for the x_k, not all of them zero. Hence the $\mathbf{b}_k$ are linearly dependent. The set is still linearly dependent if more vectors are added to it.

Theorem 2-1.6. In an n-dimensional vector space V, any set of n nonzero vectors forms a basis if and only if they are linearly independent.

Proof. To prove that the independent vectors $\mathbf{b}_1, \ldots, \mathbf{b}_n$ form a basis, let $\mathbf{v}$ be arbitrary in V. The set $\{\mathbf{v}, \mathbf{b}_1, \ldots, \mathbf{b}_n\}$ is linearly dependent, in view of Theorem 2-1.5. By Theorem 2-1.2, at least one of them is a linear combination of the rest. If $x_1\mathbf{b}_1 + \cdots + x_n\mathbf{b}_n + x_{n+1}\mathbf{v} = \mathbf{0}$, then not all $x_k = 0$, and $x_{n+1} \neq 0$ because otherwise the $\mathbf{b}_k$ would be linearly dependent. Thus $\mathbf{v} = -x_{n+1}^{-1}(x_1\mathbf{b}_1 + \cdots + x_n\mathbf{b}_n)$ is a linear combination of the $\mathbf{b}_k$. Is this representation unique? Suppose there is another such representation $\mathbf{v} = y_1\mathbf{b}_1 + \cdots + y_n\mathbf{b}_n$. Subtraction yields

$$\mathbf{0} = (y_1 + x_{n+1}^{-1}x_1)\mathbf{b}_1 + \cdots + (y_n + x_{n+1}^{-1}x_n)\mathbf{b}_n,$$

which, in view of the linear independence of the $\mathbf{b}_k$, implies

$$y_k = -x_{n+1}^{-1}x_k, \qquad \text{with} \qquad k = 1, \ldots, n,$$

and the arbitrary vector $\mathbf{v}$ has a unique representation as a linear combination of the $\mathbf{b}_k$. This makes the $\mathbf{b}_k$ a basis.

Now suppose the $\mathbf{b}_k$ to be a basis. Can they be linearly dependent? The unique representation of $\mathbf{0}$ in terms of the basis $\mathbf{b}_k$ is $\mathbf{0} = x_1\mathbf{b}_1 + \cdots + x_n\mathbf{b}_n$. But $0\mathbf{b}_1 + \cdots + 0\mathbf{b}_n = \mathbf{0}$, and hence all $x_k = 0$, and the $\mathbf{b}_k$ are linearly independent.

The concept of an n-dimensional vector space would be meaningless if not all its bases had the same number n of elements. The following theorem removes this doubt.

Theorem 2–1.7. All the bases of a finite-dimensional vector space have the same number of elements.

Proof. Suppose that there are two bases, B with n elements, C with m elements, and $m > n$. The vector space is n-dimensional, and by Theorem 2–1.5, the vectors of C are linearly dependent. Because of the basis C, the space is m-dimensional, and by Theorem 2–1.6, C has to be linearly independent, which leads to a contradiction.

Corollary. No n-dimensional vector space V can be spanned by fewer than n vectors.

The *dimension* of a vector space V is now uniquely defined as the number of vectors in any of its bases or as the maximum number of linearly independent vectors in V. The dimension of V will be denoted by dim V. If W is a subspace of V, then $\dim W \leqq \dim V$ is obvious. An n-dimensional vector space will sometimes be denoted by V^n.

Theorem 2–1.8. In a vector space V^n any set of $r(<n)$ linearly independent vectors can be completed to a basis by adjoining $n - r$ suitable vectors of V^n.

Proof. The r vectors do not span V^n; hence there is a vector in V^n which is linearly independent of the r vectors. Adjoin it and repeat the procedure successively until n linearly independent vectors are obtained. They form a basis of V^n.

The following theorem is useful for determining the dimensions of subspaces.

Theorem 2–1.9. If U and W are two subspaces of V and if $U + W$ is the subspace spanned by U and W, then

$$\dim (U + W) + \dim (U \cap W) = \dim U + \dim W.$$

Proof. Let $U \cap W$ have a basis $\mathbf{b}_1, \ldots, \mathbf{b}_r$. Now $U \cap W$ is a subspace of U as well as of W. By Theorem 2–1.8, the basis of $U \cap W$ can be completed to form a basis of U, say $\mathbf{b}_1, \ldots, \mathbf{b}_r, \mathbf{c}_1, \ldots, \mathbf{c}_s$, and of W, say $\mathbf{b}_1, \ldots, \mathbf{b}_r, \mathbf{d}_1, \ldots, \mathbf{d}_t$. The vectors $\mathbf{b}_1, \ldots, \mathbf{b}_r, \mathbf{c}_1, \ldots, \mathbf{c}_s$, and $\mathbf{d}_1, \ldots, \mathbf{d}_t$ span $U + W$. But they also form a basis of $U + W$, as will be shown in the following. Let

$$x_1\mathbf{b}_1 + \cdots + x_r\mathbf{b}_r = \mathbf{b}, \quad y_1\mathbf{c}_1 + \cdots + y_s\mathbf{c}_s = \mathbf{c}, \quad \text{and} \quad z_1\mathbf{d}_1 + \cdots + z_t\mathbf{d}_t = \mathbf{d}.$$

In order to check their linear dependence, we consider $\mathbf{b} + \mathbf{c} + \mathbf{d} = \mathbf{0}$, or $\mathbf{b} + \mathbf{c} = -\mathbf{d}$. Here $-\mathbf{d} \in W$ and $\mathbf{b} + \mathbf{c} \in U$; hence $-\mathbf{d} \in U \cap W$, and $-\mathbf{d} = w_1\mathbf{b}_1 + \cdots + w_r\mathbf{b}_r$ for some scalars w_k. Now the $\mathbf{d}_k$ and the $\mathbf{b}_k$ are linearly independent and therefore $\mathbf{d} = \mathbf{0}$, and every $z_k = 0$. The same consideration leads to $y_k = 0$, and finally, by substitution, all $x_k = 0$. Hence the $\mathbf{b}_k$, $\mathbf{c}_k$, and $\mathbf{d}_k$ are independent and form a basis of $U + W$.

This basis of $U + W$ has $r + s + t$ elements, and hence $\dim(U + W) = r + s + t$. The theorem then follows from

$$(r + s + t) + r = (r + s) + (r + t).$$

A special n-dimensional vector space is that consisting of all ordered n-tuples of scalars, written in the form $(a_1, \ldots, a_n)$, if addition is defined by

$$(a_1, \ldots, a_n) + (b_1, \ldots, b_n) = (a_1 + b_1, \ldots, a_n + b_n),$$

and scalar multiplication by $c(a_1, \ldots, a_n) = (a_1, \ldots, a_n)c = (ca_1, \ldots, ca_n)$. The null vector is $(0, \ldots, 0)$. The validity of the postulates VS is easily verified. This vector space will be designated by K^n.

A convenient basis of K^n is formed by the vectors

$$\mathbf{g}_1 = (1, 0, \ldots, 0), \qquad \mathbf{g}_2 = (0, 1, \ldots, 0), \ldots, \qquad \mathbf{g}_n = (0, 0, \ldots, 1).$$

Obviously, $(a_1, \ldots, a_n) = \sum_{j=1}^{n} a_j\mathbf{g}_j$.

Let an *isomorphism* between two vector spaces V and W be a bijective mapping α of V on W such that for all $\mathbf{v}_j \in V$ and all scalars c, $(\mathbf{v}_1 + \mathbf{v}_2)\alpha = \mathbf{v}_1\alpha + \mathbf{v}_2\alpha$ and $(c\mathbf{v}_1)\alpha = c(\mathbf{v}_1\alpha)$. Then we have:

Theorem 2–1.10. Every n-dimensional vector space V^n over K is isomorphic to K^n.

Proof. Let $\mathbf{e}_1, \ldots, \mathbf{e}_n$ be an arbitrary basis of V^n. Then let the vector $a_1\mathbf{e}_1 + \cdots + a_n\mathbf{e}_n$ of V^n correspond to the vector $(a_1, \ldots, a_n)$ of K^n. This is a one-to-one correspondence. The sum

$$(a_1\mathbf{e}_1 + \cdots + a_n\mathbf{e}_n) + (b_1\mathbf{e}_1 + \cdots + b_n\mathbf{e}_n) = (a_1 + b_1)\mathbf{e}_1 + \cdots + (a_n + b_n)\mathbf{e}_n$$

in V^n corresponds to $(a_1, \ldots, a_n) + (b_1, \ldots, b_n) = (a_1 + b_1, \ldots, a_n + b_n)$ in K^n, and in the same way the scalar products correspond. Thus the operations are preserved, and the correspondence is indeed an isomorphism. For each choice of basis in V^n there is a unique isomorphism $V^n \leftrightarrow K^n$ of this kind.

These isomorphisms cannot be considered as the same, and hence it is important to note which of the bases of V^n has been involved in the choice of any of these isomorphisms between V^n and K^n.

Exercises

1. Does the system of equations

$$x + y + z = 0, \qquad 2x + y = 0, \qquad 3x + 2y + z = 0,$$

have a nontrivial solution? Explain why.

2. Prove:
(a) $x\mathbf{0} = \mathbf{0}$ where x is real, (b) $0\mathbf{v} = \mathbf{0}$, $\mathbf{v}$ any vector,
(c) $(-1)\mathbf{v} = -\mathbf{v}$, (d) $x\mathbf{v} = \mathbf{0}$ implies $x = 0$ or $\mathbf{v} = \mathbf{0}$.

3. Prove: Two vectors (x_1, x_2) and (y_1, y_2) of K^2 are linearly dependent if and only if $x_1y_2 = x_2y_1$.

4. Which of the following sets of vectors (x, y) form subspaces of R^2, where x and y are reals?
(a) All vectors $(0, y)$.
(b) All vectors $(1, y)$.
(c) All vectors with non-negative x.
(d) All vectors with $3x - 2y = 0$.
(e) All vectors with $3x - 2y = 1$.
(f) All vectors with $xy = 0$.

5. Do the following sets of three vectors form a basis of a three-dimensional vector space over R?
(a) $(1, 0, 0)$, $(0, 1, 0)$, $(0, 0, 1)$
(b) $(1, 1, 1)$, $(2, 0, 1)$, $(-1, 1, 0)$
(c) $(1, 1, 0, 1)$, $(2, 0, 1, 1)$, $(-1, 1, 1, 0)$

6. Show that the vectors $(1, 0, 1, 2)$, $(0, 1, 1, 1)$, $(2, 1, 0, 1)$, and $(1, 2, 0, 0)$ are linearly dependent. Complete them to form a basis of R^4, after discarding one of the vectors.

7. If the basis of a space consists of the vectors $\mathbf{b}_1 = (1, 1, 0)$ and $\mathbf{b}_2 = (2, 0, 1)$, express the following vectors as linear combinations of $\mathbf{b}_1$ and $\mathbf{b}_2$.

(a) $(1, -1, 1)$ (b) $(1, 5, -2)$

Do $(1, 0, 0)$, $(0, 1, 0)$, $(0, 0, 0)$ belong to this space? Assume $K = R$.

8. If U is a subspace of V and if $\mathbf{v} \in V$ but $\mathbf{v} \notin U$, and if W is a subspace spanned by U and a vector $\mathbf{w}$ of V, does $\mathbf{v} \in W$ imply that $\mathbf{w}$ belongs to the space spanned by U and $\mathbf{v}$?

9. Can you name two different two-dimensional subspaces of R^3 which have only the null vector in common? [*Hint:* Use Theorem 2–1.9.] What is the geometric significance of your answer?

10. Let U be the subspace of R^3 consisting of all vectors of the type $(x_1, x_2, 0)$. Let W be the subspace spanned by $(0, 1, 1)$ and $(2, 0, 1)$.
(a) Describe the vectors in $U \cap W$ and in $U + W$.
(b) Find dim $(U \cap W)$ and dim $(U + W)$.

11. Let two subspaces U and W of a vector space have the same dimension.
    (a) Does $U \subseteq W$ imply $U = W$?
    (b) Does $U \cap W \neq O$ imply $U = W$?
12. The subspace U of R^4 is spanned by $(1, 2, -1, 0)$, $(2, -1, 0, 1)$, and $(1, -3, 1, 1)$, and the subspace W of R^4 by $(3, -4, 1, 2)$ and $(1, 0, 0, 0)$. Find the dimensions of U, W, $U + W$, and $U \cap W$.
13. In a V^3 over the complex field, let $\mathbf{a} = \mathbf{e}_1 + i\mathbf{e}_2$ and $\mathbf{b} = (1 + i)\mathbf{e}_2 + (1 - i)\mathbf{e}_3$.
    (a) Compute $(1 + i)\mathbf{a} - 2\mathbf{b}i$ (b) Find $\mathbf{x}$ if $\mathbf{a} + i\mathbf{x} = (1 - i)\mathbf{b}$.
14. Are the following sets fields under the usual addition and multiplication?
    (a) All the rational numbers
    (b) all the integers
    (c) all the numbers of the form $r + s\sqrt{2}$ with rational r and s,
    (d) all the fractions of the form $p/2^q$, with p and q integers.

2-2 LINEAR TRANSFORMATIONS OF VECTOR SPACES

Component transformations and matrices. If in a vector space V^n a basis $E = \{\mathbf{e}_1, \ldots, \mathbf{e}_n\}$ is given, every vector $\mathbf{v}$ in the space can be represented as

$$\mathbf{v} = x_1\mathbf{e}_1 + \cdots + x_n\mathbf{e}_n = \sum_j x_j\mathbf{e}_j.$$

Here and in the following, $\sum_j$ will stand for $\sum_{j=1}^n$, and the j will be omitted even when there is no possible ambiguity. The n real numbers $x_1, \ldots, x_n$ are called the *components* of $\mathbf{v}$ with respect to the basis E. With the basis given, the vector $\mathbf{v}$ is determined by the ordered n-tuple $(x_1, \ldots, x_n)$. With respect to another basis $F = \{\mathbf{f}_1, \ldots, \mathbf{f}_n\}$ of V^n, let the components of $\mathbf{v}$ be $y_1, \ldots, y_n$. We will now find the relations between the x's and the y's.

Since E is a basis of V^n, each $\mathbf{f}_i$ has a unique representation

$$\mathbf{f}_i = \sum_j a_{ij}\mathbf{e}_j = a_{i1}\mathbf{e}_1 + \cdots + a_{in}\mathbf{e}_n. \tag{1}$$

The n^2 scalars a_{ij}, with $i, j = 1, \ldots, n$, determine the basis F once the basis E is fixed. We arrange these numbers in an array

$$\begin{bmatrix} a_{11} & a_{12} \cdots a_{1n} \\ a_{21} & a_{22} \cdots a_{2n} \\ \vdots & \vdots \\ a_{n1} & a_{n2} \cdots a_{nn} \end{bmatrix}. \tag{2}$$

This array of numbers is a *square matrix*. It has n horizontal *rows* and n vertical *columns* and is therefore called an $n \times n$ matrix. A short notation for the matrix is (a_{ij}). The scalars a_{ij} are called the *entries* of the matrix. The rows and columns may be considered vectors of V^n because they are ordered n-tuples of scalars.

If again $\mathbf{v} = \sum_j x_j\mathbf{e}_j$ expresses $\mathbf{v}$ in the basis E, while in the basis F it is expressed as $\mathbf{v} = \sum_i y_i\mathbf{f}_i$, and if, as above, Eq. (1) holds, then we substitute and obtain

$$\mathbf{v} = \sum_i y_i\Big(\sum_j a_{ij}\mathbf{e}_j\Big) = \sum_j\Big(\sum_i a_{ij}y_i\Big)\mathbf{e}_j.$$

When we compare this with $\mathbf{v} = \sum_j x_j\mathbf{e}_j$ and recall that the representation is unique, we obtain $x_j = \sum_i a_{ij}y_i$ or, after interchanging i and j, $x_i = \sum_j a_{ji}y_j$. Comparing this expression with Eq. (1), we see that a_{ji} appears here in the same role as did a_{ij} in Eq. (1). The matrix involved here is, therefore,

$$\begin{bmatrix} a_{11} & a_{21} \cdots a_{n1} \\ a_{12} & a_{22} \cdots a_{n2} \\ \vdots & \vdots \\ a_{1n} & a_{2n} \cdots a_{nn} \end{bmatrix}. \tag{3}$$

The matrix (3) can be obtained from (2) by interchanging rows and columns, such that the entry a_{ij} in the ith row and jth column of (2) appears in (3) in the jth row and the ith column. We call (3) the *transpose* of (2). If (2) is designated A, we denote the transpose (3) by A'. Of course, $(A')' = A$. Accordingly we use the notation $a'_{ij} = a_{ji}$.

We have proved:

Theorem 2–2.1. Let $E = \{\mathbf{e}_1, \ldots, \mathbf{e}_n\}$ and $F = \{\mathbf{f}_1, \ldots, \mathbf{f}_n\}$ be two bases of a vector space V^n such that Eq. (1) holds. If the components of a vector in V^n are $x_1, \ldots, x_n$ with respect to E, and $y_1, \ldots, y_n$ with respect to F, then

$$x_i = \sum_j a'_{ij}y_j. \tag{4}$$

As an example, consider the case $n = 2$. Then $\mathbf{f}_1 = a_{11}\mathbf{e}_1 + a_{12}\mathbf{e}_2$ and $\mathbf{f}_2 = a_{21}\mathbf{e}_1 + a_{22}\mathbf{e}_2$. In F, a given vector has the representation

$$\begin{aligned} y_1\mathbf{f}_1 + y_2\mathbf{f}_2 &= y_1(a_{11}\mathbf{e}_1 + a_{12}\mathbf{e}_2) + y_2(a_{21}\mathbf{e}_1 + a_{22}\mathbf{e}_2) \\ &= (a_{11}y_1 + a_{21}y_2)\mathbf{e}_1 + (a_{12}y_1 + a_{22}y_2)\mathbf{e}_2. \end{aligned}$$

Thus $x_1 = a_{11}y_1 + a_{21}y_2$ and $x_2 = a_{12}y_1 + a_{22}y_2$, or $x_1 = a'_{11}y_1 + a'_{12}y_2$ and $x_2 = a'_{21}y_1 + a'_{22}y_2$, as stated in the theorem.

Warning. The matrix A carried E into F, while A' carried the components of a vector relative to F into those relative to E. Note this reversion of direction!

The reader may have had difficulty understanding the treatment of the summations and in particular the summation subscripts. It will be well to keep in mind two important rules. First, the index, or subscript, over which the summation is performed, is a "dummy subscript," very much like the variable of integration in a definite integral, and may therefore be replaced arbitrarily by any other letter. For instance, $\sum_j a_{ij}\mathbf{e}_j = \sum_k a_{ik}\mathbf{e}_k$ and $\sum_\alpha b_\alpha = \sum_i b_i$. However, the i in the first example is no dummy subscript, and cannot be changed in this fashion.

A second rule is the following. If the summations involve only finite numbers of summands, then the order of two or more summations may be changed without affecting the value. For example,

$$\begin{aligned}\sum_{i=1}^{2}\left(p_i\sum_{j=1}^{3} q_{ij}r_j\right) &= \sum_{i=1}^{2} p_i(q_{i1}r_1 + q_{i2}r_2 + q_{i3}r_3)\\ &= p_1(q_{11}r_1 + q_{12}r_2 + q_{13}r_3) + p_2(q_{21}r_1 + q_{22}r_2 + q_{23}r_3)\\ &= (p_1q_{11}+p_2q_{21})r_1 + (p_1q_{12}+p_2q_{22})r_2 + (p_1q_{13}+p_2q_{23})r_3\\ &= \sum_{j=1}^{3}\left(\sum_{i=1}^{2} p_iq_{ij}\right) r_j.\end{aligned}$$

In Theorem 2–2.1 we established a one-to-one mapping of the set of all n-tuples of components onto itself. Such a mapping is called a *component transformation* in V^n. How do these component transformations multiply?

Let $x_i = \sum_j a'_{ij}y_j$ and $y_j = \sum_k b'_{jk}z_k$ be two component transformations. Then,

$$x_i = \sum_j a'_{ij}\left(\sum_k b'_{jk}z_k\right) = \sum_k\left(\sum_j a'_{ij}b'_{jk}\right) z_k = \sum_k c'_{ik}z_k.$$

The product of the transformations is again a component transformation of the same form, with the matrix $(c'_{ik}) = C'$, where $c'_{ik} = \sum_j a'_{ij}b'_{jk}$.

In order to avoid the cumbrous transpose notation, let us temporarily introduce $\alpha_{ij} = a'_{ij}$, $\beta_{ij} = b'_{ij}$, and $\gamma_{ij} = c'_{ij}$. Then

$$\gamma_{ik} = \sum_j \alpha_{ij}\beta_{jk} = \alpha_{i1}\beta_{1k} + \alpha_{i2}\beta_{2k} + \cdots + \alpha_{in}\beta_{nk}. \tag{5}$$

What, then, is the matrix (γ_{ij})? When we study Eq. (5) we see that the two n-tuples $(\alpha_{i1}, \alpha_{i2}, \ldots, \alpha_{in})$ and $(\beta_{1k}, \beta_{2k}, \ldots, \beta_{nk})$ are *combined*, that is, corresponding components are multiplied and all the resulting products are added. Now the first n-tuple is the ith row of $(\alpha_{ij}) = A'$ and the second is the kth column of $(\beta_{ij}) = B'$. The result is the entry γ_{ik} in the ith row and kth column of the matrix (γ_{ij}), which we designate as the *product* $(\alpha_{ij})(\beta_{ij}) = A'B' = C'$.

Definition. The product MN of two $n \times n$ matrices M and N is the $n \times n$ matrix whose entry in the ith row and kth column is obtained by combination of the ith row of M and the kth column of N.

Thus if A' is the matrix of the component change $y \to x$, and B' is that of $z \to y$, then $A'B'$ is that of $z \to x$.

By Theorem 2–2.1, the corresponding basis transformations have the transposes of A', B', and C', that is, A, B, and C, respectively, as matrices. When we again use a, b, and c instead of α, β, and γ, then Eq. (5) becomes

$$c_{ki} = \sum_j a_{ji} b_{kj} = \sum_j b_{kj} a_{ji},$$

and therefore, in view of the definition of matrix multiplication, $C = BA$, in contrast to $C' = A'B'$. From these two equations we obtain $C' = (BA)' = A'B'$. This situation conforms exactly with the "warning" after Theorem 2–2.1: B' is the matrix of a component transformation $G \to F$, A' is that of $F \to E$, and $A'B' = C'$ is that of $G \to E$, while the respective basis changes act by the rule $A\colon E \to F$, $B\colon F \to G$, and $C = BA\colon E \to G$.

We summarize in two theorems.

Theorem 2–2.2. If M is the matrix of a component transformation L_M, and N is that of another component transformation L_N, then MN is the matrix of $L_M L_N$.

Theorem 2–2.3. $(AB)' = B'A'$.

Definition. The *main diagonal* of an $n \times n$ matrix (a_{ij}) is the ordered set $\{a_{11}, a_{22}, \ldots, a_{nn}\}$.

If to every component transformation there is a corresponding matrix, what is the matrix belonging to the identity transformation?

Theorem 2–2.4. The matrix I_n belonging to the identity transformation has entries 1 in the main diagonal and zeros elsewhere. For every $n \times n$ matrix A, $I_n A = A I_n = A$.

Proof. The transformation belonging to I_n leaves every component unchanged.

The entries of I_n are the so-called *Kronecker deltas*

$$\delta_{ik} = \begin{cases} 0 & \text{if} \quad i \neq k \\ 1 & \text{if} \quad i = k \end{cases}$$

(after the German mathematician, L. Kronecker, 1823–1891).

Under what condition does a matrix belong to a component transformation? The following theorem will answer this question.

Theorem 2-2.5. An $n \times n$ matrix A belongs to a component transformation on V^n if and only if its rows are linearly independent. In this case A has an inverse, A^{-1}.

Proof. Let E and F and the matrix A be defined as in Theorem 2-2.1, E being a basis, while the basis property of the vector n-tuple F is to be tested. Consider

$$\sum_i y_i \mathbf{f}_i = \mathbf{0}. \tag{6}$$

This is equivalent to $\sum_i y_i(\sum_j a_{ij}\mathbf{e}_j) = \mathbf{0}$ or $\sum_j (\sum_i y_i a_{ij})\mathbf{e}_j = \mathbf{0}$. In view of the linear independence of E, this holds if and only if

$$\sum_i y_i a_{ij} = 0, \qquad \text{for} \qquad j = 1, \ldots, n. \tag{7}$$

Now assume F to be a basis and therefore linearly independent. Then Eq. (6) implies that all $y_i = 0$, and since Eq. (7) is equivalent to Eq. (6), the rows of A are linearly independent.

If, on the other hand, the rows of A are linearly independent, then the vanishing of all y_i follows from Eq. (7), and consequently F is linearly independent and hence a basis.

Since every transformation, as a bijective mapping, has an inverse, a component transformation L_A, with matrix A, has an inverse which is again a component transformation. Like every component transformation, it is represented by a matrix A^{-1} with the property $AA^{-1} = A^{-1}A = I_n$. The matrix A^{-1} will be called the *inverse matrix* of A. This completes the proof.

By means of the inverse matrix we may reword Theorem 2-2.1 to say: If a matrix A carries a basis E into a basis F, then the components with respect to E are taken into those with respect to F by the matrix $(A')^{-1}$, which equals $(A^{-1})'$ (cf. Exercise 5).

Definition. A matrix whose rows are linearly independent is called *nonsingular;* otherwise it is said to be *singular*.

The component transformations have an important property; they are *linear*. By this we mean that for all scalars c and all vectors $\mathbf{v}$ and $\mathbf{w}$, every component transformation satisfies

Ln 1. $(\mathbf{v} + \mathbf{w})L = \mathbf{v}L + \mathbf{w}L$.
Ln 2. $(c\mathbf{v})L = c(\mathbf{v}L)$.

Clearly, for scalar c and d, Ln 1 and Ln 2 hold if and only if

$$(c\mathbf{v} + d\mathbf{w})L = c(\mathbf{v}L) + d(\mathbf{w}L).$$

Another consequence of Ln 1 and Ln 2 is $(\mathbf{0})L = \mathbf{0}$.

Linear transformations. In the first chapter we saw that in some instances transformations of the plane and coordinate transformations were described by the same formulas. It is, therefore, reasonable to deal here with certain transformations of vector spaces and to compare them with component transformations.

We will study only *linear transformations* of V^n, that is, those which satisfy Ln 1 and Ln 2. Let L be such a transformation. We choose a basis of V^n, say, $\mathbf{e}_1, \ldots, \mathbf{e}_n$, and represent an arbitrary $\mathbf{v} \in V^n$ as $\mathbf{v} = x_1\mathbf{e}_1 + \cdots + x_n\mathbf{e}_n$. Then, by Ln 1 and Ln 2,

$$\mathbf{v}L = (x_1\mathbf{e}_1)L + \cdots + (x_n\mathbf{e}_n)L = x_1\mathbf{e}_1L + \cdots + x_n\mathbf{e}_nL.$$

Now, for every $k = 1, \ldots, n$ there is a unique representation $\mathbf{e}_kL = a_{k1}\mathbf{e}_1 + \ldots + a_{kn}\mathbf{e}_n$ and thus, by substitution,

$$\mathbf{v}L = (a'_{11}x_1 + \ldots + a'_{1n}x_n)\mathbf{e}_1 + \ldots + (a'_{n1}x_1 + \ldots + a'_{nn}x_n)\mathbf{e}_n$$

$$= \sum_i \Big(\sum_j a'_{ij}x_j\Big)\mathbf{e}_i.$$

If y_i are the components of $\mathbf{v}L$, then

$$y_i = \sum_j a'_{ij}x_j, \tag{8}$$

which we shall express by saying that the $n \times n$ matrix $A = (a_{ij})$ carries the components of a vector into the components of the transformed vector. We can show that this matrix has to be nonsingular. For, suppose it to be singular. Then its rows are linearly dependent, and by the definition of the matrix, the $\mathbf{e}_kL$ are linearly dependent. Then there exist scalars $c_1, \ldots, c_n$, not all zero, such that $c_1\mathbf{e}_1L + \cdots + c_n\mathbf{e}_nL = \mathbf{0}$, or, by Ln 1, $(c_1\mathbf{e}_1 + \cdots + c_n\mathbf{e}_n)L = \mathbf{0}$. Since L is a transformation, it is one-to-one, and in view of $\mathbf{0}L = \mathbf{0}$, necessarily $c_1\mathbf{e}_1 + \cdots + c_n\mathbf{e}_n = \mathbf{0}$. However, this is a contradiction to the $\mathbf{e}$'s forming a basis. Thus the nonsingularity of A is proved.

Conversely, every nonsingular $n \times n$ matrix carries a basis into a basis, and we may state:

Theorem 2–2.6. The linear transformations of V^n are performed according to the rule given by Eq. (8), with a nonsingular matrix (a_{ij}).

Corollary. If two bases of V^n are given, there exists exactly one linear transformation mapping the first basis on the second basis.

Theorem 2–2.7. All linear transformations of V^n form a group.

Proof. The product of two linear transformations L_1 and L_2 is a linear transformation because

$$(c\mathbf{v} + d\mathbf{w})L_1L_2 = [c(\mathbf{v}L_1) + d(\mathbf{w}L_1)]L_2 = c(\mathbf{v}L_1L_2) + d(\mathbf{w}L_1L_2),$$

which implies that Ln 1 and Ln 2 hold for L_1L_2 if they hold for both L_1 and L_2. The inverse of a linear transformation L is linear, because from

$$(\mathbf{v}L^{-1} + \mathbf{w}L^{-1})L = \mathbf{v} + \mathbf{w}$$

it follows that $\mathbf{v}L^{-1} + \mathbf{w}L^{-1} = (\mathbf{v} + \mathbf{w})L^{-1}$, and the relation $[c(\mathbf{v}L^{-1})]L = c(\mathbf{v}L^{-1}L) = c\mathbf{v}$ implies that $c(\mathbf{v}L^{-1}) = (c\mathbf{v})L^{-1}$. The remaining group postulates are trivially satisfied.

The group of all linear transformations of V^n is called the n-dimensional *(general) linear group*, $\mathcal{GL}^n$.

Corollary. The set of all nonsingular $n \times n$ matrices forms a multiplicative group which is isomorphic to $\mathcal{GL}^n$.

Moreover we note:

Theorem 2–2.8. A linear transformation maps m-dimensional subspaces of V^n onto m-dimensional subspaces of V^n.

Proof. The transformation L maps a basis $\mathbf{e}_1, \ldots, \mathbf{e}_m$ of an m-dimensional space S onto m vectors $\mathbf{e}_jL, j = 1, \ldots, m$. If the $\mathbf{e}_jL$ are linearly dependent then there exist scalars c_j, not all of them zero, such that

$$c_1\mathbf{e}_1L + \cdots + c_m\mathbf{e}_mL = \mathbf{0}.$$

Now apply L^{-1} to this equation and obtain

$$c_1\mathbf{e}_1 + \cdots + c_m\mathbf{e}_m = \mathbf{0},$$

a contradiction. Hence the vectors $\mathbf{e}_jL$ are linearly independent. Every vector of S is a linear combination of the $\mathbf{e}_j$ and goes under L into a linear combination of the $\mathbf{e}_jL$ which are, therefore, a basis for the image space of dimension m.

Exercises

1. Find the matrix of a linear transformation which carries the subspace of R^3 generated by (1, 0, 3) and (2, 1, 0) into the subspace of vectors of the type $(x_1, x_2, 0)$.
2. In R^3 find the components of the unit vectors $\mathbf{f}_1 = (1, 0, 0)$, $\mathbf{f}_2 = (0, 1, 0)$, and $\mathbf{f}_3 = (0, 0, 1)$ with respect to the basis $\mathbf{e}_1 = (0, 1, 1)$, $\mathbf{e}_2 = (1, 0, 1)$, and $\mathbf{e}_3 = (1, 1, 0)$. What is the matrix that represents this transformation? Write down the component transformation.
3. Let $\mathbf{e}_1 = (1, 2)$, $\mathbf{e}_2 = (-1, 1)$, $\mathbf{f}_1 = (3, 3)$, and $\mathbf{f}_2 = (-2, -4)$. Find
   (a) the matrix (a_{ij}) which belongs to the basis change $\{\mathbf{e}_1, \mathbf{e}_2\} \to \{\mathbf{f}_1, \mathbf{f}_2\}$
   (b) the corresponding component transformation.
   (c) Verify your result by drawing a figure and testing the vector $\mathbf{f}_1 - \mathbf{f}_2$.

4. Let
$$A = \begin{bmatrix} 1 & 0 \\ 2 & 0 \end{bmatrix}, \quad B = \begin{bmatrix} 0 & 1 \\ 1 & 0 \end{bmatrix}, \quad C = \begin{bmatrix} 0 & 0 \\ 2 & 1 \end{bmatrix}.$$
Find
(a) A^2 (b) A^3 (c) AC (d) AB (e) BA (f) B^2 (g) B^2C

5. Prove: $(A')^{-1} = (A^{-1})'$.

6. What characterizes the 2×2 matrices
$$\begin{bmatrix} a & b \\ c & d \end{bmatrix}$$
which commute with all 2×2 matrices, that is,
$$B\begin{bmatrix} a & b \\ c & d \end{bmatrix} = \begin{bmatrix} a & b \\ c & d \end{bmatrix}B$$
for all B?

7. If
$$A = \begin{bmatrix} a & b & c \\ a & b & c \\ -a & -b & -c \end{bmatrix},$$
find A^n for each positive integer n if (a) $a + b - c = 1$, (b) $a + b - c = 0$.

8. Prove: The matrix $\begin{bmatrix} a & b \\ c & d \end{bmatrix}$ is nonsingular if and only if $ad - bc \neq 0$.

9. For what k is the matrix
$$\begin{bmatrix} 1 & 4 & 1 \\ 2 & 2 & -4 \\ 3 & 5 & k \end{bmatrix}$$
singular, and for what k is it nonsingular?

10. Find A^{-1} and B^{-1} if
$$A = \begin{bmatrix} 1 & 0 & 0 \\ 0 & 0 & 1 \\ 0 & 1 & 0 \end{bmatrix} \quad \text{and} \quad B = \begin{bmatrix} 1 & 0 & k \\ 0 & 1 & 0 \\ 0 & 0 & 1 \end{bmatrix}.$$

11. Find the inverses of the matrices
$$A = \begin{bmatrix} x & 0 & 0 \\ 0 & y & 0 \\ 0 & 0 & z \end{bmatrix} \quad \text{and} \quad B = \begin{bmatrix} a & b \\ c & d \end{bmatrix},$$
if $ad - bc = 1$, and if x, y, and z are nonzero.

12. If an *automorphism* of a vector space is an isomorphism of the space with itself, prove that every element of $\mathcal{GL}^n$ is an automorphism of V^n.
13. In the following, vectors of R^2 and R^3 and their images under linear mappings are given. Determine which mappings are linear transformations and find their matrices.
    (a) $(1, 1) \to (2, -1)$, $(2, 0) \to (-4, 2)$,
    (b) $(1, 2) \to (2, 1)$, $(0, 1) \to (1, 0)$,
    (c) $(1, 0, 0) \to (0, 2, 0)$, $(0, 1, 0) \to (3, 0, 0)$, $(0, 0, 1) \to (1, 0, 0)$.

2-3 LINEAR TRANSFORMATIONS AND DETERMINANTS. DUAL VECTOR SPACES

Determinants. A most efficient tool in dealing with actual computational problems concerning matrices is the determinant. It is assumed that most readers have at least a working knowledge of determinants, and therefore the following treatment will be something of a short review.

A *determinant* is a function of an $n \times n$ matrix. The entries of the matrix as well as the value of the function are elements of a field K. If

$$A = \begin{bmatrix} a_{11} & \cdots & a_{1n} \\ \vdots & & \vdots \\ a_{n1} & \cdots & a_{nn} \end{bmatrix},$$

then its determinant will be denoted by

$$\begin{vmatrix} a_{11} & \cdots & a_{1n} \\ \vdots & & \vdots \\ a_{n1} & \cdots & a_{nn} \end{vmatrix},$$

or $\det A$ or $\det (a_{ik})$.

The following is an inductive definition of a determinant. The determinant of the 1×1 matrix (a) is a, by definition. After the determinant of an $(n - 1) \times (n - 1)$ matrix has been defined, the determinant of an $n \times n$ matrix is defined as

$$\det A = \begin{vmatrix} a_{11} & \cdots & a_{1n} \\ \vdots & & \vdots \\ a_{n1} & \cdots & a_{nn} \end{vmatrix} = \sum_k (-1)^{k+1} a_{1k} \alpha_{1k},$$

where α_{ik} is the determinant of the $(n - 1) \times (n - 1)$ matrix which is obtained from A by deleting the ith row and the kth column.

For $n = 2$, this definition yields $\det A = a_{11}a_{22} - a_{12}a_{21}$, and for $n = 3$,

$$\det A = a_{11}(a_{22}a_{33} - a_{23}a_{32}) - a_{12}(a_{21}a_{33} - a_{23}a_{31}) + a_{13}(a_{21}a_{32} - a_{22}a_{31}).$$

Another, geometrically more meaningful, definition will be discussed later, in Section 2-9.

The following fundamental rules for computations with determinants will not be proved here. Those readers who are not familiar with these properties may verify them for the cases $n = 2$ and $n = 3$.

Dt 1. For an $n \times n$ matrix A, $\det A = \det A'$.

Dt 2. If a matrix B is obtained from A by interchange of two of its rows or by the interchange of two of its columns, then $\det B = -\det A$.

Dt 3. If two rows in A are identical, then $\det A = 0$. The same applies to two identical columns.

Dt 4. If a row or a column consists of zeros only, then $\det A = 0$.

Dt 5. If, for $A = (a_{ik})$, all $a_{ik} = 0$ whenever $i \neq k$, then

$$\det A = a_{11}a_{22}\ldots a_{nn}.$$

In particular, $\det I_n = 1$.

Dt 6. If all entries a_{ik} of one row or one column are changed to ca_{ik} (c scalar), then the value of $\det A$ changes to $c \det A$.

Dt 7. If both $\mathbf{v}$ and $\mathbf{w}$ are rows (or columns) of A, and if $\mathbf{w}$ is replaced by $\mathbf{w} + c\mathbf{v}$ (c any scalar), then the value of $\det A$ is unchanged.

Dt 8. If A and B are $n \times n$ matrices, then $\det A \det B = \det AB$.

Dt 9. If α_{ik} is the same as in the definition of the determinant, then

$$\sum_k (-1)^{k+1} a_{ik}\alpha_{jk} = \delta_{ij} \det A,$$

where δ_{ij} is the Kronecker delta.

Applications of determinants. The usefulness of determinants for matrices becomes evident in

Theorem 2–3.1. An $n \times n$ matrix is nonsingular if and only if its determinant is nonzero.

Proof. If the matrix is singular, its rows are linearly dependent. Then, by Theorem 2–1.2, there is at least one row vector $\mathbf{v}$ that is a linear combination of the remaining column vectors. Subtract this linear combination from $\mathbf{v}$ by repeated use of Dt 7, then $\mathbf{v}$ is replaced by $\mathbf{0}$. By Dt 4, the determinant is then 0.

Now assume $\det A = \det (a_{ik}) = 0$. Then, among all the matrices obtained by using any number of the rows of A and the same number of its columns, choose a nonsingular $r \times r$ matrix B; r is determined so that all such $s \times s$ matrices with $r < s$ are singular. Obviously $0 < r < n$; the case $r = 0$ would occur only if A contained nothing but zeros. Without loss of generality the columns and rows of A may be renumbered so that

$$B = \begin{bmatrix} a_{11} & \cdots & a_{1r} \\ \vdots & & \vdots \\ a_{r1} & & a_{rr} \end{bmatrix}.$$

Let C be the singular $(r+1) \times (r+1)$ matrix

$$\begin{bmatrix} a_{11} & \cdots & a_{1,r+1} \\ \vdots & & \vdots \\ a_{r+1,1} & \cdots & a_{r+1,r+1} \end{bmatrix},$$

and α_{ik} the determinant of the matrix obtained from C by deleting the ith row and the kth column. Then $\alpha_{r+1,r+1} = \det B \neq 0$. Now multiply the first row of A with $\alpha_{1,r+1}$, its second row with $-\alpha_{2,r+1}, \ldots$, its $(r+1)$-row with $(-1)^{r+2}$, and sum up. Repeated use of Dt 9, and of the fact that all $(r+1) \times (r+1)$ matrices in A are singular, yields the sum zero. Since at least one of the factors was nonzero, the linear dependence of the rows of A has been established.

Corollary 1. A matrix is nonsingular if and only if its columns are linearly independent.

Proof follows from Dt 1.

Corollary 2. The component transformations of V^n form a group which is isomorphic to $\mathcal{GL}^n$.

Proof. The nonsingular $n \times n$ matrices A form $\mathcal{GL}^n$. The component transformations have matrices $(A')^{-1}$, which are also nonsingular, in view of Corollary 1.

In "ordered" fields, where positiveness is defined, we have

Theorem 2-3.2. The linear transformations whose matrices have positive determinants form a subgroup $\mathcal{GL}^n_+$ of $\mathcal{GL}^n$. The index of $\mathcal{GL}^n_+$ in $\mathcal{GL}^n$ is 2, and $\mathcal{GL}^n_+ \triangleleft \mathcal{GL}^n$. The elements of $\mathcal{GL}^n_+$ are called *direct* linear transformations, the other elements of $\mathcal{GL}^n$ are *opposite*.

Proof. If L_A has matrix A, L_B matrix B, $\det A > 0$ and $\det B > 0$, then $L_A L_B$ has matrix AB, with $\det AB = \det A \det B > 0$. The identity transformation is direct, $\det I = 1 > 0$. The rest of the proof follows familiar lines.

If we generalize the matrix concept to cover also rectangular $m \times n$ matrices with m rows and n columns, with $m \neq n$, then a vector may be considered as a $1 \times n$ or an $n \times 1$ matrix. We will, in general, prefer the first way, that is, our vectors will be row vectors.

The rule by which matrices A and B were multiplied will be extended to rectangular matrices for the special case where A has just as many columns as B has rows. If A is $m \times n$ and B is $n \times p$, then AB is $m \times p$. On the other hand, BA in general would not exist, unless $p = m$.

This yields a convenient way of working with linear transformations. If $\mathbf{v} = (y_1, \ldots, y_n)$ is a vector, or $1 \times n$ matrix, and $A = (a_{ik})$ the $n \times n$ matrix of the linear transformation L_A, then the vector $\mathbf{v}T_A$ is the vector, or $1 \times n$ matrix $\mathbf{v}A$, where the ith component or entry x_i is the combination of the vector $\mathbf{v}$ and the ith column of A, that is,

$$x_i = y_1 a_{1i} + \cdots + y_n a_{ni} = \sum_j a_{ji} y_j = \sum_j a'_{ij} y_j,$$

in accordance with Eq. (4) of Section 2–2.

This notation will now be applied to the following question, which arises frequently in work with transformations. A linear transformation L is represented by an $n \times n$ matrix relative to a basis E of V^n. What is the matrix belonging to the same L when a different basis F of V^n is used? The answer appears in the following theorem.

Theorem 2–3.3. If a linear transformation L on V^n is represented by the matrix A relative to a basis E, and if the matrix C' takes E into a basis F, then the matrix representing L with respect to F is CAC^{-1}.

Proof. Let X be a vector (a $1 \times n$ matrix) and $XA = Y$. By Theorem 2–2.1, the component transformation resulting from the basis change has the matrix C^{-1}. Thus we have to compare the vectors XC^{-1} and YC^{-1}. Now

$$YC^{-1} = XAC^{-1} = XC^{-1}(CAC^{-1}),$$

which proves the theorem.

Another application is the solution of simultaneous systems of n linear equations with n unknowns x_k, with $k = 1, \ldots, n$. Let the equations be $XA = B$, where $X = (x_1, \ldots, x_n)$, $B = (b_1, \ldots, b_n)$, and $A = (a_{ij})$, a nonsingular $n \times n$ matrix. To solve these equations we multiply the matrix equation $XA = B$ on the right with A^{-1} and obtain uniquely $X = BA^{-1}$. We have therefore:

Theorem 2–3.4. A system $XA = B$ of n simultaneous linear equations with n unknowns has a unique solution for X if the matrix A is nonsingular.

Corollary. A simultaneous system $XA = \mathbf{0}$ of n homogeneous linear equations with n unknowns has a nontrivial solution if and only if $\det A = 0$.

Proof. If the system has a nontrivial solution, then the solution $X = \mathbf{0}$ is not unique, and by the Theorem the matrix has to be singular. Conversely, if the matrix is singular, its rows are linearly dependent. One of the rows can be expressed as a linear combination of the remaining rows. The equation in this row does not provide any information that is not contained in the other rows and is, therefore, redundant. Thus we have a homogeneous system with fewer equations than unknowns, a system which, by Theorem 2–1.1, has nontrivial solutions.

After having studied the linear transformations of V^n, we shall give a very short account of linear functionals over V^n.

Dual vector spaces. A *functional* over V^n is a one-valued mapping ϕ of V^n into the field K, satisfying the linearity requirements,

Ln 1. $(\mathbf{v} + \mathbf{w})\phi = \mathbf{v}\phi + \mathbf{w}\phi$,
Ln 2. $(c\mathbf{v})\phi = c(\mathbf{v}\phi)$,

for all $\mathbf{v}$ and $\mathbf{w} \in V^n$, and for all scalars c. Clearly, for scalars c and d,

$$(c\mathbf{v} + d\mathbf{w})\phi = c(\mathbf{v}\phi) + d(\mathbf{w}\phi),$$

and $\mathbf{0}\phi = 0$.

Theorem 2–3.5. Let $\mathbf{e}_k$, with $k = 1, \ldots, n$, be a basis of V^n and let b_k be any n scalars. Then there is exactly one functional ϕ with $\mathbf{e}_k\phi = b_k$ for all k.

Proof. For every $\mathbf{v} \in V^n$, with $\mathbf{v} = x_1\mathbf{e}_1 + \cdots + x_n\mathbf{e}_n$, we define $\mathbf{v}\phi = x_1b_1 + \cdots + x_nb_n$. This implies $\mathbf{e}_k\phi = b_k$. Moreover, if $\mathbf{w} = \sum y_k\mathbf{e}_k$, then

$$(\mathbf{v} + \mathbf{w})\phi = x_1b_1 + \cdots + x_nb_n + y_1b_1 + \cdots + y_nb_n = \mathbf{v}\phi + \mathbf{w}\phi,$$

and

$$(c\mathbf{v})\phi = cx_1b_1 + \cdots + cx_nb_n = c(\mathbf{v}\phi),$$

and ϕ is a functional. If, on the other hand, ϕ is any functional with $\mathbf{e}_k\phi = b_k$, then $\mathbf{v}\phi = x_1b_1 + \cdots + x_nb_n$ follows from Ln 1 and Ln 2.

Definition. The *sum* $\phi + \theta$ *of two functionals* ϕ and θ on V is defined by

$$\mathbf{v}(\phi + \theta) = \mathbf{v}\phi + \mathbf{v}\theta,$$

for all $\mathbf{v} \in V$. For any scalar c, the functional $\phi c = c\phi$ is defined by

$$\mathbf{v}(\phi c) = (\mathbf{v}\phi)c.$$

From these definitions it follows immediately that $\phi + \theta$ and ϕc are functionals on V, and that consequently the functionals form a vector space. This space will be called the *dual space* of V, denoted V^*. Its elements will be designated by small Greek letters, its zero by $\mathbf{o}$.

The following theorem provides a method of finding one of the bases of V^*.

Theorem 2–3.6. If V is n-dimensional, so is V^*. If $\mathbf{e}_k$, with $k = 1, \ldots, n$, form a basis of V and the functionals ϵ_k are determined by $\mathbf{e}_j\epsilon_k = \delta_{jk}$, then the ϵ_k form a basis of V^*.

Proof. The ϵ_k are uniquely defined, in view of Theorem 2–3.5. For any n given scalars $y_1, \ldots, y_n$, $\phi = \epsilon_1y_1 + \cdots + \epsilon_ny_n$ is a functional satisfying $\mathbf{e}_k\phi = y_k$, and every functional on V can be represented in this form. Thus the ϵ_k span V^*. If we could prove them to be also linearly independent, they would form a basis. Now, for the basis vectors $\mathbf{e}_k$, we have

$$\mathbf{e}_k\phi = \mathbf{e}_k\epsilon_1y_1 + \cdots + \mathbf{e}_k\epsilon_ny_n = y_k.$$

If $\phi = \epsilon_1 y_1 + \cdots + \epsilon_n y_n = o$, then $\mathbf{v}\phi = 0$ for all $\mathbf{v}$, and hence also $\mathbf{e}_i\phi = 0$ for each i, and thus all $y_k = 0$. Therefore the ϵ_k are linearly independent and form a basis of V^*, called the *cobasis* of the $\mathbf{e}_k$ basis.

Corollary. V and V^* are isomorphic.

Proof. The isomorphic mapping is $\sum x_j \mathbf{e}_j \to \sum \epsilon_j x_j$. However, this isomorphism is not unique since it depends on the choice of the basis $\mathbf{e}_k$.

An interesting result follows from Theorem 2–3.6. For every $\mathbf{v} = x_1\mathbf{e}_1 + \cdots + x_n\mathbf{e}_n$, we get $\mathbf{v}\epsilon_i = x_i$ for all $i = 1, \ldots, n$. Thus the components of any vector $\mathbf{v}$ turn out to be the values of the n cobasis functionals ϵ_i, with $\mathbf{v}$ the argument.

What, now, is the dual space of a dual space?

Theorem 2–3.7. The vector space $(V^*)^* = V^{**}$ is isomorphic to V under the mapping which makes $\mathbf{v} \in V$ correspond to the functional $\Phi_\mathbf{v}$ defined by $\phi\Phi_\mathbf{v} = \mathbf{v}\phi$.

Proof. As dual space of the n-dimensional vector space V^*, V^{**} is n-dimensional. The cobasis $\mathrm{E}_1, \ldots, \mathrm{E}_n$ of the ϵ_k is a basis of V^{**}. By Theorem 2–3.6, this means $\epsilon_j\mathrm{E}_k = \delta_{jk}$. Now E_k is one of the Φ. Assume $\mathrm{E}_k = \Phi_{\mathbf{f}_k}$, then $\phi\Phi_\mathbf{v} = \mathbf{v}\phi$ implies $\epsilon_j\mathrm{E}_k = \mathbf{f}_k\epsilon_j = \delta_{jk}$. But, together with $\mathbf{e}_k\epsilon_j = \delta_{jk}$, this results in $\mathbf{f}_k = \mathbf{e}_k$, which induces a one-to-one correspondence between the basis $\mathbf{e}_k$ of V and the basis E_k of V^{**}. The functionals Φ are linear, and hence this correspondence is an isomorphism. This isomorphism is independent of the choice of the basis, and therefore it is reasonable to identify V with V^{**}.

For notation purposes we will frequently write vectors of V as row vectors or $1 \times n$ matrices, while functionals of V^* will be written as column vectors, that is, $n \times 1$ matrices.

Exercises

1. The *sum* of two $m \times n$ matrices (a_{ij}) and (b_{ij}) is defined to be $(a_{ij} + b_{ij})$. The scalar product of a scalar c and the matrix (a_{ij}) is defined by $c(a_{ij}) = (ca_{ij})$. Show that the $m \times n$ matrices form a vector space. What is its dimension? What is the zero element? Find a basis. Do the matrices form a group under matrix multiplication?
2. What is the transformation $x_1 \to 2x_2$, $x_2 \to x_1 + x_2$ relative to the basis $\mathbf{f}_1 = (1, 1)$, $\mathbf{f}_2 = (0, 2)$?
3. Find the matrix of the transformation $x_1 \to 3x_1 - x_2$, $x_2 \to 2x_2$ relative to the basis $\mathbf{f}_1 = 3\mathbf{e}_1$, $\mathbf{f}_2 = -\mathbf{e}_1 + 2\mathbf{e}_2$.
4. Prove: If in an $n \times n$ matrix (a_{ij}), $a_{ij} = -a_{ji}$ (the matrix is *skew-symmetric*) and if n is odd, then the matrix is singular.

5. Prove: If a, b, and c are distinct, the matrix

$$\begin{bmatrix} 1 & a & a^2 \\ 1 & b & b^2 \\ 1 & c & c^2 \end{bmatrix}$$

is nonsingular.

6. Verify the following rule for 2×2 and 3×3 matrices. If $\det(a_{ij}) = \delta \neq 0$, then $(a_{ij})^{-1} = (b_{ij})$, where $b_{ij} = (-1)^{i+j}\alpha_{ji}/\delta$. Express $(a_{ij})^{-1}$ explicitly for 2×2 matrices without using the α_{ji}.

7. *Cramer's rule* (after the Swiss mathematician Gabriel Cramer, 1704–1752) states that a simultaneous system of equations $\mathbf{x}A = \mathbf{b}$ with $\det A \neq 0$ has the solution $\mathbf{x} = (x_1, \ldots, x_n)$ with $x_k = \det A_k/\det A$, where A_k is obtained from A by substitution of the kth column by $\mathbf{b}'$. Verify Cramer's rule for $n = 2$ and $n = 3$.

8. For what value of k is the following system not solvable?

$$\begin{aligned} x + y + z &= 2 \\ x + 2y &= 2 \\ kx - z &= 0 \end{aligned}$$

If the right-hand sides were replaced by zeros, would the system have nontrivial solutions for this value of k?

9. If a vector $\mathbf{x} \neq \mathbf{0}$ is taken by a linear transformation of V^n with matrix A into one of its scalar multiples, $\lambda\mathbf{x}$, then λ is called a *characteristic root* of the transformation and $\mathbf{x}$ is called a *characteristic vector*. Show that λ is a characteristic root of A if and only if $\det(A - \lambda I_n) = 0$.

10. Find three linearly independent characteristic vectors and all characteristic roots of the matrix

$$\begin{bmatrix} 2 & 0 & 0 \\ 2 & 2 & 2 \\ 4 & 2 & -1 \end{bmatrix}.$$

11. Prove:
    (a) If λ is a characteristic root of an $n \times n$ matrix and if B is a nonsingular $n \times n$ matrix, then λ is a characteristic root of BAB^{-1}.
    (b) If $\mathbf{x}$ is a characteristic vector of A, then $\mathbf{x}B^{-1}$ is a characteristic vector of BAB^{-1}.
    (c) If $\mathbf{x}$ is a characteristic vector of a matrix, so is $c\mathbf{x}$ for every scalar $c \neq 0$.

12. A square matrix where all entries outside the main diagonal are zero is called a *diagonal matrix*. What are its characteristic roots and vectors?

13. Show that the vectors $(1, 1, 0)$, $(2, 0, 1)$, and $(0, 0, 1)$ form a basis E of R^3. Find the scalars that are the values of the vector $(1, -1, 1)$ under each of the functionals of the cobasis of E. Is there a vector (x_1, x_2, x_3) whose values under the cobasis are x_1, x_2, and x_3, respectively?

14. Prove that every square matrix which has an inverse is nonsingular.

2–4 AFFINE SPACES AND THEIR TRANSFORMATIONS

Definition. An *affine n-space* A^n will be defined as a set of *points* $P, Q, R, \ldots$, which is coupled with a vector space V^n as follows.

Af 1. Every ordered pair (P, Q) of points P and Q in A^n determines a vector of V^n called $\overrightarrow{PQ}$.

Af 2. Every point $P \in A^n$ and every vector $\mathbf{v} \in V^n$ determine a point $Q \in A^n$ such that $\overrightarrow{PQ} = \mathbf{v}$.

Af 3. $\overrightarrow{PQ} = \mathbf{0}$ if and only if $P = Q$.

Af 4. If P, Q, and R are any points of A^n, then $\overrightarrow{PQ} + \overrightarrow{QR} = \overrightarrow{PR}$.

First we mention a few direct consequences of the postulates.

Theorem 2–4.1. $\overrightarrow{PQ}$ is uniquely defined by P and Q.

Proof. Suppose $\mathbf{v} = \overrightarrow{PQ}$ and $\mathbf{w} = \overrightarrow{PQ}$. Then $\overrightarrow{PQ} + \overrightarrow{QP} = \mathbf{v} + \overrightarrow{QP} = \overrightarrow{PP} = \mathbf{0}$, by Af 4 and Af 3. Hence $\overrightarrow{QP} = -\mathbf{v}$. But $\overrightarrow{PQ} + \overrightarrow{QP} = \mathbf{w} + (-\mathbf{v}) = \mathbf{0}$, and consequently $\mathbf{w} = \mathbf{v}$.

Corollary. $\overrightarrow{PQ} = -\overrightarrow{QP}$.

Theorem 2–4.2. $P \in A^n$ and $\mathbf{v} \in V^n$ uniquely define $Q \in A^n$ such that $\overrightarrow{PQ} = \mathbf{v}$.

Proof. Suppose $\overrightarrow{PQ} = \mathbf{v}$ and $\overrightarrow{PR} = \mathbf{v}$. Then, by Af 4, $\overrightarrow{PQ} + \overrightarrow{QR} = \overrightarrow{PR}$, $\mathbf{v} + \overrightarrow{QR} = \mathbf{v}$, and $\overrightarrow{QR} = \mathbf{0}$. By Af 3, then, $Q = R$.

Definition. The *translation* $T_{\mathbf{v}}$ is the mapping which takes every point P of A^n into a point Q such that $\overrightarrow{PQ} = \mathbf{v} \in V^n$.

By Theorem 2–4.1, this mapping is uniquely defined.

The reader will realize that we have met such translations in the first chapter. Also the points encountered there behaved according to our postulates Af, and this fact may to some extent (for the case $n = 2$) justify the introduction of the postulates Af. Such a justification is required because, in the absence of a model, postulates might contain contradictions.

Theorem 2–4.3. The translations on A^n form an abelian group, $\mathcal{T}^n$. This group is isomorphic to $(V^n, +)$.

Proof. Let $T_{\mathbf{v}}$ and $T_{\mathbf{w}}$ be two translations. If $(P)T_{\mathbf{v}} = Q$ and $(P)T_{\mathbf{v}}T_{\mathbf{w}} = (Q)T_{\mathbf{w}} = R$, then $\overrightarrow{PQ} = \mathbf{v}$, $\overrightarrow{QR} = \mathbf{w}$, and, by Af 4, $\overrightarrow{PR} = \mathbf{v} + \mathbf{w}$. Thus $T_{\mathbf{v}}T_{\mathbf{w}} = T_{\mathbf{v}+\mathbf{w}}$ which proves the closure and establishes the isomorphism. Since $(V^n, +)$ is an abelian group, so is $\mathcal{T}^n$.

Affine coordinates and affine subspaces. Now we choose an arbitrary point of A^n and call it O, the *origin*. Furthermore, a point P will be called $[\mathbf{v}]$ if $\overrightarrow{OP} = \mathbf{v}$. If, with respect to a particular basis of V^n, $\mathbf{v}$ has the components $x_1, \ldots, x_n$, we also write $P = [x_1, \ldots, x_n]$. In view of Theorem 2–4.2, this definition is unique. In particular, $O = [\mathbf{0}] = [0, \ldots, 0]$. The vector $\overrightarrow{OP}$ is called the *position vector* of P. Obviously to every point of A^n there belongs exactly one position vector, and every vector of V^n is the position vector of exactly one point. This provides a bijective relation between points and vectors: the vector $\mathbf{v}$ corresponds to the point $[\mathbf{v}]$ as its position vector.

If, in terms of a given basis of V^n, $P = [x_1, \ldots, x_n]$, then the scalars $x_1, \ldots, x_n$ are called the *coordinates* of P. The basis vectors of V^n, together with the point O, are referred to as a *coordinate system.* These affine coordinate systems are much more general than the usual cartesian systems. Since length and angle measure have not been mentioned in our axioms—indeed they have no place in affine geometry—we cannot restrict the basis vectors to equal and unity length, nor do we determine anything about the angles enclosed between these vectors.

In view of the one-to-one correspondence between points and their position vectors, we will speak also of translations of position vectors. If $\mathbf{v}$ is a position vector, then we define $\mathbf{v}T_{\mathbf{b}} = \mathbf{v} + \mathbf{b}$. Translations of vectors according to this definition are not linear (homogeneous) transformations and, therefore, do not belong to $\mathcal{GL}^n$. Indeed, the image of $\mathbf{0}$ under a translation is not $\mathbf{0}$ again unless the translation is the identity, while all linear transformations send $\mathbf{0}$ into itself.

The following definition introduces certain subspaces of A^n. A *linear m-variety*, α^m (or α for short), is the set of all points $Q \in A^n$ such that, for an arbitrary fixed $P \in A^n$, the vector $\overrightarrow{PQ}$ is in S^m, where S^m is an m-dimensional subspace of V^n. We say that α is *extended by* S^m *from* P. Linear m-varieties are called *planes* for $m = 2$, *straight lines* for $m = 1$, and, trivially, points for $m = 0$.

Theorem 2–4.4. The subspace S which extends a given linear m-variety α is unique.

Proof (Fig. 2–1). Suppose α is extended by S from P and by S' from P'. Any points Q and R of α satisfy $\overrightarrow{PQ} \in S$, $\overrightarrow{PR} \in S$, $\overrightarrow{P'Q} \in S'$, and $\overrightarrow{P'R} \in S'$. Then $\overrightarrow{QR} = \overrightarrow{PR} - \overrightarrow{PQ} \in S$, and $\overrightarrow{QR} = \overrightarrow{P'R} - \overrightarrow{P'Q} \in S'$. Since Q and R were arbitrarily chosen, every such vector belongs to S and S'. Hence $S = S'$.

Now we are able to define parallel varieties. Linear m-varieties are *parallel* if they are extended by the same subspace. It follows immediately that parallelism is an equivalence relation.

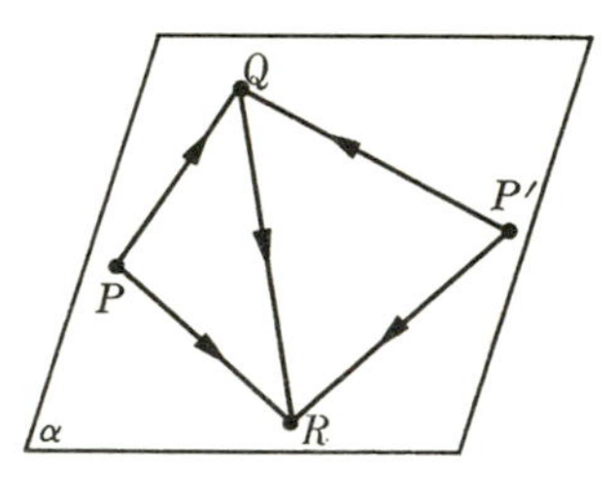

Figure 2–1

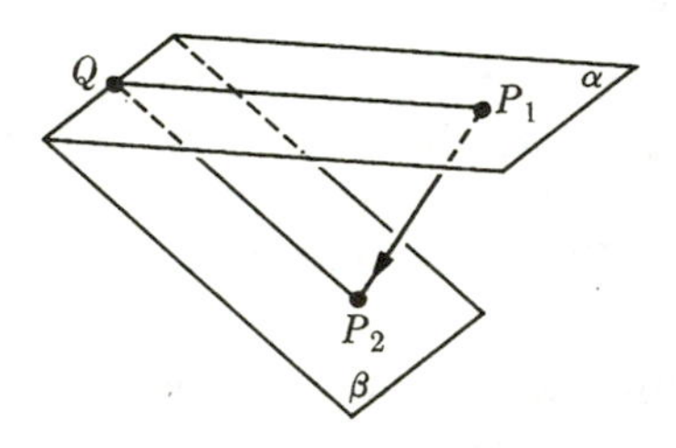

Figure 2–2

Theorem 2-4.5. Distinct parallel linear m-varieties have no points in common.

Proof (Fig. 2-2). Suppose the point Q belongs to α and β, both linear m-varieties extended by S. If α is extended from P_1, then $\overrightarrow{P_1Q} \in S$, and if β is extended from P_2, then $\overrightarrow{P_2Q} \in S$. But then also $\overrightarrow{P_1P_2} = \overrightarrow{P_1Q} + \overrightarrow{QP_2} = \overrightarrow{P_1Q} - \overrightarrow{P_2Q} \in S$, and $P_2 \in \alpha$, $P_1 \in \beta$. This would imply $\alpha = \beta$.

The following observation sheds more light on the preceding theorem. A linear variety α^m containing the origin is extended from O by a subspace $S^m \subset V^n$, and therefore the position vectors of α^m are exactly the elements of S^m. A linear variety β^m parallel to α^m has points $[\mathbf{v} + \mathbf{w}]$, where $[\mathbf{v}]$ is an arbitrary point of β^m, while $\mathbf{w}$ runs through all vectors of S^m. In other words, the position vectors of the points of β^m form a coset of V^n with respect to S^m. Theorem 2-4.5 is, therefore, a consequence of Theorem 1-3.5, which stated that cosets with respect to the same subgroup either coincide or do not have any elements in common..

This remark also implies that parallel linear varieties are taken into each other by translations. Namely, if the points of two varieties are, respectively, $[\mathbf{b} + S^m]$ and $[\mathbf{c} + S^m]$, then the translation $T_{\mathbf{c}-\mathbf{b}}$ maps the first variety onto the second.

Affine transformations. What is the analog of a component transformation in affine space? The components of the vectors of V^n are replaced here by the coordinates of points in A^n. We will attempt to study the effect on the coordinates of a change in the coordinate system of A^n. The coordinate system consists of the basis of V^n and the origin O. The basis vectors $\mathbf{e}_i$ undergo a transformation $\mathbf{e}_i \to \mathbf{e}_i(C')^{-1}$, where $C = (c_{ij})$ is a nonsingular $n \times n$ matrix. The origin O is moved to a point O' by a translation, say $TA_{\mathbf{b}}$, such that $(O)TA_{\mathbf{b}} = O'$ with $[\mathbf{b}] = [b_1, \ldots, b_n]$. Then the coordinates x_i of a point P transform in the following fashion:

$$x_i \to \sum_j c_{ij}x_j + b_i.$$

In other words, the position vector $\mathbf{x}$ of P transforms now as $\mathbf{x} \to \mathbf{x}L_CT_{\mathbf{b}}$, where L_C is the linear transformation whose matrix is C. This is an *affine coordinate transformation.*

Again, as in the case of component transformations, these coordinate transformations are, in a fixed coordinate system, also the equations of certain transformations of A^n, which are called *affinities.* Every linear transformation followed by a translation, acting on a position vector, results in an affinity operating on the corresponding point. The accurate description of the procedure is the following. The preimage for an affinity is a point in A^n. With a given origin, this point has a unique position vector which now is mapped by a linear transformation onto a position vector, to which in turn a unique point of A^n corresponds. Finally, this image point is translated to some point of A^n. Symbolically this procedure might be put into the form

$$P = [\mathbf{p}] \in A^n \leftrightarrow \mathbf{p} \xrightarrow{\mathcal{GL}^n} (\mathbf{p})L \leftrightarrow (P)L \in A^n \xrightarrow{\mathcal{T}} (P)LT \in A^n.$$

However, sometimes we will disregard the distinction between point and position vector.

What happens when the translation is performed before the linear transformation instead of after it? The following theorem provides the answer.

Theorem 2–4.6. For every $L \in \mathcal{GL}^n$ the equality $L\mathcal{T}^n = \mathcal{T}^n L$ holds.

This does not necessarily mean that $LT = TL$ for every $T \in \mathcal{T}^n$. The meaning is that, for given T and L, there is a unique $T' \in \mathcal{T}^n$ such that $LT = T'L$.

Proof. For any translation $T_\mathbf{b}$ and for any $[\mathbf{v}]$,

$$\mathbf{v}L^{-1}T_\mathbf{b}L = (\mathbf{v}L^{-1} + \mathbf{b})L = \mathbf{v}L^{-1}L + \mathbf{b}L,$$

in view of the linearity of L. But this may be written as $\mathbf{v} + \mathbf{b}L = \mathbf{v}T_{\mathbf{b}L}$, and thus $L^{-1}T_\mathbf{b}L = T_{\mathbf{b}L}$, which proves the theorem.

Now we are able to prove:

Theorem 2–4.7. The affinities of A^n form a group, the so-called *affine group* $\mathcal{A}^n$.

Proof. For proving closure, let L_1T_1 and L_2T_2 be affinities. Then $L_1T_1L_2T_2 = L_1L_2L_2{}^{-1}T_1L_2T_2 = L_1L_2T_3T_2$, where T_3 is another translation, by Theorem 2–4.6. This is of the form LT as required. The identity obviously is an affinity. The inverse of LT is $T^{-1}L^{-1} = L^{-1}LT^{-1}L^{-1} = L^{-1}T'$ for some translation T', again by Theorem 2–4.6.

Thus every affinity M has a linear part L and a translation part T. These parts are uniquely determined because, if $(O)M = B = [\mathbf{b}]$, then $T_\mathbf{b}$ is the translation part of M. The linear part is then $MT_\mathbf{b}^{-1}$.

In ordered fields we have

Theorem 2–4.8. $\mathcal{GL}^n$ is a subgroup of $\mathcal{A}^n$, and $\mathcal{T}^n \lhd \mathcal{A}^n$. The elements LT with $L \in \mathcal{GL}^n_+$ and $T \in \mathcal{T}^n$ form a normal subgroup, $\mathcal{A}^n_+$, of $\mathcal{A}^n$. The index of $\mathcal{A}^n_+$ in $\mathcal{A}^n$ is 2.

Proof. The first sentence is an immediate consequence of Theorems 2–2.6, 2–4.3, and 2–4.6. The rest follows from Theorem 2–3.2.

Let us take a closer look at the linear varieties in A^2 and A^3. We will obtain the familiar equations of lines and planes of analytic geometry.

Theorem 2–4.9. The coordinates of all points $[x_1, x_2, x_3]$ of A^3 which lie in a plane, and only these, satisfy an equation $a_1x_1 + a_2x_2 + a_3x_3 + a_0 = 0$, where not all of a_1, a_2, and a_3 are zero.

Proof (Fig. 2-3). Let a plane π be extended by a two-dimensional vector space spanned by $\mathbf{v} = (v_1, v_2, v_3)$ and $\mathbf{w} = (w_1, w_2, w_3)$ from a point $P = [p_1, p_2, p_3]$ with position vector $\mathbf{p} = (p_1, p_2, p_3)$. Then for each point $X = [x_1, x_2, x_3] = [\mathbf{x}]$ on π, $\mathbf{x} = \mathbf{p} + \rho\mathbf{v} + \sigma\mathbf{w}$, where ρ and σ are scalars. The breakdown of this vector equation into components yields $x_i = p_i + \rho v_i + \sigma w_i$, with $i = 1, 2, 3$. Now eliminate ρ and σ by multiplying the first equation with $a_1 = v_2 w_3 - v_3 w_2$, the second equation with $a_2 = v_3 w_1 - v_1 w_3$, the third equation with $a_3 = v_1 w_2 - v_2 w_1$, and then adding them. The result is

$$a_1 x_1 + a_2 x_2 + a_3 x_3 - a_1 p_1 - a_2 p_2 - a_3 p_3 = 0.$$

This equation would be meaningless only in the case $a_1 = a_2 = a_3 = 0$. But this could happen if and only if $v_1 : v_2 : v_3 = w_1 : w_2 : w_3$; that is, if $\mathbf{v}$ and $\mathbf{w}$ were linearly dependent. However, since they form a basis of the extending vector subspace S^2, this is not the case.

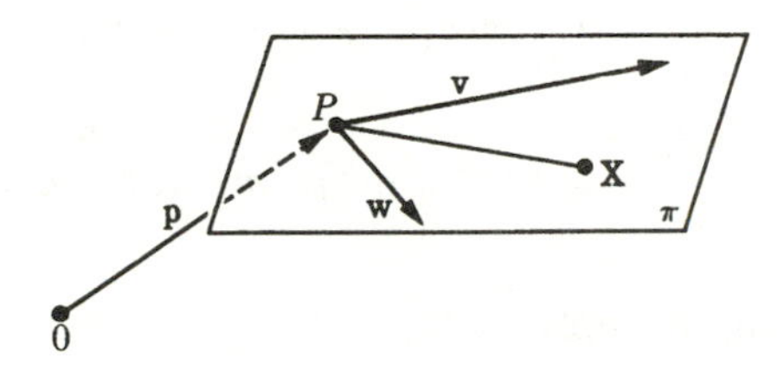

Figure 2-3

Now, conversely, consider all points $X = [x_1, x_2, x_3] = [\mathbf{x}]$ which satisfy $a_1 x_1 + a_2 x_2 + a_3 x_3 + a_0 = 0$, not all of a_1, a_2, a_3 zero. Suppose, without loss of generality, that $a_1 \neq 0$. Then $x_1 = -a_1^{-1}(a_0 + a_2 x_2 + a_3 x_3)$. The vectors $\mathbf{v} = (-a_2, a_1, 0)$ and $\mathbf{w} = (-a_3, 0, a_1)$ are linearly independent and form a basis of some $S^2 \subset V^3$. As the point from which the plane will be extended, choose $P = [-a_1^{-1}a_0, 0, 0]$. Then $\mathbf{x}$ is indeed obtained as $(-a_1 a_0, 0, 0) + \rho\mathbf{v} + \sigma\mathbf{w}$, because if we choose $\rho = a_1^{-1}x_2$ and $\sigma = a_1^{-1}x_3$, we obtain

$$\begin{aligned} -a_1^{-1}a_0 + a_1^{-1}x_2(-a_2) + a_1^{-1}x_3(-a_3) &= -a_1^{-1}(a_0 + a_2 x_2 + a_3 x_3) = x_1, \\ 0 + a_1^{-1}x_2 a_1 + a_1^{-1}x_3 0 &= x_2, \\ 0 + a_1^{-1}x_2 0 + a_1^{-1}x_3 a_1 &= x_3. \end{aligned}$$

This completes the proof.

The corresponding result for A^2 is

Theorem 2-4.10. The coordinates of all points $[x_1, x_2]$ of A^2 which lie on a straight line, and only these, satisfy an equation $a_1 x_1 + a_2 x_2 + a_0 = 0$, where a_1 and a_2 are not both zero.

The proof will be left as an exercise.

The rest of this section will be devoted to the study of the properties that are left invariant by affinities and by subgroups of $\mathcal{A}^n$.

Theorem 2–4.11. Every affinity maps a linear m-variety of A^n onto a linear m-variety of A^n.

Proof. Let $[\mathbf{x}]$ be a point of a linear m-variety extended by S^m from $P = [p_1, \ldots, p_n] = [\mathbf{p}]$. If $\mathbf{v}_1, \mathbf{v}_2, \ldots, \mathbf{v}_m$ is a basis of S^m, then for some scalars $\rho_1, \ldots, \rho_m$,

$$\mathbf{x} = \mathbf{p} + \rho_1\mathbf{v}_1 + \cdots + \rho_m\mathbf{v}_m.$$

Under an affinity $LT_\mathbf{b}$, $\mathbf{x}$ becomes

$$\begin{aligned}\mathbf{x}LT_\mathbf{b} &= (\mathbf{p} + \rho_1\mathbf{v}_1 + \cdots + \rho_m\mathbf{v}_m)LT_\mathbf{b}\\ &= (\mathbf{p}L + \rho_1\mathbf{v}_1L + \cdots + \rho_m\mathbf{v}_mL)T_\mathbf{b}\\ &= \mathbf{p}L + \mathbf{b} + \rho_1\mathbf{v}_1L + \cdots + \rho_m\mathbf{v}_mL.\end{aligned}$$

By Theorem 2–2.8, the $\mathbf{v}_jL$ form a basis of an m-dimensional subspace, say S', and all the $\mathbf{x}LT_\mathbf{b}$ then belong to the linear m-variety extended by S' from the point $[\mathbf{p}L + \mathbf{b}]$.

Corollary 1. Collinearity, coplanarity, and parallelism are preserved under affinities.

Proof. If a point belongs to a linear variety, its image under an affinity belongs to the image variety. If two varieties are parallel, they are extended by the same subspace. This subspace is mapped onto the subspace extending the image varieties which are, therefore, parallel.

Corollary 2. Affinities are collineations.

Theorem 2–4.12. Affinities preserve vector equality; that is, if $\overrightarrow{PQ} = \overrightarrow{RS}$ and if N is an affinity, then

$$\overrightarrow{(P)N(Q)N} = \overrightarrow{(R)N(S)N}.$$

Proof. If $[\mathbf{p}] = P$, $[\mathbf{q}] = Q$, $[\mathbf{r}] = R$, and $[\mathbf{s}] = S$, then $\mathbf{q} - \mathbf{p} = \mathbf{s} - \mathbf{r}$. If the affinity $LT = LT_\mathbf{b}$, then $\mathbf{q}LT = \mathbf{q}L + \mathbf{b}$, and $\mathbf{p}LT = \mathbf{p}L + \mathbf{b}$. Hence $\overrightarrow{(P)N(Q)N}$ is the vector

$$\begin{aligned}\mathbf{q}LT - \mathbf{p}LT &= \mathbf{q}L - \mathbf{p}L = (\mathbf{q} - \mathbf{p})L\\ &= (\mathbf{s} - \mathbf{r})L = \mathbf{s}L + \mathbf{b} - \mathbf{r}L - \mathbf{b} = \mathbf{s}LT - \mathbf{r}LT = \overrightarrow{(R)N(S)N}.\end{aligned}$$

This theorem enables us to define $\mathbf{v}N$. From the proof it becomes evident that $\mathbf{v}N = \mathbf{v}L$ if L is the linear part of N.

The reader may be under the mistaken impression that images of vectors under affinities have been used before and in a different way. All previous affinities acted on points, which were sometimes represented by their position vectors, but never on vectors of V^n. The difference stems from the fact that in the transformation of the vector *both* endpoints of the vector are transformed by the affinity, while in the case of the point, only the point itself is moved.

Theorem 2–4.13. If $\mathbf{w} = \rho\mathbf{v}$, then $\mathbf{w}N = \rho(\mathbf{v})N$ for any affinity N and every scalar ρ.

The significance of this theorem is that the partition ratio by which a vector is divided is an affine invariant.

Proof. If $N = LT$, then $\mathbf{w}N = \mathbf{w}L = (\rho\mathbf{v})L$ which, by Ln 2, is $\rho(\mathbf{v})L = \rho(\mathbf{v})N$.

Where K is the real field, we can say more:

Corollary 1. Affinities preserve midpoints.

Proof. This is the case $\rho = \frac{1}{2}$.

Corollary 2. If Q lies on a line with P and R and between them, then $(Q)N$ lies between $(P)N$ and $(R)N$.

In other terms, over the real field affinities preserve the "betweenness" relation.

Proof. The statement "Q lies between P and R on a straight line" can be expressed as $\overrightarrow{PQ} = \rho\overrightarrow{PR}$, with $0 < \rho < 1$. Since ρ is unchanged under an affinity, the corollary is proved.

Betweenness is not meaningful in every field: example, the complex field.

Exercises

1. Prove Theorem 2–4.10.
2. Prove: For two points P and Q in A^2 there exists exactly one line passing through them. Find an expression for the position vector of an arbitrary point on this line.
3. Prove: For three noncollinear points P, Q, and R in A^3, there exists one and only one plane passing through them. Find an expression for the position vector of an arbitrary point in this plane.
4. Prove: If two planes in A^3 have a point in common, then they have at least one other point in common.
5. Find the equation of the plane π through the points $[1, 0, 1]$, $[0, 2, -1]$, and $[3, 1, 0]$. What is the equation of the plane obtained from π by a translation through the vector $\mathbf{v} = (1, 0, -1)$?
6. If $\mathbf{v}$ and $\mathbf{w}$ are the position vectors of two points P and Q of A^3, find an equation for the position vectors $\mathbf{x}$ of all points on the line through P and Q.

7. If

$$N_1\colon x_1 \to x_1 + 2x_2 + 3, \qquad x_2 \to 2x_1 + 5x_2 - 1,$$

$$N_2\colon x_1 \to x_1 - x_2, \qquad x_2 \to x_1 + 2,$$

find

(a) N_1N_2 (b) N_2N_1 (c) N_1^{-1} (d) N_2^{-1}.

8. What affinity carries the square $ABCD$ with $A = [0, 0]$, $B = [1, 0]$, $C = [1, 1]$, and $D = [0, 1]$ into the parallelogram $A'B'C'D'$ with

(a) $A' = [1, 0]$, $B' = [3, 0]$, $C' = [4, 3]$, and $D' = [2, 3]$?
(b) $A' = [1, 0]$, $B' = [2, 3]$, $C' = [4, 3]$, and $D' = [3, 0]$?

Which of these affinities is direct? opposite? Draw a figure.

9. The planes $x_1 + x_2 = 0$ and $x_1 + x_2 = 1$ are mapped by the affinity

$$x_1 \to x_1 + x_2 + x_3 + 1$$

$$x_2 \to x_2 - x_3$$

$$x_3 \to 2x_3 - 1.$$

Show that the images are parallel planes. Is the affinity direct or opposite?

10. Is there an affinity of $\mathcal{A}^2$ carrying $[1, 2]$, $[3, 6]$, and $[-2, -4]$, respectively, into $[-1, -1]$, $[0, 0]$, and $[2, 2]$? If there is, find it. If not, explain why.

11. An affinity of $\mathcal{A}^2$ takes $[0, 0]$ and $[4, 2]$ into $[2, 1]$ and $[-2, 3]$, respectively. What is the image of $[2, 1]$? Do not compute the affinity.

12. Find an affinity that preserves the points $[-1, 0]$ and $[1, 0]$, while it carries $[0, h]$ into $[\frac{1}{2}, h]$. Show that this is possible for every $h \neq 0$. What happens when h tends toward zero? Why?

13. Find an example of an affinity which maps a right angle on an angle of 45°. Can you also find an affinity taking a right angle into an angle of 180°?

14. Use the equation of Exercise 6 for proving directly that an affinity takes a straight line of A^3 into a straight line of A^3.

15. Find the center of $\mathcal{A}^2$.

16. Let every affinity $\mathbf{x} \to \mathbf{x}A + \mathbf{b}$ (A a nonsingular $n \times n$ matrix) be expressed as an $(n + 1) \times (n + 1)$ matrix which has A in its left upper corner, $(0, \ldots, 0, 1)$ as its $(n + 1)$-column and $\mathbf{b}$ as the first n entries of its $(n + 1)$-row. Show that these matrices are nonsingular and form a group isomorphic to $\mathcal{A}^n$.

17. In the matrix representation of $\mathcal{A}^n$ (Exercise 16) which types of matrices correspond to elements of $\mathcal{GL}^n$, $\mathcal{A}^n_+$, $\mathcal{GL}^n_+$, and $\mathcal{T}^n$? Show that they form subgroups.

18. Prove:
    (a) All elements of $\mathfrak{A}^n$ that preserve some given point P of A^n form a group.
    (b) This group is isomorphic to $\mathcal{GL}^n$.
19. Show that a necessary and sufficient condition (a) for three points $[x_j, y_j]$, with $j = 1, 2, 3$, of A^2 to be collinear is

$$\begin{vmatrix} 1 & x_1 & y_1 \\ 1 & x_2 & y_2 \\ 1 & x_3 & y_3 \end{vmatrix} = 0,$$

(b) for four points $[x_j, y_j, z_j]$, with $j = 1, 2, 3, 4$, of A^3 to be coplanar is

$$\begin{vmatrix} 1 & x_1 & y_1 & z_1 \\ 1 & x_2 & y_2 & z_2 \\ 1 & x_3 & y_3 & z_3 \\ 1 & x_4 & y_4 & z_4 \end{vmatrix} = 0.$$

2–5 AFFINE PROPERTIES

Invariants under affinities. We mentioned before that a geometry deals with properties which are invariant under a certain group of transformations. In the case of affine geometry, those properties which are preserved under affinities will be studied, while other properties which might exist in one figure but not in a "congruent" one will be ignored. By a congruent figure in affine geometry we mean a figure that can be obtained from the original figure by an affinity.

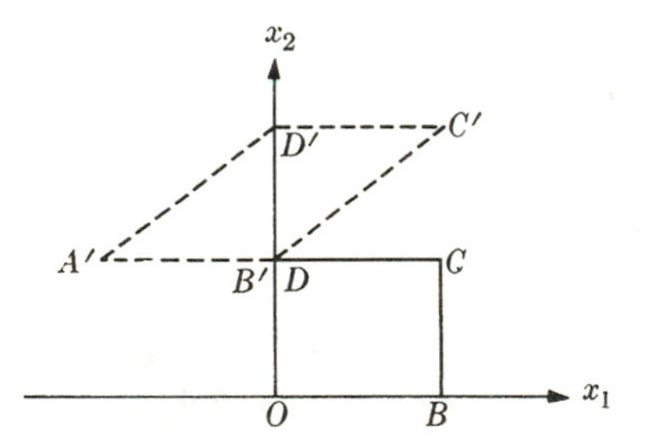

Figure 2–4

We shall discuss a few examples in A^2. The statement "$ABCD$ is a quadrangle" is affine because incidence of points and lines and the property of being a straight line are affinely invariant. "$ABCD$ is a parallelogram" is an affine statement because parallelism is preserved under affinities. "$ABCD$ is a square" is not an affinely invariant statement (Fig. 2–4) because, for instance, the square with $A = [0, 0]$, $B = [1, 0]$, $C = [1, 1]$, and $D = [0, 1]$ in a cartesian coordinate system is taken into a parallelogram, not a square, by the affinity

$$x_1 \to x_1 + x_2 - 1,$$
$$x_2 \to x_2 + 1.$$

Thus rectangularity and equality of length are not affine properties. To what extent affinities are capable of changing the shape of figures can be seen from:

Theorem 2–5.1. For any two given triangles $X_1X_2X_3$ and $Y_1Y_2Y_3$ there is a unique affinity in $\mathcal{A}^2$ mapping X_i on Y_i, with $i = 1, 2, 3$.

Proof. The vectors $\overrightarrow{X_1X_2}$ and $\overrightarrow{X_1X_3}$ are linearly independent and so are $\overrightarrow{Y_1Y_2}$ and $\overrightarrow{Y_1Y_3}$. Hence they are bases of the V^2 that contains all the vectors of A^2. By the corollary to Theorem 2–2.6, there is a unique linear transformation L taking the first pair of vectors into the second. If T is the unique translation mapping X_1 on Y_1, then LT is the desired affinity.

Of course, these considerations are not restricted to A^2. As examples of three-dimensional statements and theorems that belong to affine geometry let us mention "The solid is a parallelepiped" or "In a parallelepiped the diagonals intersect and bisect each other." A nonaffine statement would be, for instance, "$ABCD$ is a regular tetrahedron." In A^3 a theorem analogous to Theorem 2–5.1 holds.

Theorem 2–5.2. For any two given tetrahedra $X_1X_2X_3X_4$ and $Y_1Y_2Y_3Y_4$ there is a unique affinity of $\mathcal{A}^3$ mapping $X_i \to Y_i$, with $i = 1, 2, 3, 4$.

The proof is exactly the same as that of the previous theorem.

It would be easy to generalize these theorems for any A^n by using the concept of an *n-simplex*, that is, an $(n + 1)$-tuple of points such that no $k + 2$ of them lie in a linear k-variety ($k = 1, \ldots, n - 1$). A 1-simplex is a segment, and the theorem would then state that every segment can be carried into any other segment by a suitable affinity of $\mathcal{A}^1$. A 2-simplex is a triangle, a 3-simplex a tetrahedron.

Conics affinely classified. How would one go about affinely classifying conics and quadrics? By this we mean the finding of affinely different standard types such that each second-degree function in A^2 ("conics") and A^3 ("quadrics") can be transformed into exactly one of these types by an affinity or, equivalently, by an affine coordinate transformation.

For the rest of this section we will assume $K = R$.

The reader will recall from elementary analytic geometry that second-degree equations in two variables, except those that *degenerate* into sets of straight lines and isolated points, or those which do not yield any points with real coordinates, can be reduced by certain transformations to one of the following forms:

(a) circles, $x_1^2 + x_2^2 - r^2 = 0$, with $r \neq 0$,
(b) ellipses $b^2x_1^2 + a^2x_2^2 - a^2b^2 = 0$, with $a \neq 0 \neq b$,
(c) hyperbolas $b^2x_1^2 - a^2x_2^2 - a^2b^2 = 0$, with $a \neq 0 \neq b$,
(d) parabolas $x_2^2 - 2px_1 = 0$, with $p \neq 0$.

The transformations used were rotations,

$$x_1 \to x_1 \cos \alpha - x_2 \sin \alpha, \qquad x_2 \to x_1 \sin \alpha + x_2 \cos \alpha,$$

and translations, both special types of two-dimensional affinities.

Can these four types of conics be transformed into each other by further affinities? The answer is in the affirmative only with regard to cases (a) and (b). The affinity $x_1 \to ax_1/r$, $x_2 \to bx_2/r$ indeed transforms (b) into (a). Hence circles and ellipses are affinely equivalent.

The equations of type (a) can be brought into the standard form $x_1^2 + x_2^2 - 1 = 0$ by an affinity $x_1 \to rx_1$, $x_2 \to rx_2$, those of type (c) into the standard form $x_1^2 - x_2^2 - 1 = 0$ by $x_1 \to ax_1$, $x_2 \to bx_2$, those of type (*d*) into the form $x_2^2 - x_1 = 0$ by $x_1 \to x_1/2p$, $x_2 \to x_2$. It is algebraically obvious that all these equations can be transformed by affinities into second-degree equations only. Is it possible to transform the three remaining types into each other?

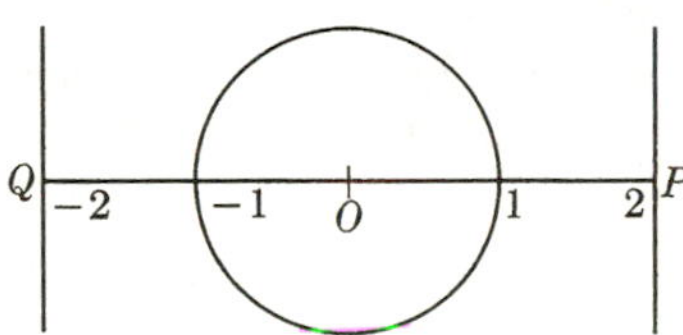

Figure 2–5

The two parallels $x_1 - 2 = 0$ and $x_1 + 2 = 0$ (Fig. 2–5) contain points $P = [2, 0]$ and $Q = [-2, 0]$, respectively, which, when connected, intersect the circle in two points between P and Q. However, the two parallels do not intersect the circle. This procedure is impossible with a parabola. Two parallels which do not meet the parabola do not have any points of the parabola between them (Fig. 2–6). Since all the concepts used here, such as straight line, parallel, incidence of a point and a line, and betweenness, are affine invariants, (a) and (d) are affinely distinct.

For the hyperbola $x_1^2 - x_2^2 - 1 = 0$ (Fig. 2–7), there exists a line, namely the x_2-axis, which does not intersect the hyperbola, but each of whose points lies between two points of the hyperbola, while such a situation is possible neither for the circle nor for the parabola. Thus case (c) is affinely distinct from the cases (a) and (d).

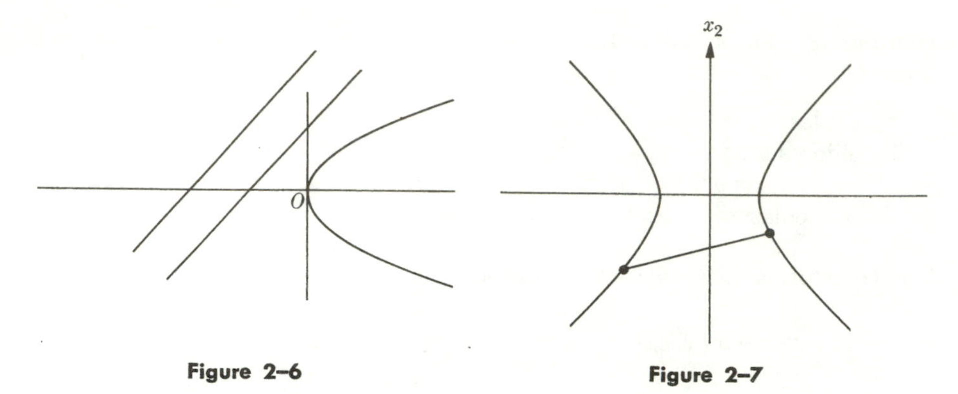

Figure 2-6

Figure 2-7

We summarize this in the following theorem.

Theorem 2-5.3. Besides the degenerate conics and those that do not contain any points with real coordinates, there are three classes of conics such that any two conics of the same class are affinely equivalent, while no two conics from distinct classes are affinely equivalent. These are the classes of all ellipses (including the circles), all hyperbolas, and all parabolas.

For A^3 an analogous treatment is possible. The standard forms will be ellipsoids, hyperboloids of one and two sheets, cones, elliptic paraboloids, hyperbolic paraboloids, elliptic, parabolic, and hyperbolic cylinders.

In this list we have omitted quadrics consisting of planes, and straight lines or isolated points only, and we have omitted quadrics without any points having real coordinates. We would have to study the theory of quadratic functions in three variables in order to show that all the classes are indeed affinely distinct. The proof would again use betweenness relations of points on a straight line and separation by parallel planes and lines.

It may appear surprising to the reader that throughout this classification special coordinate systems have been used where the coordinate axes are mutually orthogonal. However, this does not restrict the generality because every other coordinate system can be transformed into such a special system by an affinity.

We arrived at the affinities and their properties by an algebraic process avoiding any geometric intuition. This may be satisfactory for the sophisticated mathematician who despises any naïve intuitive approach. However, the more normal reader will ask himself what actually brought about the development of a geometry where only incidence, collinearity, parallelism, and partition ratio are invariant while any two simplexes (in the plane: triangles) are to be regarded as equivalent or

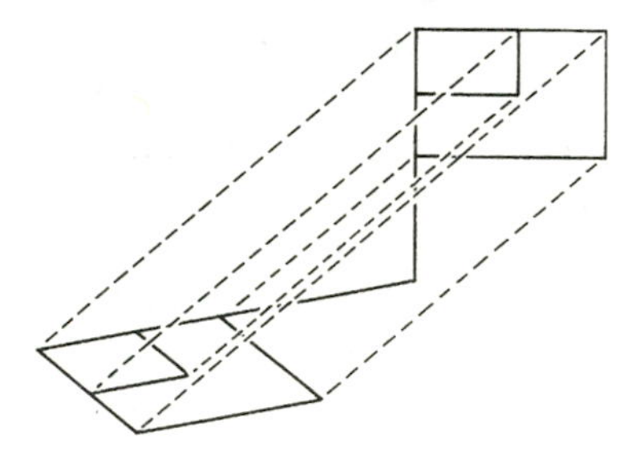

Figure 2-8

"essentially the same." In Exercise 6 we will see that parallel projection of a plane on another plane is an affinity, and naturally this fact was at the source of the development of affine geometry. Indeed, the parallel rays of the sun (Fig. 2-8) produce shadows which are affine images of the objects used, and perhaps everybody has at some time wondered about the weird shapes of shadows of perfectly ordinary objects.

Exercises

Assume $K = R$.

1. Which of the following theorems about parallelograms are affine statements?
   (a) The diagonals bisect each other.
   (b) Opposite sides are equal.
   (c) If all sides are equal, the diagonals are perpendicular to each other.
   (d) There exists a circumcircle if and only if all angles are right.
2. Which of the following properties are affine?
   (a) A quadrangle has two pairs of adjacent equal sides.
   (b) A line is tangent to an ellipse.
   (c) A quadrangle is a trapezoid.
   (d) A point is equidistant from two other points.
   (e) A solid has two parallel sides.
3. If in a triangle ABC, P is the point of intersection of
   (a) the medians
   (b) the altitudes
   (c) the angle bisectors
   (d) the perpendicular bisectors of the sides,
   does the image of P under an affinity have the same properties with regard to the image of ABC under this affinity? If the answer is negative, give counter-examples.
4. Let three affinities take the tetrahedron $OABC$ into

   (a) $OBCA$ (b) $OCAB$ (c) $OACB$,

   respectively. $A = [1, 0, 0]$, $B = [0, 1, 0]$, and $C = [0, 0, 1]$. Which of the affinities are direct? which opposite?
5. The tetrahedron whose vertices are $A = [1, 0, 0]$, $B = [2, 0, 0]$, $C = [2, 1, 0]$, $D = [1, 2, 2]$ is mapped on another tetrahedron $A'B'C'D'$ by means of an affinity. Determine this affinity for $A' = [1, 0, 0]$, $B' = [3, 0, -1]$, $C' = [2, 1, 1]$, $D' = [1, 0, 4]$,

   (a) if $A \to A'$, $B \to B'$, $C \to C'$, $D \to D'$,
   (b) if $A \to B'$, $B \to A'$, $C \to C'$, $D \to D'$.

   Which of these two affinities is direct? which opposite?

6. Prove that the mapping of a plane π onto another plane π' by parallel projection is affine. Parallel projection is to be understood such that the straight lines joining points P of π with their images $P' \in \pi'$ are parallel to each other.
7. Consider the following mapping of the points inside the unit circle of a plane into the same plane: The plane is π of Exercise 5 of Section 1-8. First map a point Q of π on a point Q' on the southern hemisphere by intersecting the sphere with the parallel to NS through Q. Then map Q' on $Q'' \in \pi$ by means of the stereographic mapping. Is the mapping $Q \to Q''$ an affinity?
8. (Pasch's axiom) Let PQR be a triangle and l a straight line in the same plane as PQR, such that l does not pass through P, Q, or R. If l meets the line PQ between P and Q, then it meets either the line PR between P and R or the line QR between Q and R. [*Hint:* Incidence and betweenness will be preserved when the triangle PQR is transformed into a conveniently located triangle.]
9. An ellipse may be considered as the affine image of a circle that has one diameter coinciding with the major axis of the ellipse. The images of two perpendicular diameters of the circle are called *conjugate diameters*. Prove:
   (a) All ellipse chords parallel to one diameter are bisected by a conjugate diameter.
   (b) The parallels to one diameter through the endpoints of the conjugate diameter are tangents of the ellipse.
10. Which surfaces are represented by $x_1^2 - x_2^2 - x_3^2 = 1$, $x_1^2 + x_2^2 + x_3^2 = 1$, and $x_1^2 + x_2^2 - x_3^2 = 1$? [*Hint:* Find their intersections with the planes $x_k = 0$ $(k = 1, 2, 3)$.] Show that they are affinely distinct.
11. The surface $x_1x_2 = x_3$ has a tangent plane $x_3 = 0$ at the origin. Show that the tangent plane separates one part of the surface from another part. Is this property affinely invariant? Test your answer with the affinity $x_1 \to x_1 - x_2$, $x_2 \to x_1 + x_2$, $x_3 \to x_3 + 2$.

2-6 SUBGROUPS OF THE PLANE AFFINE GROUP OVER THE REAL FIELD

Classification of affinities in A^2. We will try to obtain a picture of the structure of the plane affine group $\mathfrak{A}^2$ over the real field by deriving a classification of plane affinities.

We write the general plane affinity in the form

$$
\begin{aligned}
x' &= ax + by + e,\\
y' &= cx + dy + f,
\end{aligned}
\tag{1}
$$

where x and y temporarily replace our usual notation x_1 and x_2. We want to study the linear part of the transformation and, for the time being, we ignore the translation terms e and f. We divide the first equation by the second, let $z' = x'/y'$, and $z = x/y$. The result is

$$z' = \frac{x'}{y'} = \frac{ax + by}{cx + dy} = \frac{az + b}{cz + d}.$$

If $x \neq 0$ and $y = 0$, write $z = \infty$, and if $x' \neq 0$ and $y' = 0$, assume $z' = \infty$. The point [0, 0] will be disregarded.

We obtain a real bilinear transformation as in Section 1-8. In Theorem 1-9.11 we showed that by utilizing certain coordinate changes each of our transformations could be brought into exactly one of three standard forms,

(i) $z' = (az + b)/(-bz + a)$,
(ii) $z' = z + b$,
(iii) $z' = az$.

When we return to our linear transformations, these become:

(i) $x' = ax + by, \quad y' = -bx + ay$,
(ii) $x' = x + by, \quad y' = y$,
(iii) $x' = ax, \quad y' = y$.

The coordinate changes used for the standardization were the following. For (i) it was $z \to \alpha z + \beta$, with $\alpha \neq 0$, which means $x \to \alpha x + \beta y$, $y \to y$, for (ii) we had $z \to -1/(z - \gamma)$, which means $x \to -y$, $y \to x - \gamma y$, and for (iii) the change was $z \to (z - \beta)/(z - \gamma)$, with $\beta \neq \gamma$, which means $x \to x - \beta y$, $y \to x - \gamma y$. All these are linear transformations and therefore constitute coordinate transformations in the affine plane.

To every linear transformation there corresponds one well-defined real bilinear transformation. In Theorem 1-9.11, the bilinear transformations had to have a determinant $=1$ and z could be complex, whereas here we require only that the determinant not vanish, and z has to be real. Does this affect the feasibility of our procedure? If we scrutinize the proof of Theorem 1-9.11, we find that the requirement that the determinant be 1 was not used at all for the classification. On the other hand, there is no danger in our restricting ourselves only to real values of z. The bilinear transformation then becomes simply a transformation of the real axis (including the point ∞) instead of the whole plane.

Another question is whether every real bilinear transformation determines exactly one linear transformation. This, in fact, is not so. If two linear transformations L and L' yield the same bilinear transformation B, then $L'L^{-1}$ yields $BB^{-1} = I$. The only real bilinear transformations yielding I are those with $a = d = r$, any nonzero real number, and $b = c = 0$. This results in $L'L^{-1}$: $x \to rx$, $y \to ry$, which has been called a dilatation from the origin. Thus any linear transformations belong to the same bilinear transformation if and only if they can be obtained from each other by multiplication with a dilatation from the origin. This makes it possible to arrange all linear transformations in classes such that each class belongs to one bilinear transformation.

As standard form $x \to ax + by$, $y \to cx + dy$, we designate those representatives of a class which have their determinant $ad - bc = \pm 1$. Let

$$x \to a'x + b'y, \qquad y \to c'x + d'y$$

be a linear transformation. If a dilatation D_k: $x \to kx$, $y \to ky$ is performed after (or before) this transformation, the result is the transformation

$$x \to a'kx + b'ky, \qquad y \to c'kx + d'ky$$

whose determinant is $(a'd' - b'c')k^2$. If this should be in standard form with determinant ± 1, then $k = \pm|a'd' - b'c'|^{-1/2}$. Thus we have two standard forms for each class, differing from each other by a factor of -1.

We have therefore the following theorem.

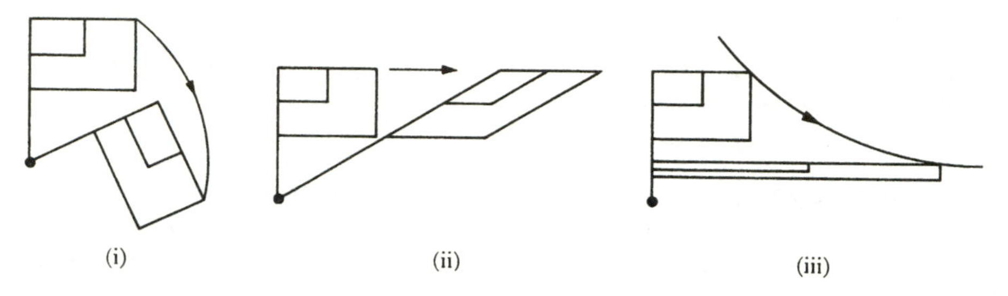

Figure 2-9

Theorem 2-6.1 (Fig. 2-9). The linear part of each affinity of $\mathfrak{A}^2$ can be brought into one of the following standard forms by a coordinate transformation followed by a dilatation from the origin.

(i) $x \to ax + by$, $y \to -bx + ay$, with a and $b \neq 0$, $a^2 + b^2 = 1$.
(ii) $x \to x + by$, $y \to y$, with $b \neq 0$.
(iii) $x \to ax$, $y \to y/a$, with $a \neq 0$.

The coefficients a, b, c, and d are real.

In the euclidean plane, type (i) would become a direct isometry as defined in Section 1-2. For, consider $z' = \alpha z + \beta$, where $\alpha = a - bi$, $\beta = e + fi$, $z = x + yi$, $z' = x' + y'i$. Then $x' + y'i = (a - bi)(x + yi) + e + fi = ax + by + e + i(-bx + ay + f)$, which is the same as (i). Moreover, $|\alpha| = a^2 + b^2 = 1$, which makes $z \to \alpha z + \beta$ a direct isometry. Since the origin is preserved in every linear transformation, this standard affinity has one invariant point and represents a rotation about 0.

In Theorem 1–9.11, the type (i) was characterized as having only $z = i$ as invariant value. Here this means that there is no real x/y invariant under a nonidentical linear direct isometry followed by a dilatation, that is, under a dilative rotation about 0. The geometrical significance is that none of the straight lines through 0 is preserved unless the dilative rotation is only a dilatation from 0.

The real bilinear transformations of type (ii) had a single real invariant point, namely ∞. Hence a linear affinity of this type has a single invariant line through O. In the case of the standard form this is the x-axis, $y = 0$. The linear affinity of this type leaves the invariant line pointwise unchanged.* Such an affinity is called a *shear transformation.*

Type (iii) of Section 1–9 had two invariant points, both real. Accordingly we have here two invariant lines through O, which in the standard case become the x- and y-axes.

In this theorem the affinities (i), (ii), and (iii) depend each on one parameter to which a second parameter has to be added in view of a possible dilatation. In order to study traces of points under a one-parameter subgroup in the case (i), we put $a = \cos t$ and $b = \sin t$. Then

$$\begin{aligned} x' &= x \cos t + y \sin t, \\ y' &= -x \sin t + y \cos t, \end{aligned} \tag{2}$$

and if $x + iy = z$ and $x' + iy' = z'$, then $z' = z(\cos t - i \sin t)$.

The affinities A_t: $z \to z(\cos t - i \sin t)$ form a group. For, by De Moivre's theorem,

$$A_{t'}A_{t''}\colon z \to z[\cos(t' + t'') - i \sin(t' + t'')] = A_{t'+t''}.$$

This group is a one-parameter group depending only on t. The trace of the constant point $[x_0, y_0]$ under this group can be obtained by putting $x = x_0$ and $y = y_0$ in Eq. (2), and by eliminating t. The result is

$$(x')^2 + (y')^2 = x_0^2 + y_0^2,$$

a circle about the origin passing through $[x_0, y_0]$, if the coordinates are considered as orthonormal.

As a further example we study the case (iii) in the same fashion. The transformations $x \to ax$, $y \to y/a$ form a one-parameter group because H_a: $x \to ax$, $y \to y/a$ implies $H_{a'}H_{a''}\colon x \to a'a''x, y \to y/(a'a'')$, and therefore $H_{a'}H_{a''} = H_{a'a''}$. The trace of a point $[x_0, y_0]$ under this group is the hyperbola $xy = x_0y_0$.

---

* In the euclidean plane this means that points outside the invariant line are moved parallel to it through a distance yb, that is, proportional to the distance from the invariant line. (This "distance" has to be measured parallel to the corresponding axis, not necessarily at right angles to the other axis!)

For the case (ii) we choose an example in which the affinity is not linear. Let J_b: $x \to x + by + b^2/4$, $y \to y + b/2$. Then the J_b form a one-parameter group. Namely,

$$J_{b'}J_{b''}\colon x \to (x + b'y + b'^2/4) + b''(y + b'/2) + b''^2/4$$
$$= x + (b' + b'')y + (b' + b'')^2/4,$$
$$y \to (y + b'/2) + b''/2 = y + (b' + b'')/2,$$

and therefore $J_{b'}J_{b''} = J_{b'+b''}$. The trace of a point $[x_0, y_0]$ under this group can be obtained by elimination of b from

$$x = x_0 + by_0 + b^2/4, \qquad y = y_0 + b/2.$$

The result is $x - y^2 = x_0 - y_0^2$, a parabola.

Thus we encountered ellipses among the traces of groups of affinities of type (i), hyperbolas among those of type (iii), and parabolas among those of type (ii). Each of these respective conics is then carried into itself by every transformation of the corresponding one-parameter group.

Affinities in A^2 preserving a conic. This provokes the question: what are the affinities that map a given conic onto itself? It is easy to see that for a given conic these affinities form a subgroup of $\mathfrak{A}^2$ because a product of such transformations preserves the conic and because the identity certainly has this property. Moreover, as the affinity maps every point of the conic on a point of the same conic, its inverse has to do the same. What, then, are the groups of affinities for each of the three affinely different types of conics?

Let us consider the circle $x^2 + y^2 = 1$ as the standard form of an ellipse. Every ellipse can be brought into this form by an affine coordinate transformation. The desired affinity has to preserve this circle. It is well known from elementary geometry that the center of the circle can be defined as the unique intersection of straight lines connecting midpoints of arbitrary parallel chords. Since this definition involves affine properties only, the origin which is the center of the circle is invariant under all affinities preserving the circle. Therefore the affinities are linear: $x \to ax + by$, $y \to cx + dy$. Then $(ax + by)^2 + (cx + dy)^2 = 1$ identically for all x, with $y^2 = 1 - x^2$. For $x = 1$, $y = 0$, we have $a^2 + c^2 = 1$. For $x = 0$, $y = 1$, we obtain $b^2 + d^2 = 1$. Substituted in $(ax + by)^2 + (cx + dy)^2 = 1$, this becomes $x^2 + y^2 + 2xy(ab + cd) = 1$, or, in view of $x^2 + y^2 = 1$, $xy(ab + cd) = 0$, which implies $ab + cd = 0$. After some computation, this, together with $a^2 + c^2 = b^2 + d^2 = 1$, results in $a = d$, $b = -c$, or $a = -d$, $b = c$. Thus the only affinities preserving the circle are $x \to ax + by$, $y \to -bx + ay$, and $x \to ax + by$, $y \to bx - ay$, with $a^2 + b^2 = 1$. In complex notation these are all the transformations $z \to \alpha z$ and $z \to \overline{\alpha z}$, respectively, with $\alpha = a - bi$, $z = x + iy$, and $|\alpha| = 1$. The group of the isometries $z \to \alpha z$ is *transitive* on the unit circle. That is, for every ordered pair of points on the circle there is an element of the group which carries the first point into the second point (cf. Exercise 11).

Every parabola can be reduced to $x = y^2$ by an affine coordinate transformation. If an affinity, (Eq. 1) sends the parabola $x - y^2 = 0$ into itself, then

$$x' - y'^2 = ax + by + e - (cx + dy + f)^2 = 0.$$

Comparison of the coefficients after a substitution $x = y^2$ yields $c = 0$, $e = f^2$, $a = d^2$, and $b = 2\,df$. The affinity then has the form

$$x' = d^2x + 2\,dfy + f^2, \qquad y' = dy + f. \tag{3}$$

Let the group of all these affinities be $\mathcal{G}$. It is a two-parameter group. Earlier we saw that the transformations J_b (which are those elements of $\mathcal{G}$ for which $d = 1$) form a subgroup of $\mathcal{G}$. In order to show that the elements of this subgroup are transitive on the points of the parabola, we let $[x, y]$ and $[x', y']$ be two arbitrary points of the parabola. Then $[x, y]J_f = [x', y']$ when f is chosen to be $y' - y$, as can be proved by substitution.

Finally, consider the hyperbola $xy = 1$ as representative of all hyperbolas. Then $(ax + by + e)(cx + dy + f) = 1$. After a substitution $y = 1/x$ we compare coefficients and obtain $e = f = 0$ and either $a = d = 0$, $bc = 1$, or $b = c = 0$ and $ad = 1$. This yields the affinities $x \to by$, $y \to x/b$, or the affinities H_a which were defined earlier and shown to form a group. This group is transitive on the points of the hyperbola (cf. Exercise 11).

Theorem 2–6.2. For every given nondegenerate conic there exists a group of all affinities which leave this conic invariant. If the conic is a parabola, this group is a two-parameter group. Each of these groups has a one-parameter subgroup which is transitive on the corresponding conic. This one-parameter subgroup is of the type (i) for ellipses, (ii) for parabolas, and (iii) for hyperbolas.

The three types (i), (ii), and (iii) are therefore sometimes called *elliptic, parabolic,* and *hyperbolic affinities,* respectively.

Exercises

Assume $K = R$.

1. An affinity has a linear part

$$L\colon\ x \to ax + by,\qquad y \to cx + dy,$$

which is an opposite transformation. Prove that it is of type (iii) and find the invariant lines of L.

2. Test the following affinities as to the types to which they belong. If applicable, find the invariant lines of their linear parts.

(a) $x \to 2x - y + 1$ (b) $x \to 2x + y$

$y \to x + 4y$ $y \to x + 5y - 5.$

3. Which of the linear transformations of the standard types are involutory? Describe them.
4. Try to classify $x \to -x + 2y + 1$, $y \to x/2 - y - 1$. What is the cause of the difficulty?
5. Determine the type and, if applicable, the invariant lines of the linear parts of

(a) $x \to x - 2y - 1$ (b) $x \to x + 2y$

$y \to 4x + y + 2$ $y \to 4x - y$

6. What is the trace of $[x_0, y_0]$ under the one-parameter group of

$$x \to r(x \cos \alpha + y \sin \alpha), \qquad y \to r(-x \sin \alpha + y \cos \alpha),$$

with α and r connected by the condition $r = \alpha$?
7. Find the affinity which maps the parabola $y^2 = 2x$ onto itself and takes O into the point [2, 2].
8. Does an affinity which maps an ellipse or a hyperbola onto itself preserve the center of the conic?
9. Are the axes of a conic preserved under affinities which map the conic on itself?
10. What happens to the asymptotes of the hyperbola $xy = 1$ under affinities which preserve the hyperbola?
11. (a) Show that the group of affinities

$$x \to ax + by, \qquad y \to -bx + ay, \qquad \text{with} \qquad a^2 + b^2 = 1$$

is transitive on the points of the circle $x^2 + y^2 = 1$.

(b) Show the same for the hyperbola $xy = 1$ and the affinities $x \to ax$, $y \to y/a$.

2-7 THE ORTHOGONAL AND EUCLIDEAN GROUPS

Euclidean vector spaces. In Section 2-3 we studied the dual space V^* of an n-dimensional vector space V and established the existence of a certain isomorphism between V and V^*. Indeed each basis of V can be mapped onto each of the bases of V^*, and any such mapping induces an isomorphism $V \leftrightarrow V^*$. When we consider such an isomorphism between V, with basis $\mathbf{f}_1, \ldots, \mathbf{f}_n$, and V^*, with basis $\phi_1, \ldots, \phi_n$, we call this relation a *pairing* of V and V^*.

Every pairing of V and V^* yields an *inner product* in the following way. Let $\mathbf{v} \in V$, $\mathbf{v} = x_1\mathbf{f}_1 + \cdots + x_n\mathbf{f}_n$, and $\theta = \phi_1 y_1 + \cdots + \phi_n y_n$. Then, in accordance with Theorems 2–3.5 and 2–3.6, we have $\mathbf{v}\theta = \sum_i \sum_j x_i y_j (\mathbf{f}_i \phi_j)$. Now the $\mathbf{f}_i\phi_j$ are well-defined scalars, say $\mathbf{f}_i\phi_j = g_{ij}$, and then

$$\mathbf{v}\theta = \sum_i \sum_j x_i y_j g_{ij}.$$

We shall now impose additional restrictions on the vector space V^n. First, the field K will be the real field R. Moreover, V^n will be *euclidean*, that is, it will have to fulfill the following requirements.

VS 4. To every $\mathbf{v}$ and $\mathbf{w}$ in V^n there corresponds a unique real number $\mathbf{vw}$ (called the *inner product* of $\mathbf{v}$ and $\mathbf{w}$) such that for all $\mathbf{u}$, $\mathbf{v}$, and $\mathbf{w}$ in V^n, and for all real r,

VS 4.1. $\mathbf{vw} = \mathbf{wv}$,
VS 4.2. $\mathbf{u}(\mathbf{v} + \mathbf{w}) = \mathbf{uv} + \mathbf{uw}$,
VS 4.3. $(r\mathbf{v})\mathbf{w} = r(\mathbf{vw})$,
VS 4.4. $\mathbf{vv} > 0$ unless $\mathbf{v} = \mathbf{0}$.

Thus if our vector space is to be euclidean, the pairing has to be such that VS 4.4 holds. The requirements VS 4.1 through 4.3 follow immediately from the linearity properties of the functionals. Now VS 4.4 in our case means that $\sum_i \sum_j x_i x_j g_{ij} > 0$ unless all $x_i = 0$. A function $\sum_i \sum_j g_{ij} x_i x_j$ is called a *quadratic form* in the x_i. We say that the quadratic form determined by the n^2 real numbers g_{ij} has to be *positive definite* if the space V with this pairing is euclidean. By positive definite we mean that the form assumes only positive values for nonzero x's.

A few simple examples should clarify this concept. The quadratic form $x_1^2 + x_1x_2 + x_2^2$ is positive definite because it can be written as

$$(x_1 + x_2/2)^2 + 3x_2^2/4,$$

which as the sum of squares can be zero only for $x_1 = x_2 = 0$. The form $x_1^2 - 2x_1x_2 + x_2^2$ (its g's are $g_{11} = 1$, $g_{12} + g_{21} = -2$, and $g_{22} = 1$) is not positive definite because it vanishes for $x_1 = x_2 = 1$, although it never becomes negative. The form $x_1^2 - x_2^3 + x_3^2$ assumes negative values.

Since we intend to use the inner product to define metrical concepts like length and distance, we say that the pairing, and concretely the g_{ij}, determine the *metric structure* of the space.

For our present purpose we choose a very simple metric structure, by pairing V, using the basis $\mathbf{e}_1, \ldots, \mathbf{e}_n$, with V^*, using the basis which is the cobasis of

$\mathbf{e}_1, \ldots, \mathbf{e}_n$, namely the $\epsilon_1, \ldots, \epsilon_n$ of Theorem 2-3.6. Then we have $\mathbf{e}_i\epsilon_j = g_{ij} = \delta_{ij}$, which makes the inner product

$$(x_1\mathbf{e}_1 + \cdots + x_n\mathbf{e}_n)(y_1\epsilon_1 + \cdots + y_n\epsilon_n) = x_1y_1 + \cdots + x_ny_n.$$

The quadratic form $x_1^2 + \cdots + x_n^2$ indeed is positive definite.

From now on we will not distinguish any more between the vectors of V and the functionals of V^* which are, of course, vectors of the space V^*. Accordingly we identify $\mathbf{e}_i$ with ϵ_i and write $y_1\mathbf{e}_1 + \cdots + y_n\mathbf{e}_n$ instead of $y_1\epsilon_1 + \cdots + y_n\epsilon_n$.

A few definitions follow. If for two vectors $\mathbf{v}$ and $\mathbf{w}$, $\mathbf{vw} = 0$, they are said to be *orthogonal*. For every $\mathbf{v}$, the product $\mathbf{vv}$ will be written $\mathbf{v}^2$ and its square root $\sqrt{\mathbf{v}^2}$ will be called the *length* of $\mathbf{v}$, denoted by $|\mathbf{v}|$. Of course, $|\mathbf{0}| = 0$ and $|\mathbf{v}| > 0$ when $\mathbf{v} \neq \mathbf{0}$. It should always be kept in mind that expressions like $\mathbf{vw}$, $\mathbf{v}^2$ or $|\mathbf{v}|$ are real numbers (scalars) and not vectors.

Euclidean spaces and isometries. An affine space A^n whose translation vectors are provided with the metric structure $g_{ij} = \delta_{ij}$ is called a *euclidean space*, E^n. Indeed E^2 turns out to be the euclidean coordinate plane and E^3 the euclidean space of three dimensions known from high school geometry. The *distance* between two points P and Q of E^n is defined as the length of the vector $\overrightarrow{PQ} = \mathbf{v}$ and denoted $|PQ|$ or $|\mathbf{v}|$. This distance meets the three requirements which usually are employed for the abstract definition of distance. For all points P, Q, and R in E^n,

Ds 1. $|PP| = 0$, $|PQ| > 0$ if $P \neq Q$.
Ds 2. $|PQ| = |QP|$.
Ds 3. $|PQ| + |QR| \geqq |PR|$, the *triangle inequality*.

The first two postulates are trivially satisfied in our case. We express the validity of Ds 3 in our space as a theorem, in a slightly different form.

Theorem 2-7.1. If $\mathbf{v}$ and $\mathbf{w}$ are two vectors in a euclidean vector space, then $|\mathbf{v} + \mathbf{w}| \leqq |\mathbf{v}| + |\mathbf{w}|$.

Proof. The theorem is trivial when $\mathbf{v}$ or $\mathbf{w} = \mathbf{0}$. Suppose that $\mathbf{v} \neq \mathbf{0}$, $\mathbf{w} \neq \mathbf{0}$. Then there exists a positive real β such that $|\mathbf{v}| = \beta|\mathbf{w}|$, and therefore $\mathbf{v}^2 = \beta^2\mathbf{w}^2 = \beta|\mathbf{v}|\,|\mathbf{w}|$. We have

$$0 \leqq (\beta\mathbf{w} - \mathbf{v})^2 = \beta^2\mathbf{w}^2 - 2\beta\mathbf{vw} + \mathbf{v}^2.$$

Consequently,

$$2\beta\mathbf{vw} \leqq \beta^2\mathbf{w}^2 + \mathbf{v}^2 = \beta|\mathbf{v}|\,|\mathbf{w}| + \beta|\mathbf{v}|\,|\mathbf{w}| = 2\beta|\mathbf{v}|\,|\mathbf{w}|,$$

and hence $\mathbf{vw} \leqq |\mathbf{v}|\,|\mathbf{w}|$. Now

$$|\mathbf{v} + \mathbf{w}|^2 = (\mathbf{v} + \mathbf{w})^2 = \mathbf{v}^2 + 2\mathbf{vw} + \mathbf{w}^2 \leqq \mathbf{v}^2 + 2|\mathbf{v}|\,|\mathbf{w}| + \mathbf{w}^2 = (|\mathbf{v}| + |\mathbf{w}|)^2.$$

We extract the positive square root on both sides and obtain the assertion.

A simple consequence from the definitions is:

Theorem 2–7.2 (Pythagoras' theorem). If in a triangle PQR, $\overrightarrow{PQ}$ and $\overrightarrow{PR}$ are orthogonal, then $|PQ|^2 + |PR|^2 = |RQ|^2$.

Figure 2–10

Proof (Fig. 2–10).

$$\overrightarrow{QR} = \overrightarrow{QP} + \overrightarrow{PR} = \overrightarrow{PR} - \overrightarrow{PQ}.$$

Hence

$$\begin{aligned} |RQ|^2 &= |\overrightarrow{PR} - \overrightarrow{PQ}|^2 \\ &= (\overrightarrow{PR} - \overrightarrow{PQ})^2 \\ &= \overrightarrow{PR^2} + \overrightarrow{PQ^2} - 2\overrightarrow{PR} \cdot \overrightarrow{PQ}. \end{aligned}$$

But $\overrightarrow{PR} \cdot \overrightarrow{PQ} = 0$ because of the orthogonality, and thus $|RQ|^2 = |PR|^2 + |PQ|^2$.

We have introduced into the affine space a metric structure which did not add or disqualify any of the points or subvarieties of A^n. It only adjoined a new relation between the existing objects. Thus every E^n is an A^n, but the converse is not true. The transformations of E^n will be affinities; however, since the metric stucture will have to be preserved, not all the affinities will be admissible. Hence the group of these transformations will form a subset of $\mathfrak{A}^n$ which is actually a subgroup.

An isometry is defined to be a distance-preserving affinity. The next theorem will show the importance of the isometries.

Theorem 2–7.3. The isometries, and they only, are the affinities that preserve inner products of vectors.

Proof. An affinity M which preserves inner products sends the inner product $\mathbf{v}^2$ into itself and is, therefore, an isometry. Conversely, let M be an isometry. Now

$$\begin{aligned} (\mathbf{v} + \mathbf{w})^2 &= \mathbf{v}^2 + 2\mathbf{v}\mathbf{w} + \mathbf{w}^2 \\ 2\mathbf{v}\mathbf{w} &= (\mathbf{v} + \mathbf{w})^2 - \mathbf{v}^2 - \mathbf{w}^2 \\ \mathbf{v}\mathbf{w} &= (|\mathbf{v} + \mathbf{w}|^2 - |\mathbf{v}|^2 - |\mathbf{w}|^2)/2. \end{aligned}$$

This expression will not be affected by any isometry, since it involves lengths of vectors, that is distances, only.

Orthonormal coordinate systems and orthogonal transformations. We will use a so-called *orthonormal* or *cartesian coordinate system*, whose basis vectors are mutually orthogonal and of length 1. We could have used such a convenient coordinate system also in our previous discussions (and actually have done so in some places). However, there was no point in doing so since length and orthogonality were not defined and certainly were not preserved in the affine or linear transformations that we had to use. In short, distance, length, and orthogonality were not affine concepts. Rather, they are euclidean in nature because they are invariant under the transformations of euclidean space, the isometries.

We are now looking for the affinities which map an orthonormal coordinate system onto an orthonormal coordinate system. We may disregard the translation part of an affinity because it preserves vectors and their lengths. A linear transformation which preserves distances and therefore also the orthonormality of a coordinate system will be called an *orthogonal transformation*, and the matrix which belongs to it an *orthogonal matrix*.

Theorem 2-7.4. For every orthogonal matrix A, $A' = A^{-1}$.

Proof. The basis vectors $\mathbf{e}_i$ are transformed by $A = (a_{ik})$ in such a way that $\mathbf{e}_i A = a_{i1}\mathbf{e}_1 + \cdots + a_{in}\mathbf{e}_n$. Now $\mathbf{e}_i\mathbf{e}_k = \delta_{ik}$, and since the transformation preserves the inner product, $\mathbf{e}_i A \cdot \mathbf{e}_k A = \delta_{ik}$, or the combination of the ith row of A with the kth column of A' is δ_{ik}. But this means $AA' = I$, and since A as the matrix of a linear transformation is nonsingular, $A' = A^{-1}$.

A converse to this theorem is provided by:

Theorem 2-7.5. If $A' = A^{-1}$, the transformation represented by A in an orthonormal coordinate system preserves the inner product and is, therefore, orthogonal.

Proof. A vector $\mathbf{v} = x_1\mathbf{e}_1 + \cdots + x_n\mathbf{e}_n$, with $|\mathbf{e}_k| = 1$, transforms into $x_1\mathbf{e}_1 A + \cdots + x_n\mathbf{e}_n A$. Now each of the $\mathbf{e}_k A$ has length 1 because $\mathbf{e}_k A = a_{k1}\mathbf{e}_1 + \cdots + a_{kn}\mathbf{e}_n$ and $a_{k1}^2 + \cdots + a_{kn}^2 = 1$, in view of $A' = A^{-1}$. Moreover, for the same reason, $\mathbf{e}_i A \cdot \mathbf{e}_k A = 0$ whenever $i \neq k$. Hence the $\mathbf{e}_i A$ form a basis for an orthonormal coordinate system. Then

$$(x_1\mathbf{e}_1 A + \cdots + x_n\mathbf{e}_n A)^2 = x_1^2 + \cdots + x_n^2 = |\mathbf{v}|^2,$$

which proves the invariance of length under the transformation belonging to A. By Theorem 2-7.3, A preserves the inner product.

Corollary 1. The columns $\mathbf{c}_i$ of an orthogonal matrix behave as do its rows, namely $\mathbf{c}_i\mathbf{c}_k = \delta_{ik}$.

Proof. If A is orthogonal, so is A', because $A' = A^{-1}$ implies

$$(A')' = A = (A^{-1})^{-1} = (A')^{-1}.$$

Corollary 2. If A is orthogonal, so is A^{-1}.

Proof. $(A^{-1})' = (A')' = A = (A^{-1})^{-1}$.

Corollary 3. The product of orthogonal matrices is orthogonal.

Proof. $(AB)' = B'A' = B^{-1}A^{-1} = (AB)^{-1}$.

From the last two corollaries the following theorem is easily obtained.

Theorem 2–7.6. The orthogonal transformations of V^n, and therefore also the orthogonal $n \times n$ matrices, form a group, the *orthogonal group*, $\mathcal{O}^n$.

Proof. The only part of the statement which has to be proved is the nonsingularity of every orthogonal matrix. Now, by definition, the matrix has an inverse, and in view of Exercise 14 of Section 2–3, it is nonsingular.

Obviously $\mathcal{O}^n$ is a subgroup of $\mathcal{GL}^n$. From Theorem 2–4.6 we see that for every $N \in \mathcal{O}^n$, $\mathcal{T}^n N = N\mathcal{T}^n$ holds. If we consider all the affinities NT with $N \in \mathcal{O}^n$ and $T \in \mathcal{T}^n$, we can show, just as in Theorem 2–4.8, that they form a subgroup of $\mathcal{A}^n$. This subgroup will be called the group $\mathfrak{M}^n$ of isometries. Again every isometry can be uniquely written as NT with a linear orthogonal part N and a translation part T.

Important subgroups of $\mathcal{O}^n$ and of $\mathfrak{M}^n$ will be obtained by studying determinants of orthogonal matrices.

Theorem 2–7.7. If a matrix A is orthogonal, then $\det A = 1$ or -1.

Proof. We have $(\det A)^2 = \det A \cdot \det A' = \det A \cdot \det A^{-1} = \det AA^{-1} = \det I = 1$, and consequently $\det A = \pm 1$.

Again the transformation is direct or opposite according to whether $\det A = 1$ or -1. Is this distinction of direct and opposite transformations dependent on the choice of the coordinate system? If A is an orthogonal matrix then, by Theorem 2–3.3, with respect to another orthonormal coordinate system it will turn into BAB^{-1}, where B is an orthogonal matrix. But then $\det BAB^{-1} = \det B \cdot \det A \cdot \det B^{-1} = \det A$.

Theorem 2–7.8. The direct orthogonal transformations of E^n form a group $\mathcal{O}^n_+ \lhd \mathcal{O}^n$, and the direct isometries form a group $\mathfrak{M}^n_+ \lhd \mathfrak{M}^n$. The only cosets are formed by the set $\mathcal{O}^n_-$ of all opposite orthogonal transformations and the set $\mathfrak{M}^n_-$ of all opposite isometries, respectively.

The proof will be left to the reader as an exercise.

EXERCISES

1. Consider a pairing of V^2 with $g_{11} = g_{22} = 0$ and $g_{12} = g_{21} = 1$. Which of the requirements for inner products are met and which are not?
2. Prove: The vector (a_1, a_2) is orthogonal to the vectors of the line whose equation in E^2 is $a_1x_1 + a_2x_2 + a_0 = 0$.
3. Prove: The vector (a_1, a_2, a_3) is orthogonal to all the vectors in the plane whose equation in E^3 is $a_1x_1 + a_2x_2 + a_3x_3 + a_0 = 0$.
4. Prove: Two lines in E^2 whose equations are $a_1x_1 + a_2x_2 + a_0 = 0$ and $b_1x_1 + b_2x_2 + b_0 = 0$ are orthogonal if and only if $a_1b_1 + a_2b_2 = 0$.

5. Two planes whose equations in E^3 are $a_1x_1 + a_2x_2 + a_3x_3 + a_0 = 0$ and $b_1x_1 + b_2x_2 + b_3x_3 + b_0 = 0$ are called orthogonal when they have only one line in common and when two vectors orthogonal to this line and each lying in either of the planes are orthogonal to each other. Prove that a necessary and sufficient condition for this orthogonality is

$$a_1b_1 + a_2b_2 + a_3b_3 = 0.$$

6. Prove: If P and Q are two points of E^3, then all points equidistant from P and Q lie on a plane. If A and B are two points of this plane then $\overrightarrow{AB} \cdot \overrightarrow{PQ} = 0$.
7. Prove: Three points P, Q, and R satisfy $|PQ| + |QR| = |PR|$ if and only if Q lies on the segment (PR), that is, if Q lies on the line PR and between P and R.
8. Construct a basis of an orthonormal coordinate system in E^3 such that one of its vectors is parallel to the vector (2, 1, 2). Is the resulting basis unique?
9. Prove: An orthogonal transformation is involutory if and only if its matrix A is *symmetric*, that is, $A = A'$.
10. If a *linear similitude* is defined as a linear transformation altering all distances by a factor k, show that every linear similitude has a matrix which can be written uniquely as kA, with A an orthogonal matrix.
11. If a *similitude* is a linear similitude followed by a translation, show that
    (a) the similitudes form a group, $\mathcal{S}^n$
    (b) $\mathcal{O}^n \lhd \mathcal{S}^n$.
12. Extend the diagram Fig. 1–10 by adding $\mathcal{A}$, $\mathcal{A}_+$, $\mathcal{GL}$, $\mathcal{GL}_+$, $\mathcal{O}$, and $\mathcal{O}_+$. Mark the normal subgroup relationship by double lines.
13. Prove Theorem 2–7.8.
14. (a) Why can one not define a euclidean vector space over the complex field?
    (b) What difficulty would arise over the rational field?

2–8 ISOMETRIES IN E^2 AND E^3

Isometries in E^2. As in Section 1–5, we will attempt to classify the isometries by studying invariant points. Our procedure will be the following. At first we will deal with the classification of the orthogonal transformations of the vectors in E^2. Then we will add translations and arrive at a classification of plane isometries. Finally we shall repeat this procedure for the three-dimensional case. The two-dimensional case is merely a review of the results of Sections 1–5 and 1–6, although the method is different.

An orthogonal transformation $\mathbf{v} \to \mathbf{v}A$, where $\mathbf{v} = x_1\mathbf{e}_1 + x_2\mathbf{e}_2$ and $A = (a_{ij})$, has an invariant vector when

$$x_1 = a_{11}x_1 + a_{21}x_2, \qquad x_2 = a_{12}x_1 + a_{22}x_2,$$

which would imply that the homogeneous equations

$$(a_{11} - 1)x_1 + a_{21}x_2 = 0 \quad \text{and} \quad a_{12}x_1 + (a_{22} - 1)x_2 = 0$$

have a nonzero solution (x_1, x_2). By Theorem 2–3.4, this is possible when and only when the matrix

$$B = \begin{bmatrix} a_{11} - 1 & a_{12} \\ a_{21} & a_{22} - 1 \end{bmatrix}$$

is singular.

First let us consider direct isometries, that is, $\det A = 1$. If there are invariant vectors, then $\det B = 0$. Now

$$\det B = \det A - a_{11} - a_{22} + 1 = 2 - a_{11} - a_{22},$$

or $a_{11} + a_{22} = 2$. However, from $a_{11}^2 + a_{21}^2 = a_{12}^2 + a_{22}^2 = 1$ we have $a_{11} \leqq 1$ and $a_{22} \leqq 1$. Hence we get $a_{11} = a_{22} = 1$, and therefore $a_{12} = a_{21} = 0$. Then $A = I$, and the only direct isometries with invariant vectors are translations.

If the direct isometry has no invariant vectors, then $\det B \neq 0$. Then

$$A^{-1} = \begin{bmatrix} a_{22} & -a_{12} \\ -a_{21} & a_{11} \end{bmatrix} = A' = \begin{bmatrix} a_{11} & a_{21} \\ a_{12} & a_{22} \end{bmatrix},$$

$a_{11} = a_{22}$, $a_{12} = -a_{21}$, and

$$A = \begin{bmatrix} a_{11} & -a_{21} \\ a_{21} & a_{11} \end{bmatrix}.$$

Here $a_{11} \neq 1$ because otherwise $\det B = 0$. The orthogonality condition now reads $a_{11}^2 + a_{21}^2 = 1$. We put $a_{11} = \cos\alpha$ and $a_{21} = \sin\alpha$, with $\alpha \neq 0$. Then we get

$$x_1 \to x_1 \cos\alpha - x_2 \sin\alpha, \qquad x_2 \to x_1 \sin\alpha + x_2 \cos\alpha,$$

a rotation about O through an angle $\alpha \neq 0$. As in Theorem 1–5.2, every nonidentical rotation about O followed by a translation is a rotation about some point.

We come now to the opposite isometries, $\det A = -1$. Again, when there are invariant vectors, we have $\det B = 0$. Then $\det B = \det A - a_{11} - a_{22} + 1 = -1 - a_{11} - a_{22} + 1 = 0$, and $a_{11} = -a_{22}$. The orthogonality of A requires that $a_{11}^2 + a_{21}^2 = 1 = a_{12}^2 + a_{22}^2$, and therefore $a_{21} = \pm a_{12}$. If $a_{21} = -a_{12}$, then

$$A = \begin{bmatrix} a_{11} & a_{12} \\ -a_{12} & -a_{11} \end{bmatrix}.$$

Since A is orthogonal, $-a_{11}a_{12} - a_{11}a_{12} = -2a_{11}a_{12} = 0$. Either $a_{11} = 0$ or $a_{12} = 0$. If $a_{11} = 0$, then $\det A = a_{12}^2 \geqq 0$, a contradiction to $\det A = -1$.

Hence $a_{12} = 0$, which may be considered as a special case of the second alternative, $a_{21} = a_{12}$; in this case we have

$$A = \begin{bmatrix} a_{11} & a_{12} \\ a_{12} & -a_{11} \end{bmatrix} = \begin{bmatrix} \cos\alpha & \sin\alpha \\ \sin\alpha & -\cos\alpha \end{bmatrix}.$$

If we let $x_1 + ix_2 = z$, then the isometry becomes $z \to \bar{z}(\cos\alpha + i\sin\alpha)$. This is, by Theorem 1–6.1, a reflection in the straight line through O which encloses with the positive x_1-axis a slope angle of $\alpha/2$.

Conversely, consider a reflection in a line not through O. It can always be handled by means of a coordinate translation such that the reflecting line passes through the new origin. Then the reflection can be represented, as in Theorem 1–6.2, as $T^{-1}UT$, where T is a translation and U a reflection in a line through the origin.

Finally, let A ($\det A = -1$) be an opposite isometry without invariant vectors ($\det B \neq 0$). Then we have

$$A^{-1} = \begin{bmatrix} -a_{22} & a_{12} \\ a_{21} & -a_{11} \end{bmatrix} = A' = \begin{bmatrix} a_{11} & a_{21} \\ a_{12} & a_{22} \end{bmatrix},$$

and $a_{11} = -a_{22}$. But then $\det B = \det A - a_{11} - a_{22} + 1 = -1 - a_{11} + a_{11} + 1 = 0$, a contradiction. Hence there are no opposite isometries without invariant vectors.

To summarize our results, we have:

Theorem 2–8.1. The direct isometries of E^2 are either translations or rotations. Every opposite isometry of E^2 is a line reflection followed by a translation (which may be I). Conversely, translations and rotations are direct isometries; line reflections are opposite isometries.

Thus we have established the analogy with our treatment of isometries in Chapter 1, and the further developments of Sections 1–5 and 1–6 may now be considered as valid in our discussion of the euclidean plane.

Angles. We have been using angles and their cosines and sines without giving a proper definition of an angle. Now this gap will be filled.

In Section 1–7 the group of two-dimensional vectors was shown to be essentially the same as the group of translations, only written additively instead of multiplicatively. Here is an analog for rotations. By Theorem 1–5.3, the rotations about any point of E^2 form a group, and all these groups (about different points) are isomorphic to $\mathcal{R}_O$, the group of rotations about O. The elements of $\mathcal{R}_O$ have the form

$$x_1 \to cx_1 - sx_2, \qquad x_2 \to sx_1 + cx_2, \qquad \text{with } c^2 + s^2 = 1. \tag{1}$$

Thus $\mathcal{R}_O$ is a one-parameter group which, moreover, is transitive on the "unit-circle," the circle about O with radius 1, by Theorem 2–6.2. A *half-line* originating from a point P (Fig. 2–11) is defined as the set of points containing with a fixed point Q all the points R on the line PQ such that either R is between P and Q, or Q is between P and R. Every half-line originating from O has one point on the unit circle, and every point on the unit circle lies on one half-line originating from O. Hence $\mathcal{R}_O$ is transitive on the set of half-lines through O. Each ordered pair of such half-lines, h_1 and h_2, determines an element of $\mathcal{R}_O$, and we call this element the *angle* $\measuredangle h_1 h_2 = \alpha$ (Fig. 2–12). If this rotation is given as Eq. (1), we furthermore define $c = \cos \alpha$ and $s = \sin \alpha$. If α_1 and α_2 are angles corresponding to rotations R_1 and R_2, respectively, then $\alpha_1 + \alpha_2$ corresponds to $R_1 R_2$. The angle that corresponds to the identity is $\alpha = 0$. The angles then form an additive group which is isomorphic to the multiplicative group $\mathcal{R}_O$. These definitions yield also the well-known rules for the cosine and the sine of sums of angles (cf. Exercise 10). It can be proved (Exercise 9) that angles are invariant under direct isometries, and this removes the impression that in the definition of the angle the origin had a privileged position. Actually, the isomorphism between the groups of rotations about different points is sufficient to take care of this doubt.

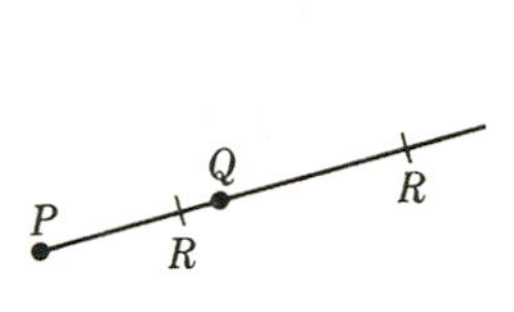

Figure 2–11

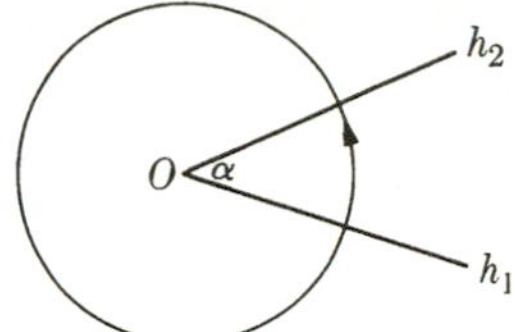

Figure 2–12

The following difficulty arises in the definition of the angle. For two given half-lines originating from a point, there is not only one rotation mapping the first onto the second, but rather an infinity of them. Thus the angle is determined by the two half-lines only up to a multiple of 2π (if angles are measured in radians). However, this multivaluedness is not reflected in the behavior of the sine and cosine functions, which are uniquely determined by the two given half-lines.

The introduction of the angle and its functions cos and sin makes it possible to express the inner product of vectors in an alternative way.

Theorem 2–8.2. If in E^2 the angle between two nonzero vectors $\mathbf{v}$ and $\mathbf{w}$ is α, then

$$\mathbf{vw} = |\mathbf{v}|\,|\mathbf{w}| \cos \alpha.$$

Proof. Let $\mathbf{v} = (x_1, x_2)$ and $\mathbf{w} = (y_1, y_2)$. A rotation of $\mathbf{v}$ through α carries $\mathbf{v}$ into

$$(x_1 \cos \alpha - x_2 \sin \alpha, x_1 \sin \alpha + x_2 \cos \alpha) = \mathbf{u}, \qquad \text{say}.$$

By hypothesis, $\mathbf{w}$ is a scalar multiple of $\mathbf{u}$, say $\mathbf{w} = c\mathbf{u}$. Then

$$\begin{aligned}\mathbf{vw} = c\mathbf{vu} &= c(x_1, x_2)(x_1 \cos \alpha - x_2 \sin \alpha, x_1 \sin \alpha + x_2 \cos \alpha) \\ &= c(x_1^2 \cos \alpha - x_1x_2 \sin \alpha + x_2x_1 \sin \alpha + x_2^2 \cos \alpha) = c(x_1^2 + x_2^2) \cos \alpha.\end{aligned}$$

But $|\mathbf{v}| = \sqrt{x_1^2 + x_2^2}$ and, since rotations are isometries, $|\mathbf{u}| = |\mathbf{v}|$, and therefore

$$|\mathbf{w}| = |c\mathbf{u}| = c|\mathbf{u}| = c|\mathbf{v}| = c\sqrt{x_1^2 + x_2^2}.$$

We substitute the obtained values for $|\mathbf{v}|$ and $|\mathbf{w}|$ and obtain

$$\mathbf{vw} = c(x_1^2 + x_2^2) \cos \alpha = c\sqrt{x_1^2 + x_2^2}\sqrt{x_1^2 + x_2^2} \cos \alpha = |\mathbf{v}|\,|\mathbf{w}| \cos \alpha,$$

as asserted.

Isometries in E^3. Let us turn to the three-dimensional case. Again we start with a study of invariant vectors of orthogonal transformations. In a somewhat more general form we look for a *characteristic vector* $\mathbf{v}$ of an orthogonal transformation L_A with matrix A which is sent by L_A into a vector $\lambda\mathbf{v}$ parallel to $\mathbf{v}$. Since L_A preserves lengths, the *characteristic root* λ has one of the values 1 or -1. If $A = (a_{ik})$, we have $\mathbf{v}A = \lambda\mathbf{v}$, or with $\mathbf{v} = (x_1, x_2, x_3)$,

$$a_{1i}x_1 + a_{2i}x_2 + a_{3i}x_3 = \lambda x_i, \qquad i = 1, 2, 3,$$

or

$$\begin{aligned}(a_{11} - \lambda)x_1 + a_{21}x_2 + a_{31}x_3 &= 0 \\ a_{12}x_1 + (a_{22} - \lambda)x_2 + a_{32}x_3 &= 0 \\ a_{13}x_1 + a_{23}x_2 + (a_{33} - \lambda)x_3 &= 0.\end{aligned}$$

This system of equations has a nontrivial solution if and only if

$$\det (A - \lambda I_3) = 0.$$

This leads to a cubic equation whose leading coefficient is -1. It has at least one real root, which, as remarked earlier, has to be 1 or -1. There exists therefore a vector $\mathbf{v} \neq \mathbf{0}$ with $\mathbf{v}L_A = \pm\mathbf{v}$. We choose a new orthonormal coordinate system in which the first basis vector is $\mathbf{e}_1 = |\mathbf{v}|^{-1}\mathbf{v}$. Let the corresponding orthogonal component transformation be L_K with matrix K. In the new coordinate system A becomes $K^{-1}AK$ and $\det K^{-1}AK = \det A$. Let $B = (b_{ij}) = K^{-1}AK$. Then $\mathbf{e}_1B = \lambda\mathbf{e}_1 = \pm\mathbf{e}_1$, or $b_{11} = \lambda$, $b_{12} = b_{13} = 0$. As a consequence of the orthogonality of B, $b_{21} = b_{31} = 0$, such that now

$$B = \begin{bmatrix}\lambda & 0 & 0 \\ 0 & b_{22} & b_{23} \\ 0 & b_{32} & b_{33}\end{bmatrix}.$$

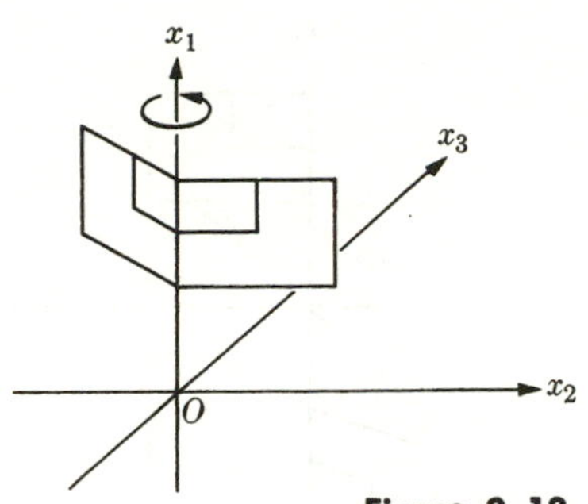

Figure 2-13

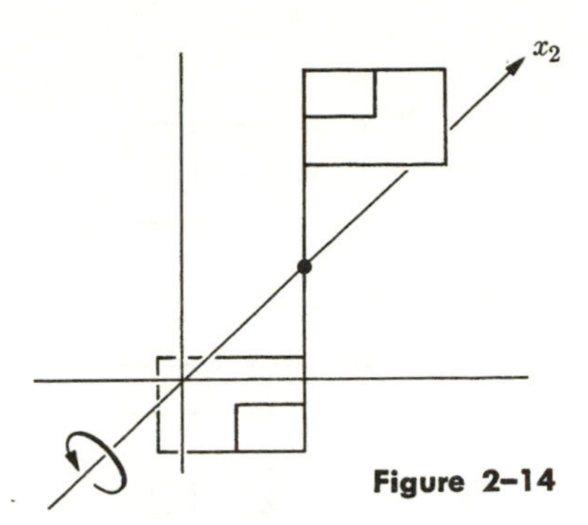

Figure 2-14

We distinguish between the Cases (1) det $A = 1$, and (2) det $A = -1$.

Case 1. Let det $A =$ det $B = 1$ and assume (a) $\lambda = 1$. Then

$$C = \begin{bmatrix} b_{22} & b_{23} \\ b_{32} & b_{33} \end{bmatrix}$$

is orthogonal and det $C = 1$. In view of our results in E^2 this isometry leaves invariant all points on the x_1-axis and only those (Fig. 2-13). A reasonable name for such an isometry is a *rotation about the x_1-axis*. In E^3 rotations about a point as used in E^2 would be meaningless and all three-dimensional rotations are assumed to be about a line. Now (b) let $\lambda = -1$. Then det $C = -1$. In the plane $x_1 = 0$ this yields a reflection in some line through the origin, according to Theorem 2-8.1. If we perform a coordinate change such that $\mathbf{e}_1$ is unchanged and $\mathbf{e}_2$ falls along the reflecting line, then B becomes

$$D = \begin{bmatrix} -1 & 0 & 0 \\ 0 & 1 & 0 \\ 0 & 0 & d \end{bmatrix}.$$

Since det $D =$ det $B = 1$, d has to be -1. Hence we obtain $x_1 \to -x_1$, $x_2 \to x_2$, $x_3 \to -x_3$, a rotation through 180° about the x_2-axis (Fig. 2-14).

We summarize Case 1 in

Theorem 2-8.3. Every direct isometry of E^3 is a rotation about a line followed by a translation (which, again, may be the identity).

For the opposite isometries:

Case 2. Let det $B =$ det $A = -1$, and assume (a) $\lambda = 1$. Then

$$B = \begin{bmatrix} 1 & 0 & 0 \\ 0 & & \\ 0 & \multicolumn{2}{c}{\mathrm{C}} \end{bmatrix}, \qquad \text{with det } C = -1.$$

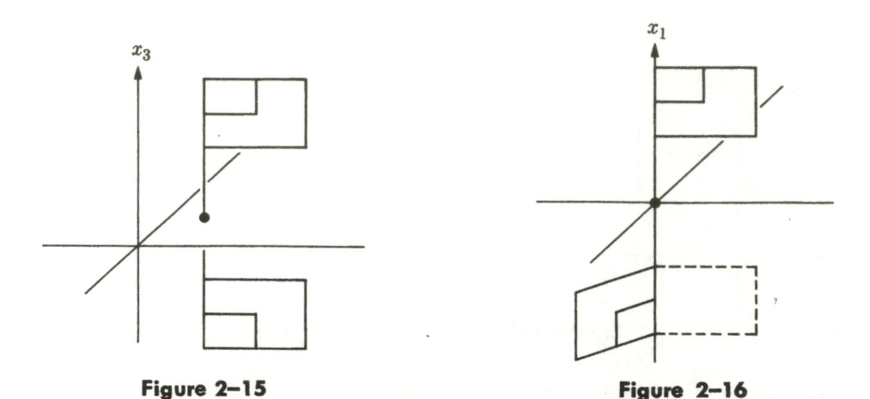

Figure 2–15

Figure 2–16

In the plane $x_1 = 0$ this is a line reflection and by the same manipulation as in Case 1 it may be brought into the form

$$\begin{bmatrix} 1 & 0 & 0 \\ 0 & 1 & 0 \\ 0 & 0 & -1 \end{bmatrix}.$$

The plane $x_3 = 0$ is pointwise invariant, and such an isometry, unless it is the identity, is called a *reflection in the plane* $x_3 = 0$ (Fig. 2–15). The motivation for this name is that this is exactly what happens when objects are reflected in a plane mirror. Again in E^3 we cannot speak of line reflections, which would be quite meaningless, and all reflections in E^3 are assumed to be either in points or in planes. Now (b) we let $\lambda = -1$, and we have

$$B = \begin{bmatrix} -1 & 0 \quad 0 \\ 0 & \\ 0 & \mathsf{C} \end{bmatrix}, \qquad \text{with } \det C = 1.$$

This (Fig. 2–16) is a reflection in the plane $x_1 = 0$ with a simultaneous rotation about the x_1-axis. Such a reflection in a plane with a simultaneous rotation about a line perpendicular to the plane is referred to as a *rotatory reflection.* In the special case of $b_{22} = -1$ we get

$$C = \begin{bmatrix} -1 & 0 \\ 0 & -1 \end{bmatrix}.$$

Then $x_1 \to -x_1$, $x_2 \to -x_2$, and $x_3 \to -x_3$, which describes an isometry which leaves invariant only the point O. If an isometry has, like this one, the property that it has only one invariant point which also serves as midpoint of

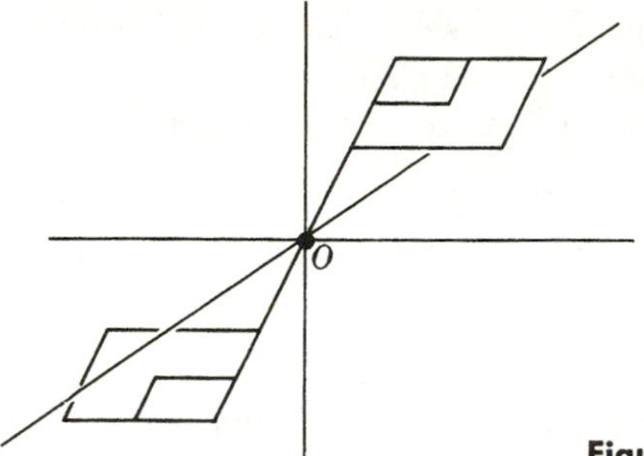

Figure 2–17

all segments joining a point and its image point, then we call it a *point reflection* (Fig. 2–17). This yields

Theorem 2–8.4. Every opposite isometry in E^3 is either a reflection in a plane followed by a translation or a rotatory reflection followed by a translation. Point reflections are a special case of rotatory reflections.

It is a remarkable fact that a point reflection in E^3 is an opposite isometry, while it is direct in E^2. In E^2 the point reflection is a rotation through 180°, while in E^3 it cannot be considered a rotation at all.

We can perform a further classification of the isometries by using invariant point properties. First we will treat the direct isometries.

Theorem 2–8.5. Every direct isometry in E^3 with an invariant point is a rotation about a line.

Proof. By Theorem 2–8.3, the isometry is a rotation followed by a translation. We choose the coordinate system so that the invariant line of the rotation is along the x_1-axis. Then

$$\begin{aligned} x_1 &\to x_1 + b_1, \\ x_2 &\to x_2 \cos\alpha - x_3 \sin\alpha + b_2, \\ x_3 &\to x_2 \sin\alpha + x_3 \cos\alpha + b_3. \end{aligned} \tag{2}$$

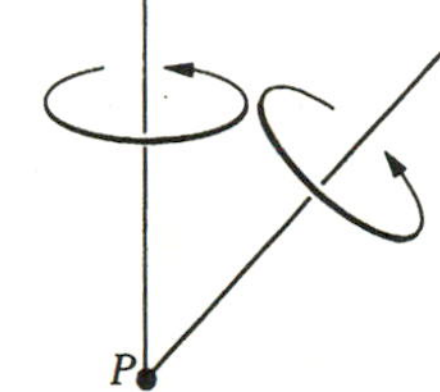

Figure 2–18

If there is an invariant point, then $b_1 = 0$. Disregarding the x_1-coordinate, we have, by Theorem 1–5.2, a rotation in each of the planes $x_1 = \text{const}$, and therefore a rotation about some line perpendicular to the plane $x_1 = 0$.

Corollary. The product of two rotations about lines through the same point P is itself a rotation about a line through P (Fig. 2–18).

Proof. The determinants of the matrices of orthogonal transformations multiply with the transformations, and hence the product of two rotations is direct. P is invariant under the product of the rotations. Thus we obtain a direct isometry preserving P which has to be a rotation about a line that contains all invariant points including P.

The following theorem is, in a sense, a converse to the corollary, above.

Theorem 2-8.6. If in E^3 two distinct lines k and l through a point P are given, then every rotation about a line through P can be generated by rotations about k and l.

The term "generated" has to be understood here as describing a combination of rotations by means of multiplications *and powers*. The rth power (r any real number) of a rotation through an angle α is the rotation about the same axis through $r\alpha$.

Proof. A euclidean coordinate transformation makes P the origin of an orthonormal system whose x_1-axis is k. A rotation about k through an angle $\alpha \neq 0$, followed by a rotation about the x_2-axis through an angle $\beta \neq 0$, is represented by

$$R = \begin{bmatrix} 1 & 0 & 0 \\ 0 & \cos\alpha & -\sin\alpha \\ 0 & \sin\alpha & \cos\alpha \end{bmatrix} \begin{bmatrix} \cos\beta & 0 & \sin\beta \\ 0 & 1 & 0 \\ -\sin\beta & 0 & \cos\beta \end{bmatrix}$$

$$= \begin{bmatrix} \cos\beta & 0 & \sin\beta \\ \sin\alpha\sin\beta & \cos\alpha & -\sin\alpha\cos\beta \\ -\cos\alpha\sin\beta & \sin\alpha & \cos\alpha\cos\beta \end{bmatrix}.$$

A point $[x_1, x_2, x_3]$ is invariant under R, that is, lies on the axis of R, when

$$x_1(\cos\beta - 1) + x_2\sin\alpha\sin\beta - x_3\cos\alpha\sin\beta = 0$$

and

$$x_2(\cos\alpha - 1) + x_3\sin\alpha = 0.$$

Solving these equations, we obtain the invariant points

$$\left[c\frac{\sin\beta}{1 - \cos\beta}, c\frac{\sin\alpha}{1 - \cos\alpha}, c\right]$$

for all real c, forming an axis through P. Thus for every given couple α and β there exists an axis of R. Are all these axes different for different choices of couples α, β? As representative of the axis points we choose the point $[x_1, x_2, x_3]$ with $x_3 = 1$. Then $x_1 = \sin\beta/(1 - \cos\beta)$. By squaring and substituting $\sin^2\beta = 1 - \cos^2\beta$, we get

$$(x_1^2 + 1)\cos^2\beta - 2x_1^2\cos\beta + x_1^2 - 1 = 0,$$

and this equation has the solutions $\cos\beta = (x_1^2 - 1)/(x_1^2 + 1)$ and 1. The second solution can be disregarded. The same argument yields $\cos\alpha = (x_2^2 - 1)/(x_2^2 + 1)$. Since

$$-1 \leqq \frac{x_i^2 - 1}{x_i^2 + 1} < 1, \qquad i = 1, 2,$$

each point $[x_1, x_2, 1]$ yields one couple α, β. Thus there is a one-to-one correspondence between all the couples α, β and all the lines through the origin as axes if the trivial values 0 and π for α and β are disregarded.

There is exactly one rotation R_α about k through α and one rotation S_β about the x_2-axis through β such that $R_\alpha S_\beta = T$ has l as axis. Let j be an arbitrary line through P. Again there is one rotation R_γ about k through γ and one S_δ about the x_2-axis through δ, such that $R_\gamma S_\delta$ has j as axis. Then

$$S_\beta = R_\alpha^{-1}T = R_{-\alpha}T, \qquad S_\delta = S_\beta^{\delta/\beta} = (R_{-\alpha}T)^{\delta/\beta}.$$

Now, $R_\gamma S_\delta = R_\gamma(R_{-\alpha}T)^{\delta/\beta}$ is a rotation about j, and every rotation about j is a power of this expression, which is thus shown to be generated by rotations about k and l.

Corollary. $\mathcal{O}_+^3$ is generated by the rotations about two distinct lines through the invariant point.

Definition. A *screw motion* is a rotation about a line followed by a translation along the invariant line (Fig. 2–19).

The trace of the continuous group generated by a screw motion is the thread line of a screw. Hence the name.

Figure 2–19

Theorem 2–8.7. A direct isometry in E^3 without invariant points is either a translation or a screw motion.

Proof. The transformation is given in Eq. (2).
Case (*i*), $b_1 = 0$. Then, in order to avoid an invariant point,

$$x_2 = x_2 \cos\alpha - x_3 \sin\alpha + b_2,$$
$$x_3 = x_2 \sin\alpha + x_3 \cos\alpha + b_3$$

must not have a solution. This means, by Theorem 2–3.4, that the determinant

$$\begin{vmatrix} \cos\alpha - 1 & -\sin\alpha \\ \sin\alpha & \cos\alpha - 1 \end{vmatrix} = 0.$$

Then $\cos^2\alpha - 2\cos\alpha + 1 + \sin^2\alpha = 2 - 2\cos\alpha = 0$ and consequently

$\cos\alpha = 1$ and $\sin\alpha = 0$. This implies

$$x_1 \to x_1, \qquad x_2 \to x_2 + b_2, \qquad x_3 \to x_3 + b_3,$$

which is a translation.

Case (*ii*), $b_1 \neq 0$. Then we have, by Theorem 1–5.2, a rotation about some line perpendicular to $x_1 = 0$ followed by the translation $x_1 \to x_1 + b_1$. This is a screw motion, which completes the proof.

Now we will attack the same questions for opposite isometries.

Theorem 2–8.8. Every opposite isometry of E^3 which has an invariant point is a rotatory reflection.

Proof. By Theorem 2–8.4, we must distinguish between two cases, (i) the isometry is a reflection in a plane followed by a translation, (ii) it is a rotatory reflection followed by a translation.

(i) If the reflecting plane is $x_1 = 0$, then we have the isometry in the form

$$x_1 \to -x_1 + b_1, \qquad x_2 \to x_2 + b_2, \qquad x_3 \to x_3 + b_3. \tag{3}$$

The existence of an invariant point implies $b_2 = b_3 = 0$, while $x_1 \to -x_1 + b_1$ means a reflection in the plane $x_1 = b_1/2$ when x_2 and x_3 are not changed. Thus, in this case, we get the special case of a rotatory reflection with a zero-rotation.

(ii) may be represented by

$$\begin{aligned} x_1 &\to -x_1 + b_1, \\ x_2 &\to x_2 \cos\alpha - x_3 \sin\alpha + b_2, \\ x_3 &\to x_2 \sin\alpha + x_3 \cos\alpha + b_3. \end{aligned} \tag{4}$$

The transformation of x_1 is independent of the transformation of the remaining two variables, so that it can be treated separately. Hence we have here a reflection in the plane $x_1 = b_1/2$, as in case (i), simultaneously with a rotation about a line perpendicular to the reflecting plane, according to Theorem 1–5.2. This is a rotatory reflection.

Definition. In E^3 a *glide reflection* is a reflection in a plane followed by a translation along a line lying in the reflecting plane (Fig. 2–20).

Theorem 2–8.9. Every opposite isometry of E^3 without invariant points is a glide reflection.

Proof (Fig. 2–21). Formula (3) is always a glide reflection because the vector $\mathbf{v} = (0, b_2, b_3)$ is parallel to the reflecting plane $x_1 = b_1/2$. If the isometry is of the type in Eq. (4) and there is no invariant point, then, as in the proof of Theorem 2–8.7, $\cos\alpha = 1$, $\sin\alpha = 0$, and Eq. (4) becomes Eq. (3), a glide reflection.

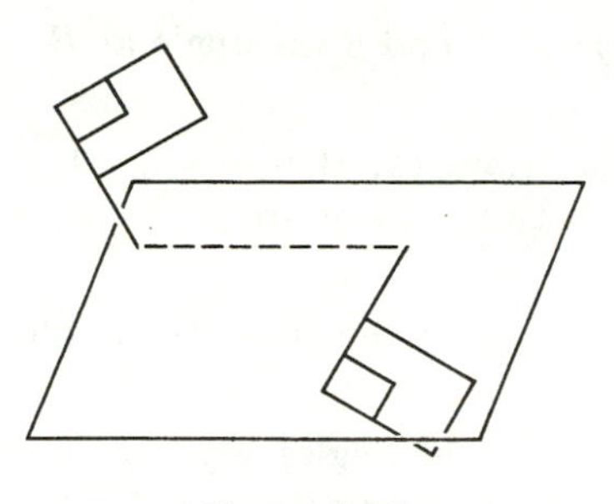
Figure 2-20

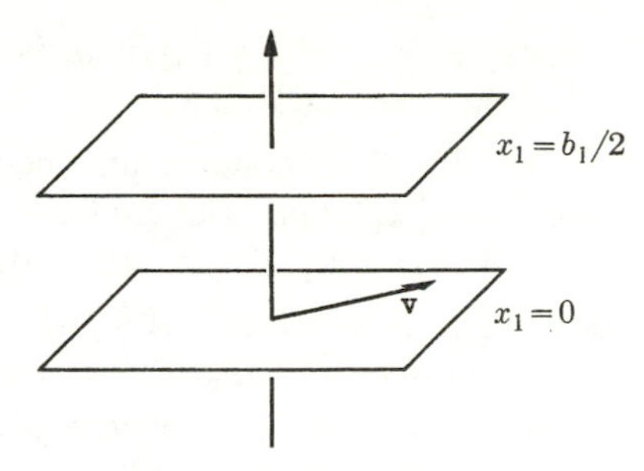

Figure 2-21

EXERCISES

1. What are the plane affinities that preserve orthogonality?
2. Prove: Every isometry of E^3 can be represented as a product of reflections in planes. The number of reflections used is even if and only if the isometry is direct.
3. What are the involutory isometries of E^3 which preserve the origin?
4. What is an isometry of E^3 if it has three noncollinear invariant points?
5. In E^3, what is the product of two reflections in planes? of reflections in three planes which have a straight line in common?
6. In E^3, what is the product of a half-turn about the x_1-axis followed by a half-turn about the x_3-axis? (A half-turn is a rotation through 180°.)
7. Write down the equations of some screw motion and find the trace of a one-parameter group generated by it.
8. $OBCD$ is a tetrahedron in E^3 with $B = [1, 0, 0]$, $C = [\frac{1}{2}, \sqrt{3}/2, 0]$, and $D = [\frac{1}{2}, \sqrt{3}/6, \sqrt{6}/3]$. Let affine mappings be determined such that O, B, C, D go, respectively, into

   (a) B, C, D, O (b) B, C, O, D (c) B, O, C, D.

   Find each transformation explicitly and check whether it is euclidean, direct, or opposite, and of what type. Find the invariant points and lines.
9. (a) Prove: Angles are invariant under direct similitudes.
   (b) What happens to angles under opposite similitudes?
10. Prove:
    (a) $\cos(\alpha + \beta) = \cos\alpha\cos\beta - \sin\alpha\sin\beta$,
    (b) $\sin(\alpha + \beta) = \sin\alpha\cos\beta + \sin\beta\cos\alpha$,
    (c) $\cos(-\alpha) = \cos\alpha$,
    (d) $\sin(-\alpha) = -\sin\alpha$.
11. Prove: In a triangle ABC with angle α between AB and AC,

$$|BC|^2 = |AB|^2 + |AC|^2 - 2|AB|\,|AC|\cos\alpha.$$

12. Prove: If in a triangle ABC α is the angle at A and β the angle at B, then $|AC| \sin \alpha = |BC| \sin \beta$.
13. Prove the "first congruence theorem" for triangles: If in a triangle ABC the sides $|AB|$ and $|BC|$ and the angle β between them are given, then the side $|AC|$ is uniquely determined.
14. Prove: If in a triangle ABC the three sides are given, then the angle β is uniquely determined (see Exercise 13).
15. Prove: If in a triangle ABC the side $|AB|$ and the angles α and β adjacent to it are given, then the remaining sides are uniquely determined.

2-9 CONTENT. ORIENTATION

Content and determinants. After having introduced the concepts of distance and length in euclidean geometry, it seems reasonable to ask about area and volume. In E^n we will use the more comprehensive term *content* which shall mean area in the two-dimensional case and volume in the three-dimensional one. The usual definition of the content, $\mu(S)$, of a set S of points in E^n refers to $\mu(S)$ as being a real-valued function of S with the properties

Ct 1. $\mu(S) \geqq 0$.
Ct 2. If S_1 and S_2 are two subsets of E^n and if $S_1 \cap S_2 = \emptyset$ (the null set), then $\mu(S_1 \cup S_2) = \mu(S_1) + \mu(S_2)$.
Ct 3. If S_0 is the set of all points represented in cartesian coordinates as $[x_1, \ldots, x_n]$ with $0 \leqq x_k \leqq 1$ for all $k = 1, \ldots, n$, then $\mu(S_0) = 1$.
Ct 4. $\mu(S)$ is independent of isometries.

It would be far beyond the scope of this treatment to discuss the content in all its generality. This is the task of other theories, e.g., measure theory and theory of integration. We will confine ourselves to a discussion of the content of the so-called *parallelotopes*, replacing Ct 2 by weaker requirements. A two-dimensional parallelotope is a parallelogram, a three-dimensional parallelotope is a parallelepiped, and an n-dimensional parallelotope is defined as a point set consisting of a point P_0 and all points obtained from P_0 by translating it through vectors $c_1\mathbf{s}_1 + \cdots + c_n\mathbf{s}_n$, where the $\mathbf{s}_k (k = 1, \ldots, n)$ are a basis of V^n and where all c_k run through the interval $0 \leqq c_k \leqq 1$. We denote the content of the parallelotope by $\mu(\mathbf{s}_1, \ldots, \mathbf{s}_n)$. This content does not depend on the choice of P_0 because, by Ct 4, it is not affected by isometries, and hence certainly not by translations which leave the $\mathbf{s}_k$ unchanged. Postulate Ct 3 now reads $\mu(\mathbf{e}_1, \ldots, \mathbf{e}_n) = 1$.

As mentioned before, Ct 2 will be replaced by the weaker requirements

Ct 2′. $\mu(\mathbf{s}_1, \ldots, \mathbf{s}_i, \ldots, \mathbf{s}_k, \ldots, \mathbf{s}_n) = \mu(\mathbf{s}_1, \ldots, \mathbf{s}_i + \mathbf{s}_k, \ldots, \mathbf{s}_k, \ldots, \mathbf{s}_n)$, $i \neq k$, $1 \leqq i, k \leqq n$.
Ct 2″. $\mu(\mathbf{s}_1, \ldots, c\mathbf{s}_i, \ldots, \mathbf{s}_n) = |c|\mu(\mathbf{s}_1, \ldots, \mathbf{s}_i, \ldots, \mathbf{s}_n)$ for all real c.

The significance of Ct 2″ is simple. If any one of the "edge vectors" $\mathbf{s}_k$ of the parallelotope is multiplied by a scalar c, then the content also is multiplied by $|c|$. The meaning of Ct 2′ will be illustrated (Fig. 2–22) for $n = 2$. The parallelograms $P_0P_1P_2P_3$, with edge vectors $\mathbf{s}_1$ and $\mathbf{s}_2$, and $P_0P_1P_4P_2$, with edge vectors $\mathbf{s}_1$ and $\mathbf{s}_1 + \mathbf{s}_2$, have equal areas by Ct 2′.

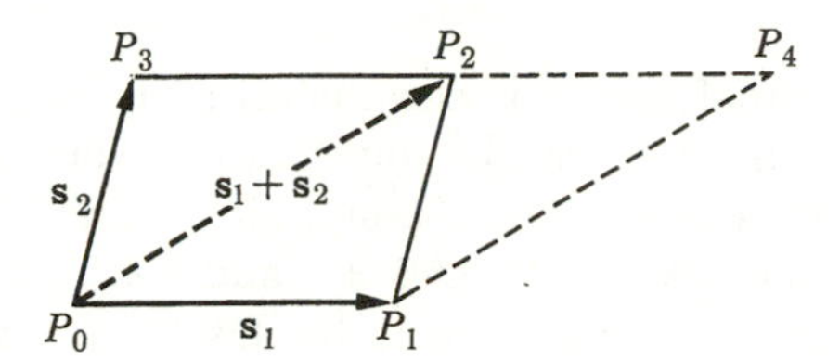

Figure 2–22

It can be proved that for parallelotopes the weakened postulates Ct 1, 2′, 2″, 3, and 4 are equivalent to the full postulates Ct 1, 2, 3, and 4. However, the proof will not be given here.

Theorem 2–9.1. If B is the $n \times n$ matrix with $\mathbf{s}_k$ $(k = 1, \ldots, n)$ as row vectors, then $\mu(\mathbf{s}_1, \ldots, \mathbf{s}_n) = |\det B|$ satisfies the postulates Ct 1, 2′, 2″, 3, 4.

Proof. Ct 1 is satisfied trivially. For Ct 3, $B = I_n$ and $\mu(\mathbf{e}_1, \ldots, \mathbf{e}_n) = |\det I_n| = 1$. Ct 2′ and Ct 2″ are the properties Dt 6 and Dt 7, respectively. In order to prove the validity of Ct 4, let the points of the parallelotope be transferred by an isometry LT whose linear part has the matrix A. Then, by Theorem 2–4.12, the vectors $\mathbf{s}_k$ are transformed by L such that

$$\mu(\mathbf{s}_1A, \ldots, \mathbf{s}_nA) = \left|\det \begin{bmatrix} \mathbf{s}_1A \\ \vdots \\ \mathbf{s}_nA \end{bmatrix}\right| = |\det BA| = |\det B|\,|\det A|$$

$$= |\det B| = \mu(\mathbf{s}_1, \ldots, \mathbf{s}_n),$$

since A is orthogonal and has $\det A = \pm 1$.

As a consequence of this theorem we adopt the definition $|\det B|$ for the content of the parallelotope.

The uniqueness of this definition is implied by the following.

In Section 2–3 we defined the determinant in an inductive and purely algebraic way. An alternative definition of the determinant

$$D = \det \begin{bmatrix} \mathbf{s}_1 \\ \vdots \\ \mathbf{s}_n \end{bmatrix}$$

as a scalar function of n n-dimensional vectors $\mathbf{s}_1, \ldots, \mathbf{s}_n$ is given by the following postulates.

Dt 1*. D remains unchanged if some $\mathbf{s}_i$ is replaced by $\mathbf{s}_i + \mathbf{s}_j$, with $i \neq j$.

Dt 2*. D changes to cD if some $\mathbf{s}_i$ is replaced by $c\mathbf{s}_i$ for every scalar c.

Dt 3*.
$$\det\begin{bmatrix}\mathbf{e}_1\\ \vdots \\ \mathbf{e}_n\end{bmatrix} = 1.$$

We will not give the proof that this new definition is indeed equivalent to that in Section 2–3. The full proof can be found in the work of O. Schreier and E. Sperner, *Modern Algebra and Matrix Theory*, New York: Chelsea, 1951.

The analogy between Dt 1*, 2*, and 3*, and Ct 2′, 2″, and 3 is obvious. Hence the definition of the determinant by Dt 1*, 2*, and 3* is sometimes called the "geometrical definition" of the determinant.

How do affinities affect the content of a parallelotope? The answer is the following theorem.

Theorem 2–9.2. Under an affinity whose linear part has the matrix A, the content of a parallelotope is multiplied by $|\det A|$.

Proof. With the same notation as in Theorem 2–9.1, the affinity sends $\mu(\mathbf{s}_1, \ldots, \mathbf{s}_n) = |\det B|$ into $|\det BA| = |\det B|\,|\det A| = \mu(\mathbf{s}_1, \ldots, \mathbf{s}_n)|\det A|$. The translation part of the affinity cannot influence this discussion because it does not affect vectors.

Corollary. An affinity preserves contents if and only if its $\det A = \pm 1$. In particular, all euclidean transformations preserve contents.

EXAMPLE. What is the area of a parallelogram whose 3 given vertices are $P_k = [x_k, y_k]$, $k = 0, 1, 2$, provided P_0P_1 and P_0P_2 are sides of the parallelogram? How does the area change under a similitude

$$x \to ax - by + e,$$
$$y \to bx + ay + f,$$

where a and b are not both zero?

It should be noted that in complex notation this similitude could be written as $z \to (a + bi)z + (e + fi)$, with $z = x + iy$.

The edge vectors of the parallelogram are

$$\mathbf{s}_1 = (x_1 - x_0, y_1 - y_0),$$
$$\mathbf{s}_2 = (x_2 - x_0, y_2 - y_0),$$

and hence the area

$$C = \left|\det\begin{bmatrix}x_1 - x_0 & y_1 - y_0\\ x_2 - x_0 & y_2 - y_0\end{bmatrix}\right|.$$

When in this equation $ax_k - by_k + e$ is substituted for x_k, and $bx_k + ay_k + f$ for y_k ($k = 0, 1, 2$), we get for the new area

$$\left|\det\begin{bmatrix} a(x_1 - x_0) - b(y_1 - y_0) & b(x_1 - x_0) + a(y_1 - y_0) \\ a(x_2 - x_0) - b(y_2 - y_0) & b(x_2 - x_0) + a(y_2 - y_0)\end{bmatrix}\right|$$

$$= \left|\det\begin{bmatrix} x_1 - x_0 & y_1 - y_0 \\ x_2 - x_0 & y_2 - y_0\end{bmatrix}\begin{bmatrix} a & b \\ -b & a\end{bmatrix}\right| = C(a^2 + b^2),$$

that is, $|\det A|$ times the original area.

Orientation. In the second part of this section we shall deal with a concept that has been mentioned in Section 1–4, the *orientation*. There we discussed the equal or different orientation of congruent triangles. However, this restriction of the orientation concept to euclidean geometry and to two dimensions is not necessary. The following more general definition will be seen to contain the treatment of Section 1–4 as a special case.

Let an arbitrary basis of V^n be given. Then for every basis of V^n there exists a well-defined n-dimensional affinity which takes the first basis into the second basis. If the transformation is direct, the bases are said to have the same orientation or to be equally oriented. If the transformation is opposite, the bases are called differently oriented. An equivalent treatment would speak about n-simplexes that are taken into each other by transformations of $\mathcal{A}^n$. Thus equally oriented simplexes are mapped on each other by direct affinities.

Theorem 2–9.3. The bases of V^n (or the n-simplexes of A^n) are divided into 2 classes such that any two elements of the same class are equally oriented, and that any two elements from different classes are differently oriented. Any interchange of two of the vectors in the basis (simplex) changes the orientation. (The ground field is assumed to be ordered.)

Proof. Since the direct transformations form a group, $\mathcal{GL}_+^n$ (or $\mathcal{A}_+^n$, respectively), the relation "equally oriented" is an equivalence relation. That is, if B_1, B_2, and B_3 are arbitrary bases (n-simplexes), and if $\sim$ stands for "equally oriented," then (i) $B_1 \sim B_1$, (ii) $B_1 \sim B_2$ implies $B_2 \sim B_1$, and (iii) $B_1 \sim B_2$ and $B_2 \sim B_3$ imply $B_1 \sim B_3$. It follows that if B is any basis (n-simplex), all the images of B under direct transformations, and only these, form the first class. Now, interchanging two vectors in B is achieved by a transformation whose matrix is the identity matrix with two lines interchanged. By Dt 2 the determinant is then -1, and the original basis (simplex) B and the modified basis (simplex), B', are differently oriented. Hence B' does not belong to the first class. All the bases (simplexes) that are equally oriented with B' form a second class. Every basis (simplex) is obtained from B either by a direct or an opposite affinity and thus belongs to either of the classes.

Let us study the significance of orientation in the cases $n = 1$, 2, and 3. For $n = 1$, a segment has two possible orientations.

For $n = 2$, a triangle has two possible orientations, and coordinate systems might be equally or differently oriented (Figs. 2-23 and 24).

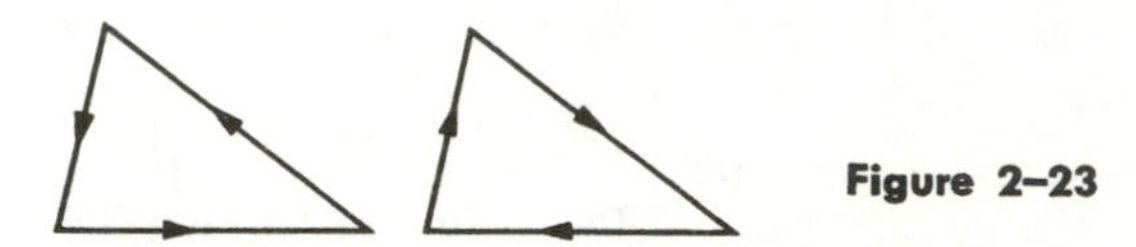

Figure 2-23

For $n = 3$ we distinguish between a right-hand coordinate system and a left-hand coordinate system which are differently oriented and therefore cannot be transformed into each other by direct affinities (Fig. 2-25).

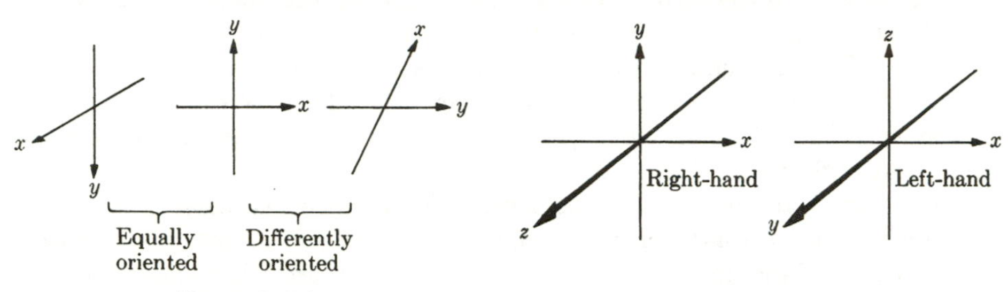

Figure 2-24

Figure 2-25

It should not be overlooked that we have merely defined *equal* orientation and *different* orientation. We also know that these two relations are preserved by linear transformations. Theorem 1-4.8 is a special case of this. However, we have not worked out any way of defining one orientation as, say, positive, and it is characteristic that nonmathematical concepts have to be used in order to make this possible. The name "right-hand system," meaning that the x-, y-, and z-axes resemble the first three fingers of the right hand, or the usage of "counterclockwise" in reference to the vertices of a triangle are examples.

EXERCISES

Assume $K = R$.

1. Prove: The volume of a tetrahedron in E^3 with vertices $[x_i, y_i, z_i]$ and with $i = 1, 2, 3, 4$, is the absolute value of

$$\frac{1}{6}\begin{vmatrix} x_1 & y_1 & z_1 & 1 \\ x_2 & y_2 & z_2 & 1 \\ x_3 & y_3 & z_3 & 1 \\ x_4 & y_4 & z_4 & 1 \end{vmatrix}.$$

2. Prove: The area of a triangle in E^2 with vertices $[x_i, y_i]$, with $i = 1, 2, 3$, is the absolute value of

$$\frac{1}{2}\begin{vmatrix} x_1 & y_1 & 1 \\ x_2 & y_2 & 1 \\ x_3 & y_3 & 1 \end{vmatrix}.$$

3. Prove: The equation of a line through $[x_i, y_i]$, with $i = 1, 2$, in E^2, is

$$\begin{vmatrix} x & y & 1 \\ x_1 & y_1 & 1 \\ x_2 & y_2 & 1 \end{vmatrix} = 0.$$

The equation of a plane through $[x_i, y_i, z_i]$ with $i = 1, 2, 3$, in E^3 is

$$\begin{vmatrix} x & y & z & 1 \\ x_1 & y_1 & z_1 & 1 \\ x_2 & y_2 & z_2 & 1 \\ x_3 & y_3 & z_3 & 1 \end{vmatrix} = 0.$$

4. The content of an n-simplex can be defined as being $1/n!$ of the content of the parallelotope spanned by the same n vectors. Discuss this for $n = 1$, 2, and 3.

5. Prove:
(a) The content-preserving n-dimensional affinities form a group, say $\mathcal{G}^n$.
(b) $\mathcal{G}^n \cap \mathcal{A}^n_+$ is normal in $\mathcal{G}^n$. It is called the *equiaffine* group.

6. Which of the following bases of V^2 have the same orientation as the orthonormal basis $\mathbf{e}_1$, $\mathbf{e}_2$, and which have a different orientation?

(a) $\mathbf{e}_2, -\mathbf{e}_1$ (b) $\mathbf{e}_1 + \mathbf{e}_2, -\mathbf{e}_2$ (c) $\mathbf{e}_2, \mathbf{e}_1 - 3\mathbf{e}_2$.

Draw figures and test the smallest positive angle leading from the first basis vector of each basis to the second vector.

7. Prove: A basis $\mathbf{f}_1 = a\mathbf{e}_1 + b\mathbf{e}_2$, $\mathbf{f}_2 = c\mathbf{e}_1 + d\mathbf{e}_2$ of V^2 has the same orientation as the orthonormal basis $\mathbf{e}_1$, $\mathbf{e}_2$ if and only if $\sin \sphericalangle \mathbf{f}_1\mathbf{f}_2 > 0$.

8. Prove: A simplex spanned by $\mathbf{v}_1, \ldots, \mathbf{v}_n$ has the same orientation as the simplex spanned by the ordered basis vectors if and only if the determinant having $\mathbf{v}_1, \ldots, \mathbf{v}_n$ as ordered rows is positive.

9. Two points P and Q are defined to lie on different sides of a line l (in A^2) or a plane p (in A^3) if the line PQ meets l (or p) in a point R such that R lies between P and Q. Prove (a) that P and Q lie on different sides of a line

AB if and only if the triangles ABP and ABQ have different orientation, (b) that P and Q lie on different sides of a plane determined by three noncollinear points A, B, and C if and only if the tetrahedra $ABCP$ and $ABCQ$ have different orientation.

10. Let P be the vertex of an angle $\measuredangle kl$ in E^2, and let Q and R be points lying on the two sides of the angle. If the orientation of PQR is like that of $O[1, 0][0, 1]$, call the angle positive. Show that the sine of a positive angle is positive.

2-10 FINITE AFFINE PLANES

In this section we shall discuss affine planes over certain special finite fields.

Theorem 2-10.1. For p a prime integer, the integers mod p form a field which has p elements.

This field is called $GF(p)$, as a special case of the more complicated $GF(p^n)$ for integers $n \geq 1$. (The abbreviation is short for *Galois field*, another term for a finite field, named for the French mathematician Évariste Galois, 1811–1832).

Proof. In order to prove the validity of the field properties, consider the p numbers $0, 1, \ldots, p - 1$ as representatives of their classes which form the set S. Then it is easy to show that the postulates Fd 1, 2, 4, and 5 are satisfied. For Fd 3, the only difficulty arises regarding the existence of the multiplicative inverse. Let a be nonzero, that is, not divisible by p. Consider all ab where b ranges through all the $p - 1$ nonzero elements. All these ab are distinct because if $ab = ac$, then $a(b - c)$ is divisible by p. But since a is not divisible by p, $b - c$ must be a multiple of p. But then $b = c$, mod p. Hence $ab = ac$ if and only if $b = c$, and all the ab are distinct. There are $p - 1$ of them and none of them is 0 because a and b are nonzero. But then ab assumes as values all $p - 1$ elements of S, and among them 1, and in this case b is the inverse of a.

Remark. Had we taken a composite number instead of p, this proof would have been impossible. Namely, modulo a composite number, say $m = pq$, there are at least two integers p and q, both positive and $<m$, whose product is m. On the other hand, there are finite fields with a composite number of elements, but they will not be discussed here.

EXAMPLE. In $GF(2)$ the addition and multiplication tables are

+	0	1
0	0	1
1	1	0

and

·	0	1
0	0	0
1	0	1

Now consider the two-dimensional vector space over $GF(p)$ and call it $V^2(p)$. Each ordered couple of elements of $GF(p)$ is a vector of $V^2(p)$ or a point of the affine plane $A^2(p)$, which sometimes is designated $AG(2, p)$, for "affine geometry of dimension 2 over $GF(p)$." Since $GF(p)$ has p elements, there are p^2 vectors and therefore p^2 points. By Theorem 2–4.10, straight lines have the equations

$$a_1x_1 + a_2x_2 + a_0 = 0. \tag{1}$$

If $a_1 \neq 0$, we multiply with a_1^{-1} and obtain

$$x_1 + a_1^{-1}a_2x_2 + a_1^{-1}a_0 = 0.$$

There are p^2 possibilities of choosing the two coefficients in this equation. If $a_1 = 0$ and $a_2 \neq 0$, there are p possibilities for

$$x_2 + a_2^{-1}a_0 = 0.$$

There is no straight line for $a_1 = a_2 = 0$. Hence there is a total of $p^2 + p$ lines.

In Eq. (1) all p elements of $GF(p)$ can be substituted for x_1, and for each of them a value of x_2 is obtained. Hence every line contains p points. On the other hand, through every point $[c_1, c_2]$ there is a line $x_1 = c_1$ and p lines $x_2 - c_2 = m(x_1 - c_1)$, one for each possible value of m. All these $p + 1$ lines through $[c_1, c_2]$ are distinct.

We summarize.

Theorem 2–10.2. For every $GF(p)$ there is an affine plane whose points are the p^2 ordered couples $[x_1, x_2]$ with $x_1, x_2 \in GF(p)$. This plane has $p^2 + p$ distinct lines. Through each point pass $p + 1$ lines, and on each line lie p points.

How can we visualize these finite geometries? Let us deal with the simple cases $p = 2$ and $p = 3$. For $p = 2$ there are 4 points, O, B, C, and D. There are 6 lines, OB, BC, CD, OD, OC, and BD. If $\|$ means "parallel," then $OB\|CD$, $OD\|BC$, and, surprisingly, $OC\|BD$. The two diagonals of the quadrangle $OBCD$ are, therefore, parallel. If $O = [0, 0]$, $B = [1, 0]$, $C = [1, 1]$, and $D = [0, 1]$, then the equations of OC and BD are, respectively, $x_1 + x_2 = 0$ and $x_1 + x_2 + 1 = 0$, and it is analytically obvious that they are parallel. Indeed they do not have any point in common. In Fig. 2–26 all the points which bear the same name have to be considered as abstractly identical. These relations will be recognized more clearly if we draw the plane on the surface of a doughnut-shaped "torus" (Fig. 2–27).

In the case $p = 3$, there are 9 points, $O, B, C, D, E, F, G, H, J$. The 12 lines, arranged in four groups of parallel triples, are $OBC\|DEF\|GHJ$, $ODG\|BEH\|CFJ$, $OEJ\|BFG\|CDH$, and $CEG\|OFH\|BDJ$. In order not to confuse the figure, the lines in Fig. 2–28 have not been drawn in all possible positions. Once more a clearer understanding could be obtained by drawing the figure on a torus.

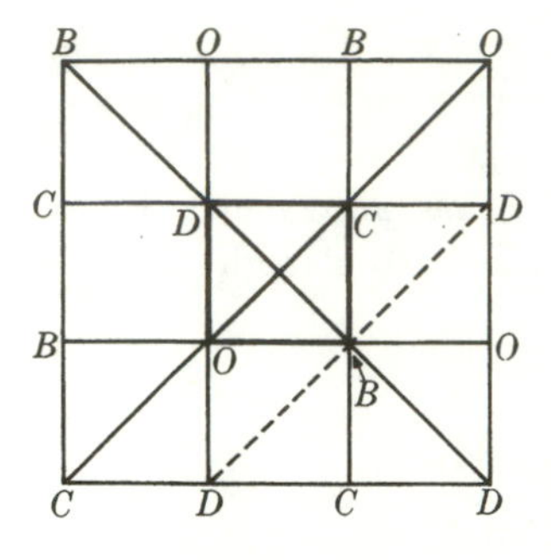

Figure 2–26

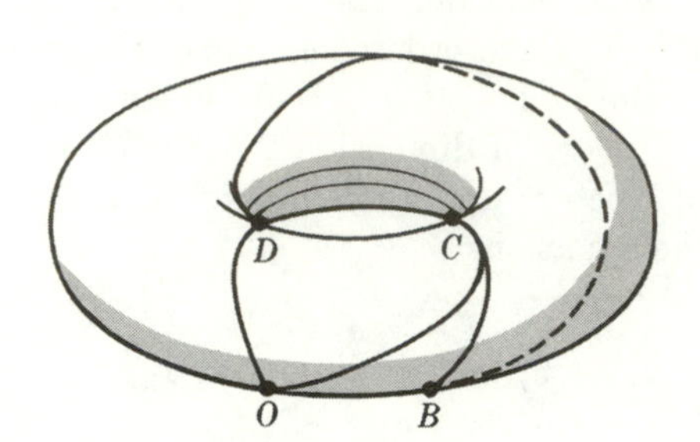

Figure 2–27

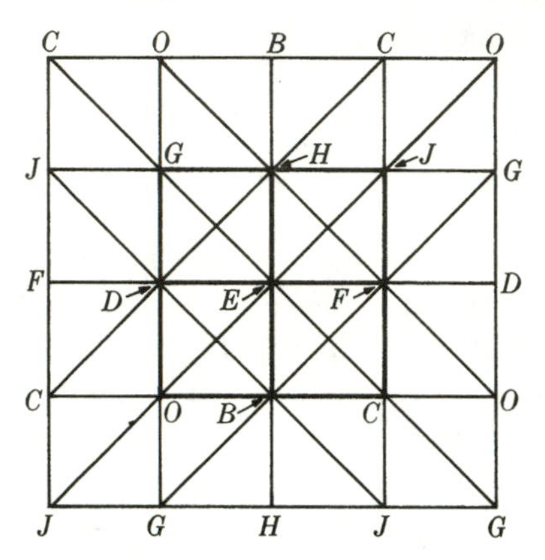

Figure 2–28

There, for instance, the segment ODG could be extended beyond G to reach O from the other side, and no abstract identification of points would be necessary.

The planes exhibited here are not all possible finite planes over Galois fields. There is a Galois field of order q whenever q is a power of a prime number. However, when q is not a prime, the geometry of the finite plane is more complicated than the simple case studied by us. As an example, here are the addition and multiplication tables of $GF(2^2)$.

+	0	1	2	3
0	0	1	2	3
1	1	0	3	2
2	2	3	0	1
3	3	2	1	0

·	0	1	2	3
0	0	0	0	0
1	0	1	2	3
2	0	2	3	1
3	0	3	1	2

An interesting geometrical property of this plane will be explored in the exercises.

Is a euclidean vector space possible over $GF(p)$? The answer is no. The length of a vector exists only if the square root can be extracted from every non-negative element of the field. This is impossible for Galois fields because positiveness of an element cannot be properly defined ($2 = -1$, mod 3, is it positive?). Moreover, there exist elements in every Galois field that do not have a square root, e.g., mod 3, 2 does not have a square root.

In the same manner as affine planes over Galois fields, affine n-spaces over Galois fields can be constructed. This will not be attempted here.

EXERCISES

1. Which of the 7 elements of $GF(7)$ have square roots in the field?
2. How many points, lines, and planes has $A^3(2)$?
3. Do the 3 lines $3x_1 + 2x_2 + 2 = 0$, $2x_1 + x_2 = 0$, and $x_1 - 3x_2 + 1 = 0$ have a common point in $A^2(R)$? in $A^2(5)$?
4. Find the intersection point of the 3 medians of the triangle [0, 0][1, 0][3, 1] in $A^2(5)$.
5. Prove: In $A^2(2)$ the two diagonals of every parallelogram are parallel to each other. The same holds in $A^2(2^2)$ and for any other field that has *characteristic* 2, that is, $b + b = 0$ for all b in the field.
6. Prove: In $A^2(2)$ and $A^2(2^2)$ midpoints do not exist.
7. In the figure describing $A^2(3)$, let $O = [0, 0]$, $B = [1, 0]$, $C = [2, 0]$, $D = [0, 1]$, $G = [0, 2]$, etc.
   (a) What is the midpoint of E and J? of O and H?
   (b) Choose any parallelogram and show that the diagonals bisect each other.
8. How many points of $A^2(R)$ satisfy the equation $x_1^2 + x_2^2 = 0$? How many points of $A^2(5)$ do?
9. (a) How many affinities exist in $A^2(2)$?
   (b) How many of these are linear?
   (c) Can you distinguish between direct and opposite affinities?
   (d) Which of the linear transformations have invariant lines?
10. In $A^2(5)$ find all affinities leaving invariant all points on $2x_1 + x_2 + 3 = 0$.
11. Which of the results of Sections 2–4 and 2–5 do not carry over when the vector spaces over the reals are replaced by vector spaces over
    (a) $GF(p)$, p a prime (b) the complex field?

    Which property of the real numbers is required by these results and absent in the cases (a) and (b)?

CHAPTER 3

Projective Geometry and Noneuclidean Geometries

In this chapter we continue the introduction to various geometric spaces which are coordinatized by means of vector spaces. Projective spaces, like affine spaces, are not vector spaces, and a procedure has to be found which adapts vector spaces to the task of coordinatization. In the case of projective spaces this is done by considering the one-dimensional vector subspaces rather than the vectors themselves. Some of the most interesting aspects and invariants of projective geometry are then developed. Of particular interest is the way in which affine and euclidean concepts appear as specializations of notions in projective geometry. Most often, the coordinate field is the field of reals, but occasionally a glimpse into the possible generalizations for other fields is permitted, in particular for the complex field. A projective model is defined for noneuclidean, that is, hyperbolic and elliptic geometries, a model which shows that these, as well as euclidean geometry, can be considered as projective geometry with a special metric imposed.

The idea of the transformation group belonging to a geometry serves as the basic tool for all these investigations. Finally an outline of possible further developments is presented, along with a list of books for suggested reading.

3–1 PROJECTIVE SPACES

Basic facts. When the concept of affine n-space was introduced, the starting point was an abstract n-dimensional vector space. In order to define projective n-space, we start from an $(n + 1)$-dimensional vector space V^{n+1} over a field K. Every two vectors $\mathbf{v}$ and $\mathbf{w}$ of V^{n+1} will be considered as "projectively equivalent" if they are linearly dependent, that is, if there is a scalar $r \neq 0$ such that $\mathbf{w} = r\mathbf{v}$. Obviously this equivalence satisfies Ev 1 through 3 of page 12, and the vectors of V^{n+1} are thus partitioned into equivalence classes. The null vector forms a class by itself. If a class contains a vector $\mathbf{v}$, it will be called $[\mathbf{v}]$. Evidently $[\mathbf{v}] = [r\mathbf{v}]$ for all nonzero scalars r. Each class distinct from $[\mathbf{0}]$ is called a *projective point*. The set of all projective points in V^{n+1} is said to be the *projective n-space* over K, denoted $P^n(K)$ or, for short, P^n. Thus the points of P^n are the one-dimensional subspaces of V^{n+1}. The two-dimensional subspaces of V^{n+1} are the *projective lines*, the three-dimensional subspaces the *projective planes*, and the n-dimensional subspaces the *projective hyperplanes*, if $n > 3$. If $[\mathbf{u}]$, $[\mathbf{v}]$, and $[\mathbf{w}]$ are distinct points of P^n, then the set of all $r\mathbf{u} + s\mathbf{v}$, for all scalars r and s, represents a projective line, designated by $[\mathbf{u}] + [\mathbf{v}]$. Similarly, $r\mathbf{u} + s\mathbf{v} + t\mathbf{w}$ is a projective plane, denoted $[\mathbf{u}] + [\mathbf{v}] + [\mathbf{w}]$.

The following theorem and its corollary will be helpful for visualizing basic relations between points and lines in projective geometry.

Theorem 3–1.1. In P^2, (i) any two distinct points belong to exactly one line, (ii) any two distinct lines have exactly one point in common.

Proof. Two one-dimensional subspaces of V^{n+1} are distinct if and only if a nonzero vector of $[\mathbf{v}]$ and a nonzero vector of $[\mathbf{w}]$ are linearly independent. However we choose from the two subspaces $\mathbf{v}$ and $\mathbf{w}$ as linearly independent vectors, we always obtain the same two-dimensional subspace of V^{n+1} spanned by them. Thus indeed the line is uniquely determined by the two points. Now let S_1 and S_2 be two distinct lines, that is, two-dimensional subspaces. Consider $S_1 + S_2$. This subspace cannot be two-dimensional because in that case S_1 and S_2 could not be distinct. Hence necessarily $\dim (S_1 + S_2) > 2$. But P^2 itself was formed by V^3 and thus $\dim (S_1 + S_2)$ cannot exceed 3 and therefore is $= 3$. By Theorem 2–1.9,

$$\dim (S_1 + S_2) + \dim (S_1 \cap S_2) = \dim S_1 + \dim S_2.$$

Here we have $3 + \dim (S_1 \cap S_2) = 2 + 2$, and consequently $\dim (S_1 \cap S_2) = 1$. This means that S_1 and S_2 share a common one-dimensional subspace, that is, one common point.

Corollary. In every $P^n (n > 0)$, any two distinct points belong to exactly one line.

Homogeneous coordinates and duality. We need a suitable system of coordinates for P^n. First we choose a basis $\mathbf{e}_0, \ldots, \mathbf{e}_n$ of V^{n+1}. Then every point of P^n can be represented in the form $[\mathbf{x}] = [\sum_0^n x_j\mathbf{e}_j]$. We call the point $[\mathbf{x}]$ or $[x_0, \ldots, x_n]$ if $x_0, \ldots, x_n$ are the components of $\mathbf{x}$. The numbers $x_0, \ldots, x_n$ are the *homogeneous coordinates* of the point $[\mathbf{x}]$. These coordinates are not determined uniquely because multiplication of all coordinates with any fixed nonzero scalar would yield the same point. Thus a point is represented by the ratio of the coordinates $x_0 : x_1 : \cdots : x_n$. Conversely, the ratio of every $(n + 1)$-tuple of scalars, not all of which are zero, determines a unique point of P^n.

Let $[\mathbf{x}] = [x_0, \ldots, x_n], \ldots, [\mathbf{z}] = [z_0, \ldots, z_n]$ be $n + 1$ points of P^n which lie in the same hyperplane. This means that $[\mathbf{x}], \ldots, [\mathbf{z}]$ are all contained in the same n-dimensional subspace of V^{n+1}. In other words, $\mathbf{x}, \ldots, \mathbf{z}$ are linearly dependent, which happens if and only if

$$\begin{vmatrix} x_0 & \cdots & x_n \\ \vdots & & \vdots \\ z_0 & \cdots & z_n \end{vmatrix} = 0. \tag{1}$$

We have, therefore,

Theorem 3–1.2. In P^n, $n+1$ points $[\mathbf{x}] = [x_0, \ldots, x_n], \ldots, [\mathbf{z}] = [z_0, \ldots, z_n]$ belong to the same hyperplane if and only if Eq. (1) holds.

Corollary 1. The equation of a straight line in P^2 is of the form $u_0x_0 + u_1x_1 + u_2x_2 = 0$, where not all the u's are zero.

Proof. If a line is given, then it contains at least two distinct points, $[\mathbf{y}] = [y_0, y_1, y_2]$ and $[\mathbf{z}] = [z_0, z_1, z_2]$. For every point $[\mathbf{x}] = [x_0, x_1, x_2]$ on the line, we have

$$\begin{vmatrix} x_0 & x_1 & x_2 \\ y_0 & y_1 & y_2 \\ z_0 & z_1 & z_2 \end{vmatrix} = 0,$$

or

$$(y_1z_2 - y_2z_1)x_0 + (y_2z_0 - y_0z_2)x_1 + (y_0z_1 - y_1z_0)x_2 = 0.$$

All the coefficients u_j vanish only in the case when $\mathbf{y}$ and $\mathbf{z}$ are linearly dependent, that is, when $[\mathbf{y}]$ and $[\mathbf{z}]$ coincide. On the other hand, if $[\mathbf{x}]$ is not on the line, $\mathbf{x}$, $\mathbf{y}$, and $\mathbf{z}$ could not be linearly dependent and the determinant would not vanish. This completes the proof.

The corollary implies that there is a one-to-one correspondence between the lines of P^2 and the nonzero ordered triples of scalars, if proportional triples are considered identical. The same was true about the points of P^2. Just as a point $[x_0, x_1, x_2]$ is determined by the ratio of the three homogeneous coordinates x_0, x_1, and x_2, a line is given when the ratio of its three *homogeneous line coordinates* u_0, u_1, and u_2 is known. In complete analogy to the homogeneous point coordinates, the line will be denoted by $\langle \mathbf{u} \rangle = \langle u_0, u_1, u_2 \rangle$. So far as incidence of points and lines in P^2 is concerned, points and lines play exactly the same roles, and every statement about incidence of points and lines in the plane will remain true if the words "point" and "line" are interchanged. This is the so-called *duality principle* in the projective plane, and a statement after the interchange will be called the *dual* statement.

When a theorem involving only incidence has been found, the dual theorem, that is, that theorem in which the words "point" and "line" have been interchanged, is automatically true. This might require simple grammatical changes. For instance, "Two distinct points lie on exactly one line" might sound unwieldy in the translation, "Two lines lie on exactly one point," and we might replace "lie on" by "pass through" or "have . . . in common." The advantage of the term "incident" is that it expresses a mutual relation between point and line and does not have to be replaced when an interchange is desired. The two parts of Theorem 3–1.1 can then be joined and rephrased:

"In P^2, two distinct $\begin{Bmatrix}\text{points}\\ \text{lines}\end{Bmatrix}$ are incident with exactly one $\begin{Bmatrix}\text{line}\\ \text{point}\end{Bmatrix}$."

Corollary 2. The equation of a plane in P^3 is of the form

$$u_0x_0 + u_1x_1 + u_2x_2 + u_3x_3 = 0.$$

The proof will be left as an exercise.

Here again a duality principle exists. In P^3, planes and points play the same roles so far as mere incidence statements are concerned. This is the duality principle for P^3. For instance, it will be proved in the exercises that, if two points lie in a plane q, then every point of their straight line join lies in q. Since this statement involves only incidences in P^3, its dual is automatically true. It reads, "If two planes pass through a point Q, then every plane passing through the line common to the two planes also passes through Q." Note that in this duality lines correspond to lines. It might seem that there is a difference between the two statements. While for the first statement, Theorem 3–1.1 ensured that there is a line joining the two points, the reader might be in doubt about the existence of the line common to the two planes in the dual statement. However, this doubt will be removed in Exercise 1.

The algebraic background for the duality principle and its application lies in the theory of the dual vector spaces. The ordered triples $\langle u_0, u_1, u_2\rangle$ representing lines in P^2 and the quadruples $\langle u_0, u_1, u_2, u_3\rangle$ representing planes in P^3 stand for the one-dimensional subspaces of linear functionals of V^3 and V^4, respectively. By Theorem 2–3.6 and its corollary these functionals form vector spaces isomorphic to V^3 and V^4, respectively. The duality principle is a geometric expression of this isomorphism between a vector space and its dual space. In general there exists in P^n a duality between points and hyperplanes. When we use homogeneous coordinates in P^n, we have to consider points $[\mathbf{x}] = [x_0, \ldots, x_n]$ characterized by the ratio of their point coordinates and hyperplanes $\langle \mathbf{u}\rangle = \langle u_0, \ldots, u_n\rangle$ determined by the ratio of their *hyperplane coordinates*. The isomorphism between the vector space and its dual was determined after the bases of the two spaces had been chosen. The same is true for the geometric duality; only after the choice of the bases can it be said which hyperplane corresponds to a given point, and vice versa.

Projective space and affine space. How is a projective space P^n related to an affine space A^n? Suppose that the basis of the V^{n+1} that is used for the coordinatization of P^n consists of $\{\mathbf{e}_0, \ldots, \mathbf{e}_n\}$ and that of the V^n which gives rise to A^n consists of $\{\mathbf{e}_1, \ldots, \mathbf{e}_n\}$. We distinguish between two categories of points $[\mathbf{x}]$ of P^n: those with $x_0 \neq 0$, and those with $x_0 = 0$. Each point of the first kind can be written as $[x_0, \ldots, x_n] = [1, x_1/x_0, \ldots, x_n/x_0]$. Thus we have a bijective mapping of all these points of P^n onto all points of A^n, namely $[1, y_1, \ldots, y_n] \in P^n$ corresponds to $[y_1, \ldots, y_n] \in A^n$. There are no points in A^n which correspond to the points of the second kind in P^n.

In order to show that the described mapping is meaningful, we will prove that it preserves incidence. We perform the proof for P^2 and A^2. For higher dimensions the proof is analogous.

Suppose the three points $[x_0, x_1, x_2]$, $[y_0, y_1, y_2]$, and $[z_0, z_1, z_2]$ with nonzero x_0, y_0, and z_0 are collinear. Then, by Theorem 3–1.2,

$$0 = \begin{vmatrix} x_0 & x_1 & x_2 \\ y_0 & y_1 & y_2 \\ z_0 & z_1 & z_2 \end{vmatrix} = x_0 y_0 z_0 \begin{vmatrix} 1 & x_1/x_0 & x_2/x_0 \\ 1 & y_1/y_0 & y_2/y_0 \\ 1 & z_1/z_0 & z_2/z_0 \end{vmatrix},$$

and therefore

$$\begin{vmatrix} 1 & x_1/x_0 & x_2/x_0 \\ 1 & y_1/y_0 & y_2/y_0 \\ 1 & z_1/z_0 & z_2/z_0 \end{vmatrix} = 0.$$

This is precisely the condition for the collinearity of the corresponding affine points, in view of Exercise 19 of Section 2–4.

No duality should be expected in A^n. Since A^n was obtained from P^n by removal of only *one* hyperplane and *many* points in this hyperplane, the balance between points and hyperplanes, which made duality in P^n possible, does not exist in A^n. Thus in A^n any two points are incident with a line, while parallel lines are not incident with any common point.

Since the points of the first kind in a projective space contain the images of all the points of the corresponding affine space, and in addition there are also the points of the second kind, it is worth while to look for a model consisting of an affine space extended by the latter, so-called *improper* points. Since all improper points correspond to projective points with $x_0 = 0$, these projective points lie in a projective hyperplane and their model images all belong to an affine linear $(n - 1)$-variety, that is, a straight line in the case of A^2 and a plane in the case of A^3.

Another property of the improper points is provided by using parallelism. If $a_1x_1 + a_2x_2 + a_0 = 0$ is the equation of a straight line l in A^2, all its parallels can be written in the same form, with the same a_1 and a_2, but with varying a_0. Conversely, this equation with any scalar replacing a_0 will always represent a parallel of l. In the projective plane the corresponding equation is $a_1x_1/x_0 + a_2x_2/x_0 + a_0 = 0$ or $a_0x_0 + a_1x_1 + a_2x_2 = 0$. The only point with $x_0 = 0$ lying on it is $[0, a_2, -a_1]$. Since a_0 does not appear in its coordinates, all the lines corresponding to parallels of l contain this improper point. Therefore, in the affine model of the projective plane, all parallel lines meet in exactly one improper point, and every improper point is the meeting point of a line and all its parallels. Thus every two lines intersect in exactly one point, in accordance with Theorem 3-1.1. The analog of this situation in A^3 is that all planes parallel to a fixed plane contain exactly one improper line, and every improper line is common to a set of parallel planes.

We have proved the following:

Theorem 3-1.3. A model of P^1 is provided by A^1 augmented by one improper point. A model of P^2 is provided by A^2 augmented by a line called the *improper line* all of whose points are improper points. Every proper line of the augmented A^2 contains exactly one improper point. Proper lines of the augmented A^2 have an improper point in common if and only if they are parallel. An affine model of P^3 consists of A^3 augmented by a so-called *improper plane* whose points are the improper points. Every proper plane of A^3 contains exactly one improper line and every proper line of the augmented A^3 contains exactly one improper point. Two proper planes (lines) of the augmented A^3 have an improper line (point) in common if and only if they are parallel.

The augmented affine plane and the augmented affine three-space are coordinatized by homogeneous coordinates which are identical with the corresponding homogeneous coordinates of P^2 or P^3. The usual affine coordinates are sometimes called *nonhomogeneous coordinates.*

The construction of the augmented affine plane is independent of the coordinate system. It is performed by assigning an improper point to each of the possible slopes of lines, including the "slope ∞" of the lines $x_1 = \text{const}$, and by assuming that all the improper points lie on an improper line. On the other hand, when P^2 is given, the corresponding affine plane and its improper line do depend on the choice of the coordinate system. In P^2 all lines are equivalent, in that each can be chosen arbitrarily as the line $x_0 = 0$, so that in the model its image becomes the improper line. This explains the often-heard statement that "P^2 becomes A^2 after the removal of one line." An analogous situation holds in higher dimensions.

The word "augmented" will sometimes be omitted when no ambiguity results. Thus we might mention improper points in an affine space although, properly speaking, there is no such thing unless the space is augmented.

The real projective plane. We will attempt to study the augmented affine plane over the real field in order to obtain information about the real projective plane, denoted $P^2(R)$. Let a straight line in the plane be given by two proper points on it, say $[x_0, x_1, x_2]$ and $[y_0, y_1, y_2]$. Then every point on the line except $[\mathbf{y}]$ can be represented as $[x_0 + ry_0, x_1 + ry_1, x_2 + ry_2]$ by suitable choice of the real number r. For $r = -x_0/y_0$, we obtain the improper point of the line,

$$[0, x_1 - x_0y_1/y_0, x_2 - x_0y_2/y_0] = [\mathbf{z}].$$

For

$$r = -x_0/y_0 + \epsilon \quad \text{and} \quad r = -x_0/y_0 - \epsilon,$$

with ϵ a small positive number, we get (Fig. 3–1)

$$[\mathbf{z}'] = [\epsilon y_0, x_1 - x_0y_1/y_0 + \epsilon y_1, x_2 - x_0y_2/y_0 + \epsilon y_2],$$
$$[\mathbf{z}''] = [-\epsilon y_0, x_1 - x_0y_1/y_0 - \epsilon y_1, x_2 - x_0y_2/y_0 - \epsilon y_2],$$

respectively. These are two proper points, and by using nonhomogeneous coordinates we can see that these two points lie in different quadrants and very far from the origin, if ϵ is made sufficiently small. Thus, with decreasing ϵ, the points move to opposite sides farther and farther from the origin, to the limit $\epsilon = 0$ where they coincide and become improper. In other words, the proper lines of the affine model of $P^2(R)$ are closed, and every two proper points determine two segments between them, one of which contains an improper point. Improper points of this model are "points at infinity."

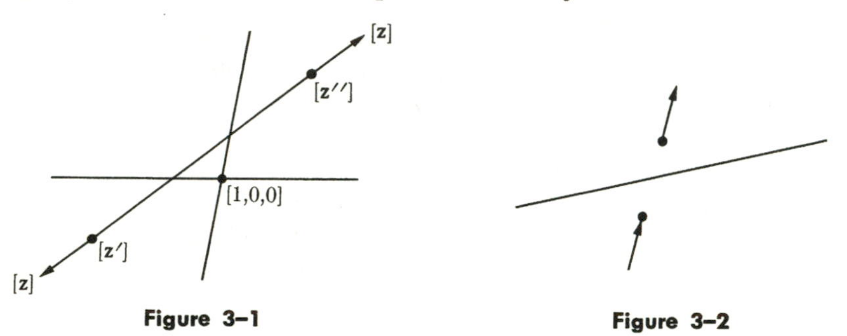

Figure 3–1 **Figure 3–2**

We will conclude this section with a few topological observations which should serve to clarify some concepts, although no proofs will be given. The euclidean plane is separated into two half-planes by every straight line, and it is impossible to get from one half-plane to the other on a continuous curve without crossing the line. In $P^2(R)$ this is not so (Fig. 3–2). No line in $P^2(R)$ separates $P^2(R)$. That is, given a line in $P^2(R)$ and two points not on it, there is a continuous path, namely a straight line segment, containing both points but not intersecting the line. In other words, a line has only one side in $P^2(R)$. In order to cut this projective plane into two disjoint pieces we therefore need, in addition to the first straight line, one other line. If on a surface at most k closed curves are possible which do not cut the surface into disjoint pieces, we say that the surface is $(k + 1)$-*ply connected.* Thus $P^2(R)$ is doubly connected ($k = 1$). The euclidean plane, however, is simply connected ($k = 0$). If a disk (that is, a circle and all its interior points) is cut out of the projective plane, it can be continuously mapped on, or in topological language is *homeomorphic* to, a *Möbius strip* (Fig. 3–3) in which the two ends of a rectangular strip are pasted together after being twisted through 180°. In Fig. 3–4 points with the same number could be pasted together along the narrow sides of the rectangle, to produce the Möbius strip in Fig. 3–3. The side 1–1 does not divide the strip surface into separate halves because there exists the path 3–4 via 2. The Möbius strip is a *one-sided* surface; an ant could run around the entire surface without crossing its edge, that is, the boundary is a simple closed curve.

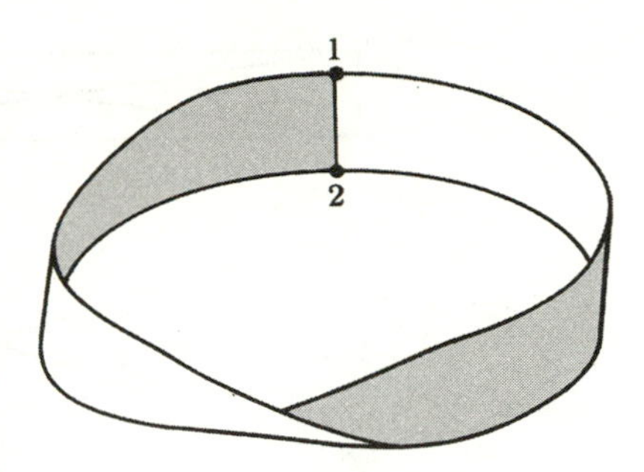

Figure 3–3

An important observation is that on the real projective plane orientation becomes meaningless, which can be seen by tracing a quandrangle when it is pushed across the improper line (Fig. 3–5). A surface is said to be *orientable* if it can be entirely covered, with neither interstices nor overlapping, by a finite number of triangles, so that an orientation can be assigned to each triangle such that, if two triangles $\triangle_1$ and $\triangle_2$ share a side, the orientation of this side in $\triangle_1$ is the opposite of that in $\triangle_2$. A sphere is orientable (Fig. 3–6). The Möbius strip is not (Fig. 3–7), nor is the projective plane which is obtained from the Möbius strip by "closing the hole," that is, pasting together (in imagination only!) the sides 1–5–2 so that equal numbers again fall on each other. The real projective plane thus is one-sided and nonorientable. In fact, a one-sided surface is never orientable.

Where does the name "projective" stem from? And where do we encounter situations like those of $P^2(R)$ in everyday life? If a plane π (Fig. 3–8) is reproduced on a plane π', this is done by joining every point Q of π to the eye of the observer. Let the eye be at E. The line QE intersects π' at Q', say, and then Q is said to be *projected* on Q' from E. It can be shown that this gives rise to a projective plane on π'. Let us check the behavior of parallel lines in π, when projected on π' (Fig. 3–9). The dotted line in π', the horizon, contains the points in which the images of the parallels of π meet. In our notation this is the improper line, and K and L are improper points.

Another kind of model for $P^1(R)$ and $P^2(R)$ is the following (Fig. 3–10). Consider all the straight lines in E^2 through O. Each of them, except the x-axis, intersects the line $y = 1$ in exactly one point. Thus we obtain a one-to-one correspondence between the lines of the euclidean bundle, except the line $y = 0$, and the points of the line $y = 1$. If we want the x-axis also to be represented, we have to augment the image line by an improper point "at infinity." Thus we obtain the bundle as a model of $P^1(R)$. If we replace the bundle by the three-dimensional bundle of lines through O and the image line by the plane $y = 1$ in E^3, then this procedure yields a model of $P^2(R)$. All the lines in the plane

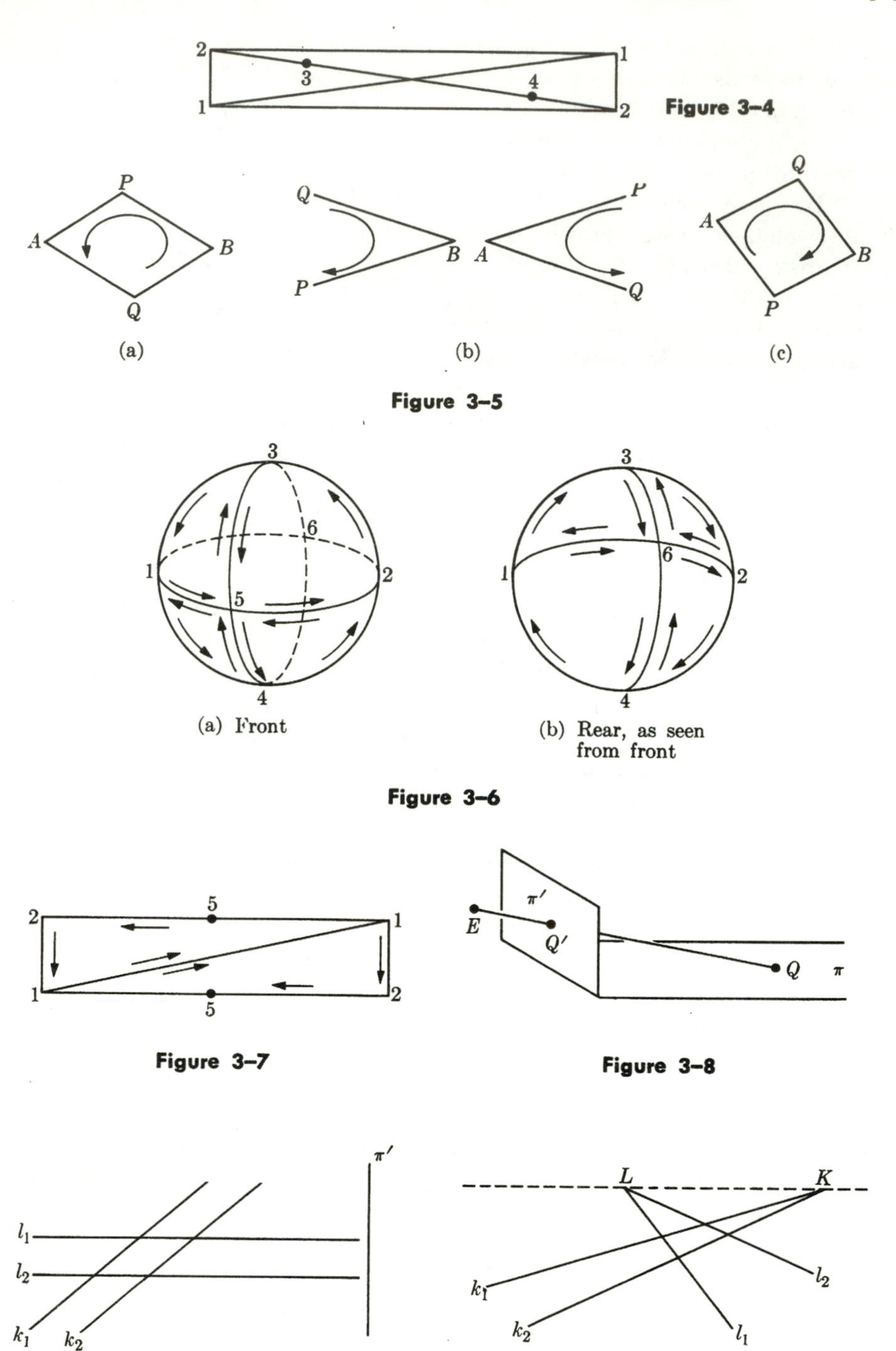

Figure 3–4

(a) (b) (c)

Figure 3–5

(a) Front

(b) Rear, as seen from front

Figure 3–6

Figure 3–7

Figure 3–8

(a) Preimage in π

(b) Projection in π'

Figure 3–9

$y = 0$ are mapped on improper points, and the plane $y = 0$ meets the plane $y = 1$ in the improper line. This is the so-called *bundle model* of $P^2(R)$.

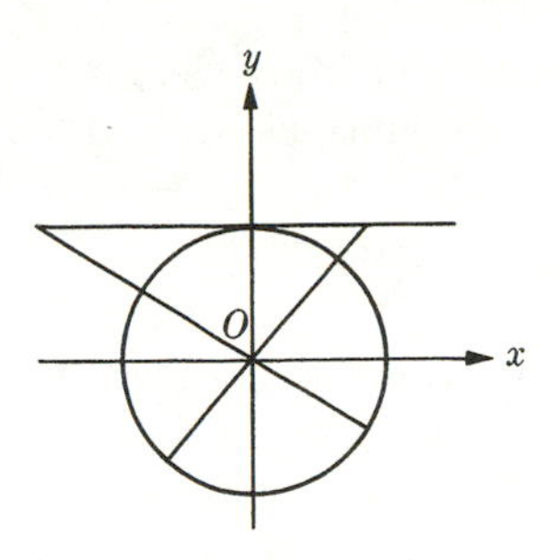

Figure 3-10

The bundle model can be clarified by an abstract identification of antipodal points of a circle or a sphere. If the bundle is intersected by the unit circle or the unit sphere, respectively, the points of intersection on the circle (in pairs) represent the points of $P^1(R)$, and the pairs of antipodal points on the sphere represent the points of $P^2(R)$. A homeomorphic modification of this model for $P^2(R)$ would be a hemisphere with only the points on its boundary antipodally identified; or equivalently, a rectangle with antipodally identified boundary points. The result is again Fig. 3-4 or 3-7.

Exercises

1. Prove: Every two distinct projective planes in P^3 have precisely one line in common.
2. Prove: If two points of P^3 lie in a plane, then every point of their join also lies in this plane.
3. Prove Corollary 2 of Theorem 3-1.2.
4. Find
   (a) the equation of the projective line joining the points $[2, 1, 3]$ and $[3, 0, 1]$ of $P^2(R)$
   (b) the point on this line which corresponds to an improper point in the augmented $A^2(R)$.
5. Let $[\mathbf{x}] = [1, 3, 1]$ and $[\mathbf{y}] = [0, 2, 1]$.
   (a) Show that $[\mathbf{z}] = [1, -1, -1]$ lies on the join of $[\mathbf{x}]$ and $[\mathbf{y}]$.
   (b) Find a representation such that $[\mathbf{z}] = [\mathbf{x} + \mathbf{y}]$, and another such that $[\mathbf{x}] = [\mathbf{y} + \mathbf{z}]$.
6. Find the equation of the plane through the points $[0, 0, 1, 0]$, $[0, 1, 0, 1]$, and $[1, 1, 1, 1]$ in $P^3(R)$.
7. Write each of the following equations of conics in $A^2(R)$ in homogeneous coordinates. Which improper points does each conic contain when $A^2(R)$ is augmented?
   (a) $x^2 + y^2 - 2x = 3$ (b) $x^2 - 3y^2 - 2xy = 0$
   (c) $3x^2 + 4y^2 - 2x + 1 = 0$ (d) $x^2 - 2xy + y^2 - y = 1$.
8. Find the homogeneous coordinates of the following points of $P^3(R)$.
   (a) the origin
   (b) the point on the x_2-axis which corresponds to an improper point
   (c) the point with nonhomogeneous coordinates $[1, 1, 1]$
   (d) the point where the join of the origin and the point of (c) intersects the improper plane of $A^3(R)$.

9. Which are the improper points of the point sets described by the following equations when $A^3(R)$ is augmented?
   (a) $x - 2y + 3z = 4$ (b) $x^2 + y^2 + z^2 - 2x = 0$
   (c) $xy = z$ (d) $x^2 + y^2 - z^2 = 1$.
   If the improper points form a figure, describe it.
10. What is the plane dual of
    (a) a triangle (b) all points on a line
    (c) four lines no three of which are concurrent?
11. What is the three-dimensional dual of
    (a) the vertices of a triangle (b) all points on a line
    (c) all points in a plane (d) the sides of a tetrahedron?
12. Prove: Through three noncollinear projective points there is exactly one plane. What is the dual statement?
13. In $P^3(R)$, does the point $[x_0, x_1, x_2, x_3]$ lie in the plane $\langle x_0, x_1, x_2, x_3 \rangle$?
14. In P^2, find a necessary and sufficient condition for three lines $\langle \mathbf{u} \rangle$, $\langle \mathbf{v} \rangle$, and $\langle \mathbf{w} \rangle$ to be concurrent.
15. In P^2 a line is represented by an equation in point coordinates.
    (a) How is a point represented in line coordinates?
    (b) A triangle in $P^2(R)$ is given by its sides $\langle 1, 0, 2 \rangle$, $\langle 2, 0, 1 \rangle$, and $\langle 0, 1, 1 \rangle$. Find its vertices in point coordinates and in line coordinates.
16. What is the condition for
    (a) the two lines $\langle u_0, u_1, u_2 \rangle$ and $\langle v_0, v_1, v_2 \rangle$ to intersect in an improper point?
    (b) the three planes $\langle u_0, u_1, u_2, u_3 \rangle$, $\langle v_0, v_1, v_2, v_3 \rangle$, and $\langle w_0, w_1, w_2, w_3 \rangle$ to intersect in an improper point?
17. In P^3, how is a line represented
    (a) in point coordinates (b) in plane coordinates?
18. In the real projective plane a straight line turned out to be a "closed curve." What, then, is the difference between a projective straight line and a closed curve (for instance, a circle) of the real affine plane imbedded into the real projective plane? Use the bundle model as well as the Möbius strip model for your answer.

3–2 PROJECTIVITIES AND THEIR GROUPS. PROJECTIVE COORDINATES

Projectivities. Our next objective is the introduction of certain transformations of P^n. As a starting point we choose the linear transformations of V^{n+1} and their group $\mathcal{GL}^{n+1}$. Every linear transformation of V^{n+1} will bring about a transformation of P^n because, by Theorem 2–2.8, the one-dimensional subspaces of V^{n+1}, that is, the points of P^n, will be carried by it into one-dimensional subspaces of V^{n+1}. If proportional nonsingular $(n+1) \times (n+1)$ matrices A and cA, for every nonzero scalar c, are considered as "projectively equivalent," then equivalent matrices carry equivalent vectors of V^{n+1} into equivalent vectors of V^{n+1}. In other terms, equivalence classes of such matrices map points of P^n onto points of P^n.

We have proved the following:

Theorem 3–2.1. Two linear transformations of V^{n+1} represent the same transformation of P^n if and only if their matrices are proportional. That is, they can be obtained from each other by composition with a transformation $x_j \to cx_j$ $(j = 0, \ldots, n;\ c \neq 0)$.

The transformations of P^n thus obtained are called *projectivities.*

The above theorem raises the following question. A multiplication of all the coefficients of the $(n+1) \times (n+1)$ matrix B with -1 represents the same projectivity as before. When we dealt with euclidean transformations or similitudes, the sign of $\det B$ determined whether the transformation was direct or opposite. But multiplication of B with -1 changes the sign of $\det B$ if and only if n is even. Thus we should not expect the concept of orientation to be meaningful in any P^n with even n, and indeed we saw that in the case $n = 2$ the real projective plane is not orientable.

Theorem 3–2.2. The projectivities of P^n form a group. This group is called the *projective group,* $\mathcal{P}^n$.

Proof. To every element of $\mathcal{GL}^{n+1}$ there is exactly one projectivity of P^n. This correspondence is a single-valued (but not necessarily one-to-one) mapping, say α. Now the mapping satisfies $(L_1L_2)\alpha = L_1\alpha L_2\alpha$ according to the definition of projectivity. This implies the closure immediately. Also $I\alpha = I$, and $L\alpha L^{-1}\alpha = (LL^{-1})\alpha = I\alpha = I$ yields the existence of the inverse.*

Since we have introduced hyperplane coordinates, it is reasonable to ask about the fate of a hyperplane $\langle \mathbf{u} \rangle$ under a projectivity. This question could be answered by direct reference to Theorem 2–2.1. We will develop the result again. If $\mathbf{x} = (x_0, \ldots, x_n)$ and $\mathbf{u} = (u_0, \ldots, u_n)$ are considered as $1 \times (n+1)$ matrices, then $\mathbf{xu}' = 0$. If the hyperplane coordinates after the transformation A are $\mathbf{v} = (v_0, \ldots, v_n)$, we have $(\mathbf{x}A)\mathbf{v}' = \mathbf{x}(A\mathbf{v}') = 0$. Combination of the two equations yields $\mathbf{u}' = A\mathbf{v}'$ and finally $\mathbf{v} = (A^{-1}\mathbf{u}')' = \mathbf{u}(A^{-1})'$.

As a consequence we have:

Theorem 3–2.3. Under a projectivity with matrix A the hyperplane $\langle \mathbf{u} \rangle$ goes into $\langle \mathbf{u}(A^{-1})' \rangle$.

Corollary. The projectivities of P^2 are collineations.

---

* The mapping is a so-called *homomorphism* of groups.

In Section 3-1 we saw that A^n can be considered as P^n with one hyperplane removed. What, now, is the relation between the groups $\mathcal{A}^n$ and $\mathcal{P}^n$? The following theorem provides an answer that is perfectly in agreement with the relationship between A^n and P^n.

Theorem 3-2.4. $\mathcal{A}^n$ consists exactly of all projectivities of $\mathcal{P}^n$ which preserve the hyperplane $x_0 = 0$.

Proof. First it should be made clear that the projectivities preserving $x_0 = 0$ form a subgroup of $\mathcal{P}^n$. Obviously the product of two such projectivities leaves $x_0 = 0$ invariant, as does I and the inverse of such a projectivity. Now suppose that $rx_j' = \sum_k a_{jk}x_k$ $(j, k = 0, \ldots, n)$ is a projectivity, with nonzero r. If it preserves $x_0 = 0$, then $0 = \sum_k a_{0k}x_k$ for all x_k, with $k = 1, \ldots, n$. Hence

$$a_{01} = a_{02} = \cdots = a_{0n} = 0, \qquad \text{and} \qquad rx_0' = a_{00}x_0.$$

Since $\det(a_{ik}) \neq 0$, necessarily $a_{00} \neq 0$. What happens to those points of P^n which can be considered also as proper points of A^n? None of these becomes an improper point because $x_0' = a_{00}x_0/r \neq 0$. For all proper points we may write

$$x_j'/x_0' = \sum_k (a_{jk}x_k)/(a_{00}x_0).$$

If we consider the x_j/x_0 and x_j'/x_0', with $j = 1, \ldots, n$, as affine coordinates, this is an affinity if indeed the $n \times n$ matrix (a_{jk}/a_{00}) is nonsingular. Now

$$0 \neq \begin{vmatrix} a_{00} & 0 & \cdots & 0 \\ a_{10} & a_{11} & \cdots & a_{1n} \\ \vdots & & & \vdots \\ a_{n0} & a_{n1} & \cdots & a_{nn} \end{vmatrix} = a_{00} \begin{vmatrix} a_{11} & \cdots & a_{1n} \\ \vdots & & \vdots \\ a_{n1} & \cdots & a_{nn} \end{vmatrix}$$

$$= a_{00}^{n+1} \begin{vmatrix} a_{11}/a_{00} & \cdots & a_{1n}/a_{00} \\ \vdots & & \vdots \\ a_{n1}/a_{00} & \cdots & a_{nn}/a_{00} \end{vmatrix},$$

and therefore $\det(a_{jk}/a_{00}) \neq 0$. Thus all the projectivities which leave $x_0 = 0$ invariant are affinities and form a group. Conversely, every affinity can be obtained from a projectivity by replacing the homogeneous coordinate x_j by x_j/x_0 and by putting $x_0' = x_0$. Hence the group is $\mathcal{A}^n$.

Theorem 3-2.5. In P^n, let two ordered $(n + 2)$-tuples of points be given such that in neither of them $n + 1$ points lie in the same hyperplane. Then there exists one and only one projectivity which takes one $(n + 2)$-tuple into the other.

Corollary. If in P^n a projectivity has $n + 2$ invariant points, no $n + 1$ of which lie in a hyperplane, then the projectivity is the identity.

Proof. At first we will show that there is a unique projectivity taking the $n + 2$ special points

$$Q_0 = [1, 0, \ldots, 0], \qquad Q_1 = [0, 1, \ldots, 0], \ldots,$$

$$Q_n = [0, 0, \ldots, 1], \qquad \text{and} \qquad U = [1, 1, \ldots, 1]$$

into the arbitrary points $[\mathbf{a}_0] = [a_{00}, \ldots, a_{0n}]$, $[\mathbf{a}_1] = [a_{10}, \ldots, a_{1n}], \ldots,$ $[\mathbf{a}_n] = [a_{n0}, \ldots, a_{nn}]$, and $[\mathbf{b}] = [b_0, \ldots, b_n]$, respectively, such that no $n + 1$ of the image points are in the same hyperplane. Considering only the first $n + 1$ points and their images, we obtain a projectivity with matrix

$$L = \begin{bmatrix} a_{00}r_0 & a_{01}r_0 & \cdots & a_{0n}r_0 \\ \vdots & & & \vdots \\ a_{n0}r_n & a_{n1}r_n & \cdots & a_{nn}r_n \end{bmatrix},$$

such that $(Q_j)L = [r_j\mathbf{a}_j]$.

The coefficients r_j are not determined because, for instance, $[a_{00}, \ldots, a_{0n}] = [a_{00}r_0, \ldots, a_{0n}r_0]$ for all nonzero r_0. The matrix is nonsingular because the points $[\mathbf{a}_j]$ do not lie in a hyperplane. When we apply L to the additional points U and $[\mathbf{b}]$, we get

$$a_{0j}r_0 + \cdots + a_{nj}r_n = b_j, \quad \text{with} \quad j = 0, \ldots, n.$$

This is a system of $n + 1$ nonhomogeneous linear equations in the unknowns r_j, whose determinant does not vanish. Hence, by Theorem 2-3.4, it has a unique solution. None of the r_j can be zero because otherwise $n + 1$ of the image points would be collinear. Thus a unique projectivity has been obtained as required. Call it T.

Let $[\mathbf{a}_0']$, $[\mathbf{a}_1'], \ldots, [\mathbf{a}_n']$, and $[\mathbf{b}']$ be $n + 2$ other image points subject to the same independence restriction, and let the projectivity carrying $Q_0, Q_1, \ldots, Q_n$, and U into these points be T_1. Then $T^{-1}T_1$ maps $[\mathbf{a}_j] \to [\mathbf{a}_j']$, $[\mathbf{b}] \to [\mathbf{b}']$, $j = 0, \ldots, n$. Thus we have found a projectivity as asserted in the theorem.

Is $T^{-1}T_1$ the only projectivity that does so? Suppose that there is another such projectivity W. In view of the uniqueness of T, there is only one projectivity which leaves the $n + 2$ points Q_j and U unchanged. The projectivity $T_1W^{-1}T^{-1}$ preserves these points and so does the identity, and consequently $T_1W^{-1}T^{-1} = I$, which implies $T^{-1}T_1 = W$. This completes the proof.

Projective coordinates. After we have studied transformations of P^n, it is reasonable to ask about coordinate transformations. However, this requires an introduction of appropriate general coordinate systems.

Let $n + 2$ points $Q_0, \ldots, Q_n$ and U be given such that no $n + 1$ of them lie in a hyperplane. We call them the *reference points* of a general *projective coordinate system.* Let $Q_j = [\mathbf{q}_j]$, with $j = 0, \ldots, n$, where all $\mathbf{q}_j \in V^{n+1}$. From each of the $[\mathbf{q}_j]$ we choose arbitrarily one representative vector $\mathbf{q}_j$ and henceforth keep it fixed. The $n + 1$ vectors $\mathbf{q}_j$ are linearly independent and hence form a basis of V^{n+1}. For every point $[\mathbf{x}]$ of P^n there is a representation $\mathbf{x} = \alpha_0\mathbf{q}_0 + \cdots + \alpha_n\mathbf{q}_n$. If $U = [\mathbf{u}]$, then such a representation exists also for $\mathbf{u}$, say, $\mathbf{u} = \epsilon_0\mathbf{q}_0 + \cdots + \epsilon_n\mathbf{q}_n$. None of the ϵ_j is zero, otherwise U would lie in a hyperplane with n of the Q's. If for a point $[\mathbf{x}]$ of P^n, $\mathbf{x} = \alpha_0\mathbf{q}_0 + \cdots + \alpha_n\mathbf{q}_n$, then we say that its *general projective coordinates* in the coordinate system with the reference points $Q_0, \ldots, Q_n$ and U are $\alpha_0/\epsilon_0, \ldots, \alpha_n/\epsilon_n$. Is the representation of $\mathbf{x}$ unique? Suppose another representation is $\beta_0\mathbf{q}_0 + \cdots + \beta_n\mathbf{q}_n$. Then

$$\alpha_0\mathbf{q}_0 + \cdots + \alpha_n\mathbf{q}_n = r(\beta_0\mathbf{q}_0 + \cdots + \beta_n\mathbf{q}_n),$$

for some nonzero r, and consequently, in view of the linear independence of $\mathbf{q}_0, \ldots, \mathbf{q}_n$, one gets $\alpha_j - r\beta_j = 0$ for $j = 0, \ldots, n$. Thus the representation of every point in a projective coordinate system is unique up to multiplication of all the coordinates with a nonzero r. Similarly, had we chosen different, that is, proportional ϵ_j's for U, the resulting change in the coordinates would not have affected their ratio.

What are the projective coordinates of the reference points? From $\mathbf{x} = \alpha_0\mathbf{q}_0 + \cdots + \alpha_n\mathbf{q}_n$ we obtain the α_j of Q_k to be $r\delta_{jk}$, where r is an arbitrary nonzero real and δ_{jk} the Kronecker delta. All the ϵ_j are equal and nonzero, and so we get, for the projective coordinates of Q_k, $\alpha_j/\epsilon_j = s\delta_{jk}$, with some $s \neq 0$.

A special coordinate system is that in which the projective coordinates coincide with the homogeneous coordinates, that is, where the projective coordinates of a point $[\alpha_0, \ldots, \alpha_n]$ are $\alpha_0, \ldots, \alpha_n$. For this system we have to choose $Q_j = [\mathbf{e}_j]$ and $U = [\sum_j \mathbf{e}_j]$. Then we see that in the augmented A^n all the reference points, except Q_0 and U, are improper, while Q_0 and U must be proper. The point $Q_0 = [1, 0, \ldots, 0]$ is the origin of the affine system and $U = [1, 1, \ldots, 1]$ is the affine point whose coordinates are all 1. The point Q_j is the improper point on the x_j-axis of the affine coordinate system because, for all points on this axis, except the origin, the j-coordinate is the only nonzero nonhomogeneous coordinate. When this axis intersects the improper hyperplane $x_0 = 0$, all the homogeneous coordinates vanish except the j-coordinate, which may be assigned the value 1.

So we see that the homogeneous coordinates of the augmented A^n are only a special case of projective coordinates.

The use of the general projective coordinates makes it possible for us to avoid the reference to an augmented A^n every time we want to discuss a projective space. We should always keep in mind that P^n has an existence in its own right and that the augmented A^n merely served as a model of P^n.

We summarize in the following theorem.

Theorem 3–2.6. If $n + 2$ reference points $Q_0, \ldots, Q_n$, U are given in P^n such that no $n + 1$ of them lie in a hyperplane, then they determine a general projective coordinate system in which every point of P^n has $n + 1$

projective coordinates whose ratio determines this point uniquely. These projective coordinates coincide with the special homogeneous coordinates of an augmented A^n if and only if $Q_j = [\mathbf{e}_j]$ and $U = [\sum_j \mathbf{e}_j]$ for $j = 0, \ldots, n$.

Theorem 3–2.7. Under any projectivity of the space, the ratio of the projective coordinates of a point with respect to a coordinate system is equal to the ratio of the coordinates of the image point with respect to the image of the coordinate system. In short, projective coordinates are invariant under projectivities.

Proof. If the point is $[\mathbf{y}] = [x_0\mathbf{q}_0 + \cdots + x_n\mathbf{q}_n]$ in the system with reference points $Q_0, \ldots, Q_n$, U, and if π is a projectivity, then $Q_j\pi$ and $U\pi$ form the reference points for a new system. Then

$$[\mathbf{y}]\pi = [x_0\mathbf{q}_0 + \cdots + x_n\mathbf{q}_n]\pi = x_0[\mathbf{q}_0]\pi + \cdots + x_n[\mathbf{q}_n]\pi.$$

A simple consequence of Theorem 3–2.5 is:

Theorem 3–2.8. For every two projective coordinate systems of P^n there exists a unique projectivity taking one into the other.

The last two theorems enable us to introduce coordinate transformations by using projectivities. This provides a further example of the dual role of transformation formulas in a geometry, namely, as transformations of the space as well as coordinate changes. Thus every linear transformation of a vector space V^{n+1} induces a projectivity of P^n and can be considered a projective coordinate transformation of P^n.

Exercises

1. The ratio of segments on a line is not always preserved under projectivities. Find an example.
2. Use the bundle model for answering:
   (a) Is betweenness of points on a line a projectively meaningful property?
   (b) Is the property of two point couples on a line separating each other projectively invariant? In case of a negative answer to either (a) or (b), find an example in which a projectivity destroys the property.
3. Write the equation of the circle $x^2 + y^2 = 1$ in homogeneous coordinates, thus imbedding the circle in P^2. Then let the x_2-axis be a new improper line, and at the same time make the former improper line into the x_2-axis. What happens to the circle now if the improper line is removed?
4. (a) Find the plane projectivity mapping $[1, 0, 1]$, $[-1, 1, 0]$, $[1, 0, 0]$, and $[-1, 1, -1]$, respectively, on $[3, 0, 5]$, $[2, -1, 3]$, $[2, 0, 3]$, and $[3, -1, 5]$.
   (b) Find the line coordinates of the join of the first two points and verify that they transform according to Theorem 3–2.3.

5. What is the matrix of the projectivity which takes the special reference points Q_0, Q_1, Q_2, Q_3, U of Theorem 3–2.6 into $[1, 1, 0, 0]$, $[0, 0, 1, 1]$, $[0, 1, 1, 0]$, $[1, 0, 0, 0]$, and $[1, 2, 3, 4]$, respectively?
6. (a) Show that the plane projectivities which preserve the special reference points Q_0, Q_1, and Q_2 are affinities.
   (b) Describe these affinities.
   (c) Generalize the result for n dimensions.
7. Find all involutory projectivities of P^1.
8. Find the two-dimensional projectivities that leave the line $\langle 1, 1, 1\rangle$ pointwise invariant.
9. $Q_0 = [2, 3]$, $Q_1 = [1, 1]$, and $U = [0, 1]$ are the reference points of a coordinate system in $P^1(R)$.
   (a) Find the projective coordinates of $P = [4, 5]$.
   (b) What are the projective coordinates of P in a system with $Q_0 = [1, 0]$, $Q_1 = [0, 1]$, and $U = [1, 1]$?
10. Let $Q_0 = [1, -1, 2]$, $Q_1 = [1, 0, 1]$, $Q_2 = [0, 2, 1]$, and $U = [1, 0, 4]$ be the reference points of a coordinate system. Find the projective coordinates of $[-1, 1, -1]$.
11. Dualize Theorem 3–2.6 with particular attention to the cases $n = 2$ and $n = 3$.
12. Dualize Theorem 3–2.5. In particular, formulate the dual statement for $n = 2$ and $n = 3$.

3–3 THE CROSS RATIO AND ITS APPLICATIONS

Cross ratio and coordinates in P^1. In order to better understand the concept of projective coordinates, we will establish a connection between them and the cross ratio. The cross ratio was introduced in Section 1–8, but there we used the concept of distance, which has no meaning in projective geometry. We will define the cross ratio without using distances, and it will be shown that the definitions are equivalent in the pertinent cases. The cross ratio of four collinear points will turn out to be a projective invariant.

Let Q_0, Q_1, and U be the reference points of a projective coordinate system in P^1; that is, on a projective line. Let W be a point on the line, not coinciding with Q_1. Let $Q_0 = [\mathbf{q}_0]$, $Q_1 = [\mathbf{q}_1]$, $U = [\mathbf{u}]$, $W = [\mathbf{w}]$, $\mathbf{u} = \epsilon_0\mathbf{q}_0 + \epsilon_1\mathbf{q}_1$, and $\mathbf{w} = \alpha_0\mathbf{q}_0 + \alpha_1\mathbf{q}_1$. The projective coordinates of W are α_0/ϵ_0 and α_1/ϵ_1, and their ratio is invariant under all projectivities of P^1. This ratio

$$\frac{\alpha_1}{\epsilon_1} : \frac{\alpha_0}{\epsilon_0} = \frac{\alpha_1\epsilon_0}{\alpha_0\epsilon_1}$$

will be called the *cross ratio* $(Q_0Q_1|WU)$. Since $W \neq Q_1$, α_0 cannot vanish, and hence the cross ratio always exists.

Theorem 3–3.1. The cross ratio of four collinear points is invariant under projectivities.

Proof. Theorem 3–2.7.

Consider the special case in which all four points Q_0, Q_1, W, and U are proper points of the augmented A^1. Then we may put $\mathbf{q}_0 = [1, q_0]$, $\mathbf{q}_1 = [1, q_1]$, $\mathbf{u} = [1, u]$, and $\mathbf{w} = [1, w]$. We have then

$$[\mathbf{u}] = \epsilon_0[1, q_0] + \epsilon_1[1, q_1] = [\epsilon_0 + \epsilon_1, \epsilon_0 q_0 + \epsilon_1 q_1] = [1, u],$$
$$[\mathbf{w}] = \alpha_0[1, q_0] + \alpha_1[1, q_1] = [\alpha_0 + \alpha_1, \alpha_0 q_0 + \alpha_1 q_1] = [1, w].$$

By comparison of the corresponding coefficients, we get

$$\epsilon_0 + \epsilon_1 = s, \quad \alpha_0 + \alpha_1 = r, \quad su = \epsilon_0 q_0 + \epsilon_1 q_1, \quad \text{and} \quad rw = \alpha_0 q_0 + \alpha_1 q_1,$$

for some nonzero r and s. This yields

$$\epsilon_1 = s\,\frac{u - q_0}{q_1 - q_0}, \qquad \alpha_1 = r\,\frac{w - q_0}{q_1 - q_0}.$$

Now we substitute the values for ϵ_1 and α_1 and obtain

$$\frac{\alpha_1/\epsilon_1}{\alpha_0/\epsilon_0} = \frac{r(w - q_0)(s - \epsilon_1)}{s(u - q_0)(r - \alpha_1)} = \frac{(w - q_0)[(q_1 - q_0) - (u - q_0)]}{(u - q_0)[(q_1 - q_0) - (w - q_0)]}$$
$$= \frac{(w - q_0)(q_1 - u)}{(u - q_0)(q_1 - w)}.$$

Hence

$$(Q_0Q_1|WU) = \frac{(q_0 - w)(q_1 - u)}{(q_0 - u)(q_1 - w)}. \tag{1}$$

On the euclidean line this expression involves four segments and coincides exactly with the definition of the cross ratio in Section 1–8, when applied to collinear points.

Another case we will discuss is that in which $Q_0 = [1, 0]$ is the origin, $U = [1, 1]$ the unit point, and $W = [1, w]$ a proper point of the augmented A^1, while $Q_1 = [0, 1]$ is its improper point. Then, with ϵ_0, ϵ_1, α_0, and α_1 having the same meaning as above, we obtain $\epsilon_0 = \epsilon_1 = \alpha_0 = 1$ and $\alpha_1 = w$, and therefore

$$(Q_0Q_1|WU) = \frac{\alpha_1/\epsilon_1}{\alpha_0/\epsilon_0} = w.$$

Thus in this special case the cross ratio turns out to be the nonhomogeneous affine coordinate of W. Since every triple Q_0, Q_1, U, of points of P^1 can be mapped onto the points $[1, 0]$, $[0, 1]$, and $[1, 1]$ by a projectivity, the cross ratio $(Q_0Q_1|WU)$, which is invariant under this projectivity, can be defined as the nonhomogeneous coordinate of W after this projectivity has been performed.

Thus we have proved the following.

Theorem 3-3.2. The nonhomogeneous affine coordinate of a proper point W of P^1 is the cross ratio $([1, 0],[0, 1]|W[1, 1])$.

Corollary. If three distinct collinear points and a scalar c are given, then there is a unique fourth point on the same line such that their cross ratio is c.

For the three given points can be transformed into $Q_0 = [1, 0]$, $Q_1 = [0, 1]$, and $U = [1, 1]$, respectively, and there is a point for every value of the nonhomogeneous coordinate. In particular, we obtain $c = 0$ for $W = Q_0$ and $c = 1$ for $W = U$. The only point on the line that does not have a nonhomogeneous coordinate is Q_1 and, indeed, in the case $W = Q_1$, Eq. (1) does not yield any result for c. In order to take care of this case, we say that

$$(Q_0Q_1|Q_1U) = \infty.$$

This is a purely symbolic notation which, on the other hand, becomes meaningful in $P^1(R)$ when we recall that $\lim_{w\to\infty} [1, w] = [0, 1] = Q_1$.

Harmonic quadruples. Of special importance for real projective geometry is the case $(Q_0Q_1|WU) = -1$. The four points are then called a *harmonic* point quadruple. If we again assume $q_1 \to \infty$ and $w \neq q_1$, then for a harmonic quadruple Eq. (1) yields $q_0 - w = u - q_0$; that is, Q_0 lies midway between W and U. Thus the harmonic cross ratio constitutes a projective generalization of the concept of a midpoint.

A useful property of the cross ratio is the following.

Theorem 3-3.3. If $c = (B_1B_2|B_3B_4)$ and if B_1, B_2, B_3, and B_4 are distinct, then $(B_2B_1|B_3B_4) = (B_1B_2|B_4B_3) = 1/c$.

Proof. If $B_1 = [\mathfrak{q}_0]$, $B_2 = [\mathfrak{q}_1]$, $B_3 = [\alpha_0\mathfrak{q}_0 + \alpha_1\mathfrak{q}_1]$, and $B_4 = [\epsilon_0\mathfrak{q}_0 + \epsilon_1\mathfrak{q}_1]$, then $c = (\alpha_1/\epsilon_1)/(\alpha_0/\epsilon_0)$. In the cross ratio $(B_2B_1|B_3B_4)$, $\mathfrak{q}_0$ and $\mathfrak{q}_1$ are interchanged, and therefore so are their coefficients in the representation of B_3 and B_4. Hence $(B_2B_1|B_3B_4) = (\alpha_0/\epsilon_0)/(\alpha_1/\epsilon_1) = 1/c$. This value is finite because $c = 0$ only if two of the four points coincide. The same applies to $(B_1B_2|B_4B_3)$.

Corollary. If $(B_1B_2|B_3B_4) = -1$, then $(B_1B_2|B_3B_4) = (B_1B_2|B_4B_3)$.

If $(B_1B_2|B_3B_4) = -1$, then B_4 is called the *fourth harmonic* of B_1, B_2, and B_3. In view of the corollary, B_3 is the fourth harmonic of B_1, B_2, and B_4 if and only if B_4 is the fourth harmonic of B_1, B_2, and B_3.

Applications in the projective plane. All the discussion of the cross ratio was carried out in P^1 only. It could have been done for an arbitrary projective line in any P^n, $n \geqq 1$. The projective lines are subspaces of P^n, and since the cross ratio is invariant under projectivities of P^1, it will be invariant also under a higher-dimensional projectivity π. For P^1 will be carried by π into some one-dimensional projective subspace of P^n and π will induce a projectivity in this subspace.

Theorem 3-3.4. If Q_1, Q_2, and Q_3 are three points on a line l in P^n and if R_1, R_2, and R_3 are three points on a line k, then there exists a projectivity α such that $Q_j\alpha = R_j$, with $j = 1, 2, 3$. If β is another projectivity with $Q_j\beta = R_j$, and if Q is an additional point on l, then $Q\alpha = Q\beta$.

Proof. Obviously there exists a projectivity, say π, which takes l into k. Let $Q_j\pi = R'_j$. By Theorem 3-2.5, there is a unique projectivity π' from $\mathcal{P}^1$ such that $R'_j\pi' = R_j$. Hence $Q_j\pi\pi' = R_j$ and $\alpha = \pi\pi'$. By Theorem 3-3.1,

$$(Q_1Q_2|Q_3Q) = (Q_1\alpha, Q_2\alpha|Q_3\alpha, Q\alpha) = (Q_1\beta, Q_2\beta|Q_3\beta, Q\beta).$$

By the corollary to Theorem 3-3.2, $Q\alpha = Q\beta$.

Our next concern is the dualization of the concepts developed so far. We will restrict ourselves to the study of P^2, where the cross ratio of four collinear points will now be replaced by that of four, concurrent lines. The general treatment in P^n would have to deal with four hyperplanes intersecting in a projective linear $(n - 2)$-variety. In P^2 we introduce the notation $\delta[a, b, c] = \langle a, b, c\rangle$ and $\delta\langle a, b, c\rangle = [a, b, c]$. If D_1, D_2, D_3, D_4 are collinear points, then δD_1, δD_2, δD_3, and δD_4 are concurrent lines, and we formally define $(\delta D_1, \delta D_2 \mid \delta D_3, \delta D_4) = (D_1D_2 \mid D_3D_4)$. Let (Fig. 3-11) $Q_0 = [1, 0, 0]$, $Q_1 = [0, 1, 0]$, $U = [1, 1, 0]$, and $W = [1, w, 0]$; these points lie on the line $x_2 = 0$. The four lines δQ_0, δQ_1, δU, and δW pass through the point $Q_2 = [0, 0, 1]$, and their cross ratio is $(\delta Q_0, \delta Q_1 \mid \delta W, \delta U) = (Q_0 Q_1 \mid WU) = w$.

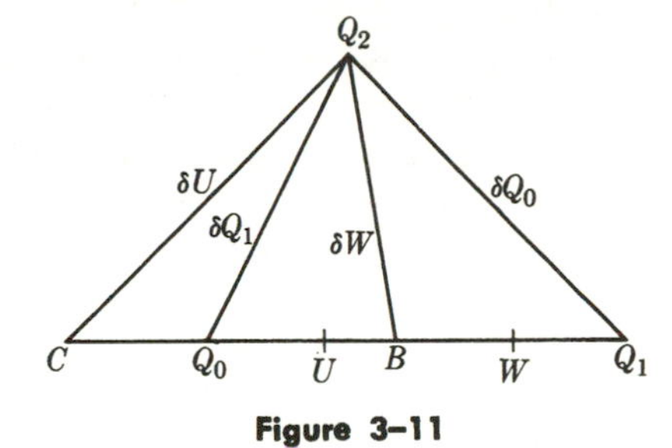

Figure 3-11

Theorem 3-3.5. If in P^2, four lines through a point S are intersected by a line l which does not contain S, then the cross ratio of the four lines is equal to the cross ratio of their corresponding points of intersection with l.

Proof (Fig. 3-11). We may assume that the four lines are δQ_0, δQ_1, δW, and δU, as defined above, and that l is $x_2 = 0$. In any other case, a projectivity exists which maps any four concurrent lines with cross ratio w on δQ_0, δQ_1, δW, and δU, and any l on $x_2 = 0$, as a consequence of the dual of Theorems 3-2.5 and 3-3.4. A simple computation shows that the four lines intersect the line $x_2 = 0$ in the points $Q_1 = [0, 1, 0]$, $Q_0 = [1, 0, 0]$, $B = [w, -1, 0]$, and $C = [1, -1, 0]$, respectively. In order to compute $(Q_1Q_0 \mid BC)$, we may ignore the third coordinate for all points. The vectors of Q_1 and Q_0 are, respectively, $\mathbf{e}_1$ and $\mathbf{e}_0$. The vector of B is $w\mathbf{e}_0 - \mathbf{e}_1$, and that of C is $\mathbf{e}_0 - \mathbf{e}_1$. Consequently, $\alpha_0 = -1, \alpha_1 = w, \epsilon_0 = -1, \epsilon_1 = 1$, and the cross ratio is

$$(Q_1Q_0|BC) = \frac{\alpha_1/\epsilon_1}{\alpha_0/\epsilon_0} = \frac{w/1}{-1/-1} = w.$$

Corollary 1. (Fig. 3–12). *Let four lines k_j, with $j = 1, 2, 3, 4$, through a point S be intersected by two distinct lines l and l', both of which do not contain S. If k_j intersects l in K_j and l' in K_j', then $(K_1K_2|K_3K_4) = (K_1'K_2'|K_3'K_4')$.*

A mapping of points K_j of one line onto points K_j' of another line such that all lines joining K_j and K_j' pass through a point S is called a *point perspectivity*. The corollary can now be expressed: The cross ratio of four collinear points is preserved under point perspectivities.

Corollary 2. *Every point perspectivity in P^2 is induced by a projectivity.*

With the help of the above theorem we are able to describe a construction for finding the fourth harmonic to a given triple of collinear points in $P^2(R)$. The following convention will be used: BC is the join of the points B and C, and $b \times c$ is the point of intersection of the lines b and c.

Construction of the fourth harmonic (Fig. 3–13). Let B_1, B_2, and B_3 be distinct collinear points and S any point not on the line B_1B_2. Choose any point C on B_1S distinct from B_1 and S. Let $B_2C \times SB_3 = D$, $B_1D \times SB_2 = E$, and $CE \times B_1B_2 = B_4$. Then B_4 is the desired fourth harmonic.

Proof. All the points and lines constructed lie in one plane. Hence Theorem 3–1.1 applies, and the points of intersection and the joins exist unless two intersecting lines or two points to be joined coincide. This cannot happen, as can be verified. For instance, $B_2 \neq C$ because S is not on B_1B_2; therefore $SB_1 \neq B_1B_2$ and $SB_1 \times B_1B_2 = B_1$ is the only common point of the two lines. But $C = B_2$ would be another common point. As another example, check $B_2C \neq$

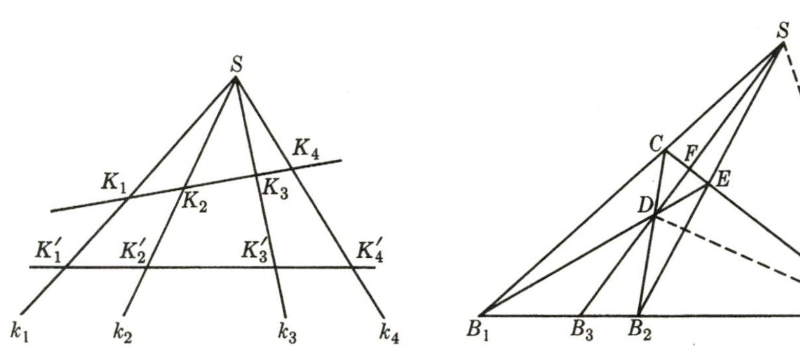

Figure 3–12

Figure 3–13

SB_3. We know that $B_2 \neq B_3$. If $B_2C = SB_3$, then B_2, B_3, C, and S are collinear; that is, S lies on B_2B_3, which is a contradiction.

Now we approach the actual proof. By repeated use of Theorem 3–3.5, we obtain

$$\begin{aligned}(B_1B_2|B_3B_4) &= (SB_1, SB_2|SB_3, SB_4) = (SC, SE|SF, SB_4) \\ &= (CE|FB_4) = (DC, DE|DF, DB_4) = (DB_2, DB_1|DB_3, DB_4) \\ &= (B_2B_1|B_3B_4) = c, \text{ say.}\end{aligned}$$

By Theorem 3–3.3, $c = 1/c$, and hence $c = \pm 1$. The value $c = 1$ cannot occur unless $B_3 = B_4$. But this would imply $FB_3 = FB_4 = FC = FS$, and hence $C = S$, a contradiction to the construction rule. The remaining possibility is $c = -1$, which makes B_4 the fourth harmonic as required.

A direct consequence of the construction of the fourth harmonic in $P^2(R)$ is

Theorem 3–3.6 (Fig. 3–13). If $CDES$ is a quadrangle in $P^2(R)$ and if $SC \times ED = B_1$, $SE \times CD = B_2$, $B_1B_2 \times SD = B_3$, and $B_1B_2 \times CE = B_4$, then $(B_1B_2|B_3B_4) = -1$.

Four distinct lines, no three of which are concurrent, and all six points of intersection form a *complete quadrilateral.* The six points are called its vertices, the four lines its sides. In Theorem 3–3.6, the vertices are C, D, E, S, B_1, and B_2; SD and CE are two of the diagonals of the complete quadrilateral.

The above theorem implies Theorem 3–3.7, which will involve only incidences of points and lines and will not use any other concept, such as the value of a cross ratio. Such a theorem is called an *incidence theorem.*

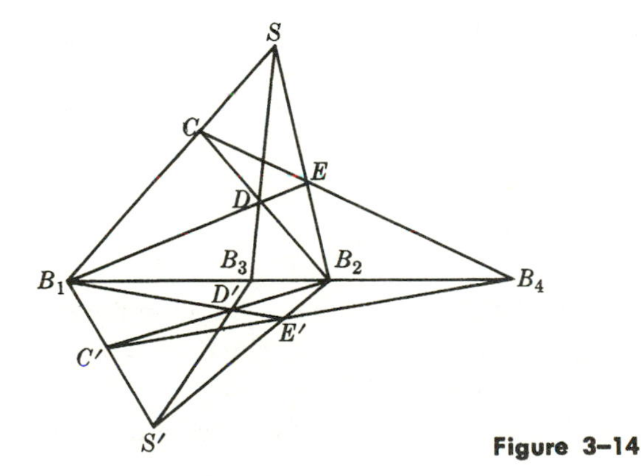

Figure 3–14

Theorem 3–3.7 (The complete quadrilateral theorem, Fig. 3–14). If $CDESB_1B_2$ and $C'D'E'S'B_1B_2$ are two complete quadrilaterals in $P^2(R)$, with $SC \times ED = S'C' \times E'D' = B_1$, $SE \times CD = S'E' \times C'D' = B_2$, $B_1B_2 \times SD = B_1B_2 \times S'D' = B_3$, then B_1B_2, CE, and $C'E'$ meet in one point.

Proof. By Theorem 3–3.6, B_4 is the fourth harmonic for B_1, B_2, and B_3 when we use the quadrilateral $CDESB_1B_2$. There is only one such fourth harmonic. Thus when the second quadrilateral is used, the line $C'E'$ must pass through B_4.

Whenever the harmonic cross ratio was mentioned, we restricted the discussion to the special case of $K = R$, the real field. This was done, among other reasons, in order to avoid the complication arising in fields in which $1 = -1$, that is, whose characteristic is 2. (The *characteristic* p of a skew-field was defined as the minimum number of times that 1 has to be added to itself such that $1 + \cdots + 1 = 0$. If this equality does not hold for any positive integer p, we say that the characteristic is 0. Cf. Exercise 5 of Section 2–10.) Most of the statements would remain true if $K = C$, the complex field.

Many of the developments in this section are useful in dual interpretation. We will mention some of these.

Theorem 3–3.5 is dual to itself. The definition of a perspectivity which maps points on points can be dualized in the following way.

Definition (Fig. 3–15). A mapping of lines k_j, which meet in a point, on lines k'_j, which meet in another point, such that all points common to k_j and k'_j lie on a line, is called a *line perspectivity*.

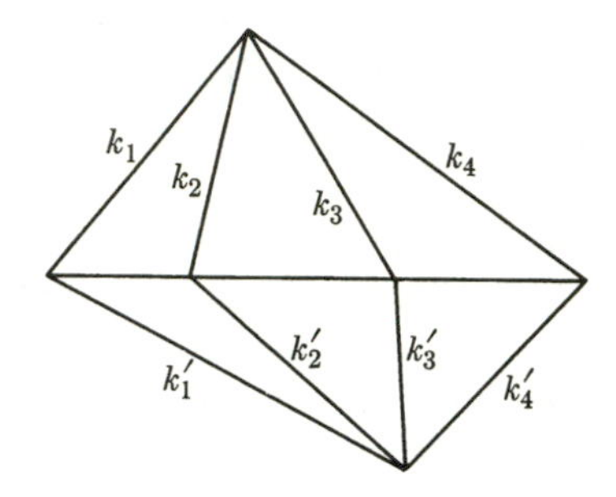

Figure 3–15

The dual of the corollary to Theorem 3–3.5 is: The cross ratio of four concurrent lines is preserved under line perspectivities.

The dualization of the theorems about the complete quadrilateral will be left for the exercises.

EXERCISES

1. (a) Show that $(AB|CD) = c$ assumes the six different values c, $1/c$, $1 - c$, $1/(1 - c)$, $1 - 1/c$, $c/(c - 1)$ under all the possible permutations of A, B, C, and D, and list the 24 cases.

   (b) Which 8 of these permutations preserve a harmonic cross ratio?

2. Prove: The value of $(AB|CD)$ in $P^1(R)$, with distinct collinear points A, B, C, and D, is negative when the point couples A, B and C, D separate each other, and positive when they do not.

3. Prove: If A, B, C, D, and E are distinct collinear points, then

   (a) $(AB|CD)(AB|DE)(AB|EC) = 1$, (b) $(AB|CD)(AB|DE) = (AB|CE)$.

4. Let $A = [1, 0, -3]$, $B = [0, 2, 1]$, $C = [1, 6, 0]$, and $D = [2, 2, -5]$. Evaluate $(AB|CD)$.
5. The following lines in $P^2(R)$ are given by their equations, l_1: $x_1 + x_2 = 0$, l_2: $x_0 = 0$, and l_3: $3x_0 + 2x_1 + 2x_2 = 0$. Find the equation of l_4 such that $(l_1 l_2|l_3 l_4) = -3$.
6. Generalize the cross-ratio concept so as to define a cross ratio of planes. What is the statement corresponding to Theorem 3–3.5?
7. (a) If A, B, and C are proper collinear points and if D is the improper point of the line $A^1(R)$, prove that C is the midpoint of the segment AB if and only if $(AB|CD) = -1$.
   (b) Dualize the statement for the plane.
8. (a) What becomes of the cross ratio $(AB|CD)$ of four collinear points when D is the improper point of the line AB?
   (b) Use your answer to (a) for showing that Theorem 2–4.13 is but a special case of Theorem 3–3.1.
9. (a) Prove that the cross ratio of four concurrent lines in E^2 is equal to the cross ratio (in the sense of Section 1–8) of their slopes with respect to an orthonormal system.
   (b) Find $(l_1 l_2|l_3 l_4)$ if in an orthonormal system the equations of the lines are l_1: $y = 0$, l_2: $x = 0$, l_3: $y = mx$, and l_4: $y = x$.
   (c) How would you define the "slope" of a line in P^2 with respect to a projective coordinate system?
10. In the proof for the construction of the fourth harmonic, elaborate all the possible coincidences of points and lines.
11. Describe the construction of the fourth harmonic line when three distinct concurrent lines are given.
12. Dualize Theorems 3–3.6 and 3–3.7. (The dual of a complete quadrilateral is called a *complete quadrangle*.)
13. What is the affine specialization of Theorem 3–3.6 which results from choosing $CDES$ such that SB_4 is the improper line?
14. If in $P^1(R)$, for a nonidentical one-dimensional involutory projectivity π, $A\pi = A$ and $B\pi = B$, and if A, B, and C are distinct, prove that $(A, B|C, C\pi) = -1$.

3–4 PROJECTIVITIES IN P^1 AND P^2

Invariant points of projectivities in $P^1(R)$ and $P^2(R)$. We shall attempt a classification of the one- and two-dimensional projectivities over the real field.

In $P^1(R)$, we ask first about the invariant points of projectivities. This means solving the equations $rx_0 = ax_0 + bx_1$ and $rx_1 = cx_0 + dx_1$, with $ad - bc \neq 0$ and nonzero r. If we put $x_0/x_1 = z$, the equations are equivalent to

$z = (az + b)/(cz + d)$ or $cz^2 + z(d - a) - b = 0$. By distinguishing the possible values of the discriminant of this equation we obtain the following theorem.

Theorem 3-4.1. A projectivity $x_0 \to ax_0 + bx_1$, $x_1 \to cx_0 + dx_1$ on the projective line has two, one, or no invariant real points according to whether $(a - d)^2 + 4bc$ is positive, zero, or negative, respectively.

As in the case of plane affinities (page 97), these types of one-dimensional projectivities are sometimes called *hyperbolic, parabolic,* and *elliptic,* respectively.

In the projective plane we have to study the three equations

$$rx_j = a_{j0}x_0 + a_{j1}x_1 + a_{j2}x_2,$$

with $j = 0, 1, 2$, where $\det (a_{jk}) \neq 0$. This means solving the simultaneous equations

$$(a_{00} - r)x_0 + a_{01}x_1 + a_{02}x_2 = 0,$$

$$a_{10}x_0 + (a_{11} - r)x_1 + a_{12}x_2 = 0,$$

$$a_{20}x_0 + a_{21}x_1 + (a_{22} - r)x_2 = 0.$$

This is possible if and only if $\det [(a_{jk}) - rI_3] = 0$. As we said in Sections 2-3 and 2-8, this involves finding the characteristic root r of the matrix (a_{jk}). Now

$$\begin{vmatrix} a_{00} - r & a_{01} & a_{02} \\ a_{10} & a_{11} - r & a_{12} \\ a_{20} & a_{21} & a_{22} - r \end{vmatrix} = 0$$

is a cubic equation in r with real coefficients. Such an equation has one real solution or two or three real solutions. It cannot happen that it has no real solutions at all. There is, therefore, at least one characteristic vector with components x_0, x_1, x_2 or, in other words, an invariant point $[x_0, x_1, x_2]$. Thus, in contrast to projectivities of the line, we have:

Theorem 3-4.2. A projectivity in $P^2(R)$ always has at least one invariant point.

In view of Theorem 3-2.3, the theorem can be dualized to:

Theorem 3-4.3. A projectivity in $P^2(R)$ always has at least one invariant line.

Perspectivities. We introduce a special type of plane projectivity in which we have a particular interest. A *perspectivity* in P^2 is a projectivity such that each line through its invariant point C is an invariant line. C is called the *center* of the perspectivity. The perspectivities of Section 3-3 were of this kind.

Theorem 3–4.4. Every perspectivity, not the identity, has exactly one center and exactly one pointwise-invariant line, called its *axis*.

Proof. In view of applications in Chapter 4, the proof will not use any properties of projectivities beyond the fact that they are collineations.

The first step of the proof will deal with the uniqueness of the center and the axis. Suppose that the perspectivity has a center C and two invariant lines k_1 and k_2 not through C. Then every point on k_1 or on k_2 lies on some line k_3 through C, and hence is an invariant point as the intersection of the two invariant lines k_1 and k_3 or k_2 and k_3. Let j be an arbitrary line not through $k_1 \times k_2$. Then $j \times k_1$ and $j \times k_2$ are invariant, which makes j an invariant line because it contains two invariant points. If h is an arbitrary line containing the invariant point $k_1 \times k_2$, then $h \times j$ is a second invariant point on h, and h is invariant. Thus every line is invariant and $\alpha = I$. Therefore the assumption of two invariant lines not through C led to a contradiction, and a nonidentical perspectivity can have at most one invariant line not through C. Is a second center C' possible? In this case, every point P not on CC' would lie on two invariant lines PC and PC' and be invariant. This would yield at least two invariant lines not through C, a contradiction.

If indeed there exists an invariant line l not through C, then every point P on l lies on some line through C. Hence P is invariant because it is the intersection of the two invariant lines l and PC. Thus l is an axis and, as explained before, the single axis of α.

In the case where no invariant lines exist other than those through C, let k be a line not through C, and therefore $k \neq k\alpha$. Let $(k \times k\alpha)C = l$. Then $l\alpha = l$, because l passes through C, and

$$(k \times l)\alpha = k\alpha \times l\alpha = k\alpha \times l = k \times l.$$

Thus every line not through C has precisely one invariant point, namely its intersection with its image under α. The join of the invariant point of k with the invariant point of another such line is an invariant line and must pass through C. This join is, therefore, l. Conversely, every point of l is the invariant point of some line through it. Hence l is the axis. A second axis through C is impossible, because then α would be the identity.

Corollary 1. A nonidentical perspectivity in P^2 has no invariant points besides its center and the points on its axis.

The dual of Theorem 3–4.4 yields the following.

Corollary 2. A projectivity has an axis (is a "line perspectivity") if and only if it is a perspectivity.

As we saw in the proof of Theorem 3–4.4, one has to distinguish between two types of perspectivities. A *minor perspectivity* or an *elation* has its center on its axis. For a *major perspectivity* or a *homology*, the center is not on the axis.

Theorem 3–4.5. The minor perspectivities of P^2 with a fixed axis form a group.

Proof. When two minor perspectivities α and β with the same axis and centers C and D are multiplied, all the points of the axis are preserved. The factors do not have any invariant points outside the axis. If the product has such an invariant point Q, then $Q\alpha\beta = Q$, $Q\alpha = Q\beta^{-1}$. But $Q\alpha$ lies on CQ and $Q\beta^{-1}$ on QD, which implies $Q\alpha = Q\beta^{-1} = Q$, a contradiction unless $\alpha = \beta = I$. This proves the closure. The inverse π^{-1} of a minor perspectivity π obviously preserves all points of the axis. If it had an invariant point Q outside the axis, then $Q = Q\pi^{-1}\pi = Q\pi$, a contradiction. Hence it is a minor perspectivity. Finally, the identity is a minor perspectivity.

We shall make the dual statement as follows.

Corollary 1. The minor perspectivities of P^2 with a fixed center form a group.

A combination of the theorem and the corollary yields:

Corollary 2. The minor perspectivities of P^2 with a fixed center and a fixed axis form a subgroup of the groups mentioned in the theorem and in Corollary 1.

The translations of A^2 are an example of minor perspectivities. The improper line is the axis, and the traces of all points under a given translation and all its powers meet in one improper point on the axis. In this case the group mentioned in Theorem 3–4.5 is $\mathcal{T}$. One fixed translation and all its powers form a group, as in Corollary 2.

All the perspectivities do not form a group. It is easy to construct examples of perspectivities whose product is not a perspectivity. On the other hand, it can be shown (cf. any elementary text in projective geometry) that every plane projectivity can be represented as the product of perspectivities. In other terms, the plane perspectivities generate $\mathcal{P}^2$.

Further examples of perspectivities will be discussed in the exercises.

The theorems of Desargues and Pappus. We encountered an incidence theorem before: the theorem of the complete quadrilateral. In this section two further incidence theorems will be studied—those of Gérard Desargues, French mathematician and architect (1593–1662), and of Pappus of Alexandria (*c.* 300 A.D.).

Theorem 3–4.6 (Desargues). In P^2, let two triangles (Fig. 3–16) ABC and $A'B'C'$ be given such that $A \neq A'$, $B \neq B'$, and $C \neq C'$. If the joins of corresponding vertices of the triangles are concurrent, then the points of intersection of corresponding sides of the triangles are collinear.

Proof. The lines SAA', SBB', and SCC' are distinct. (If two coincided, the theorem would be trivially true.) Neither of the quadruples $SABC$ and $SA'B'C'$ contains collinear triples, and there exists a unique projectivity α such that

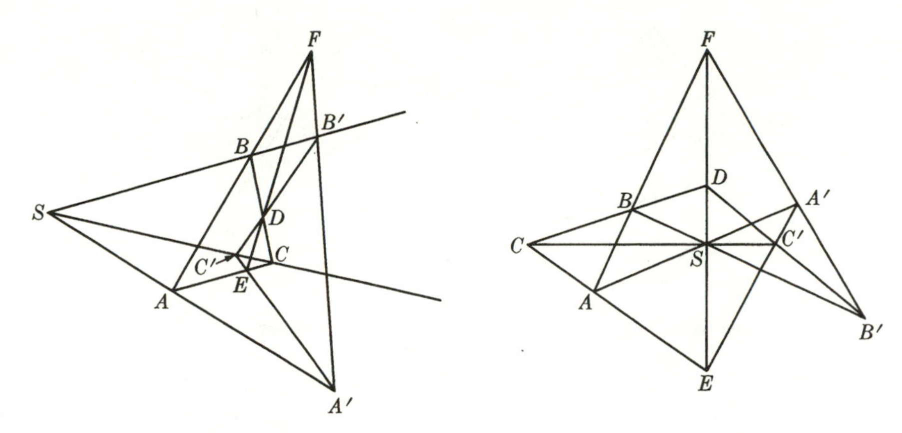

Figure 3–16 **Figure 3–17**

$A\alpha = A'$, $B\alpha = B'$, $C\alpha = C'$, and $S\alpha = S$. Under α, each of the lines SA, SB, and SC goes into itself. Then also every other line through S is invariant because the cross ratio of the four lines is invariant under α. Hence α is a perspectivity with center S. The points

$$F = AB \times A'B', \qquad E = AC \times A'C', \qquad \text{and} \qquad D = BC \times B'C'$$

are invariant under α. Namely, for instance, $F\alpha$ must be on SF as well as on $A'B'$ and must therefore coincide with F. By Corollary 1 to Theorem 3–4.4, D, E, and F must lie on the axis of α and thus are collinear. The case that any of the three points D, E, and F coincides with S can be dismissed, because this would cause two vertices of each of the triangles to coincide.

It should be noted that the case of S lying on the axis DEF has not been excluded. This (Fig. 3–17) is indeed a specialization of Desargues' theorem, called the *minor Desargues theorem*, in view of the perspectivity α being minor in this case.

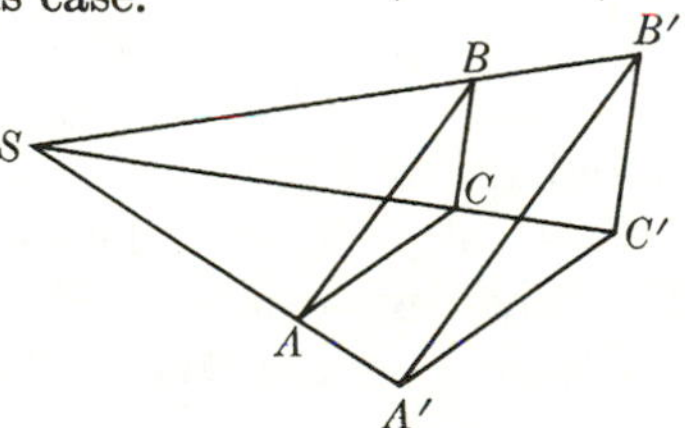

Figure 3–18

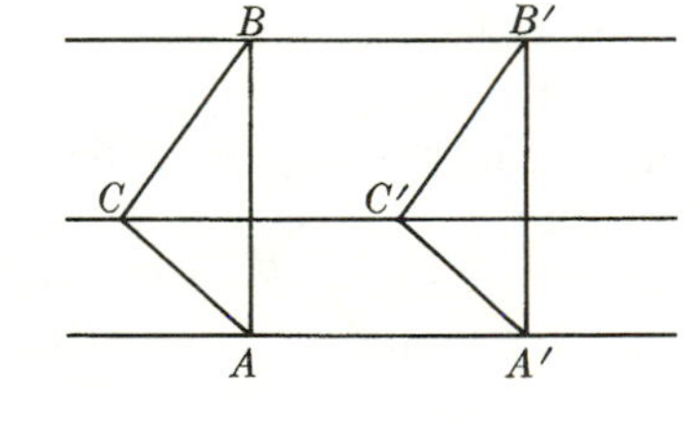

Figure 3–19

A special position for Desargues' theorem is the one in which the axis of α is the improper line of an affine plane. This is the so-called *affine Desargues theorem* (Fig. 3–18). If in this case α is major, it is a dilatation. If it is minor, so that the center also is improper, then α is a translation and we obtain the *minor affine Desargues theorem* (Fig. 3–19).

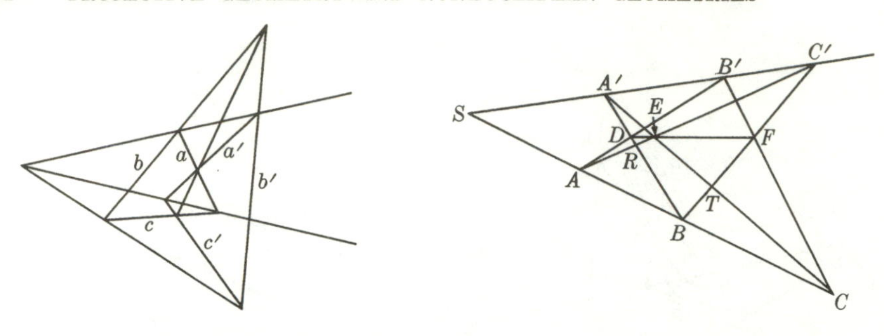

Figure 3–20

Figure 3–21

Let us study the dual statement of Desargues' theorem. A dualization of the statement, substituting the words "line" and "point" for each other, would have approximately the following form (Fig. 3–20): "Let two triangles be given by their sides a, b, c, and a', b', c', such that $a \neq a'$, $b \neq b'$, and $c \neq c'$. If the points of intersection of corresponding sides of the triangles are collinear, then the joins of corresponding vertices are concurrent." But this is just the converse of Desargues' theorem, where the hypothesis "joins of corresponding vertices are concurrent" and the resulting statement "points of intersection of corresponding sides are collinear" have been interchanged. As we pointed out before, the dual of a correct statement involving only incidences is itself a correct statement. We have therefore:

Corollary. Desargues' theorem and its converse are equivalent statements.

Theorem 3–4.7 (Pappus). In P^2, if A, B, and C (Fig. 3–21) lie on a line and A', B', C' on another line, then $AB' \times A'B$, $AC' \times A'C$, and $BC' \times B'C$ are collinear.

Proof. Let α be the perspectivity with center A which takes the points of $A'B$ into those of $A'B'$. If $S = AB \times A'B'$, $R = AC' \times A'B$, and $T = BC' \times A'C$, then we have α: $B \to S$, $R \to C'$, $D \to B'$, and $A' \to A'$. Now let β be the perspectivity with center C, mapping the points of $A'B'$ on those of BC'. Then β: $S \to B$, $C' \to C'$, $B' \to F$, and $A' \to T$. The projectivity $\alpha\beta$ therefore maps B on B, R on C', D on F, and A' on T. Now consider the perspectivity π with center E carrying the points of $A'B$ into those of BC'. It yields π: $B \to B$, $R \to C'$, and $A' \to T$. When we compare $\alpha\beta$ and π, Theorem 3–3.4 implies $D\pi = D\alpha\beta = F$, and since π is a perspectivity with center E, the points D, E, and F have to be collinear.

The dual of Pappus' theorem will be discussed in the exercises. Both Desargues' and Pappus' theorems are of great importance ir the axiomatic theory of the foundations of geometry, and we will meet them again in the next chapter.

Collineations and projectivities. After studying various types of projectivities in $P^2(R)$, we may ask the following question. Every projectivity is a collineation, but is every collineation in $P^2(R)$ a projectivity? In other words, are projectivity and collineation in $P^2(R)$ just synonyms? The answer will be in the affirmative. However, this result will be seen to be correct under the restriction $K = R$.

In order to prepare the main theorem we prove two lemmas.

Lemma 3–4.8. The real field has no automorphism besides the identity. The complex field has at least one nonidentical automorphism.

Proof. For an automorphism α of a field, $(a + b)\alpha = a\alpha + b\alpha$ and $(ab)\alpha = (a\alpha)(b\alpha)$. For the reals we have $0\alpha = 0$ because $a\alpha = (0 + a)\alpha = 0\alpha + a\alpha$, and $1\alpha = 1$ because $a\alpha = (a \cdot 1)\alpha = (a\alpha)(1\alpha)$. Now $2\alpha = (1 + 1)\alpha = 1 + 1 = 2$, and, by repetition, $n\alpha = n$ for all positive integers n. Also $(-1)\alpha = -1$, because $0 = [1 + (-1)]\alpha = 1 + (-1)\alpha$, and consequently $(-n)\alpha = -n$. For rationals we have $(m/n)\alpha = m/n$ because $m = (m/n \cdot n)\alpha = (m/n)\alpha \cdot n$. For a real $b > 0$, $\sqrt{b}\sqrt{b} = b$, and hence $\sqrt{b}\alpha \cdot \sqrt{b}\alpha = b\alpha$, and as a nonzero square, $b\alpha > 0$. It follows that α preserves positiveness.

In order to prove that for every real number b, $b\alpha = b$, suppose that $b\alpha = c > b$. Then there is a rational number r such that $c > r > b$ and, therefore, $r - b > 0$. Consequently $(r - b)\alpha = r - b\alpha = r - c < 0$, a contradiction to the preservation of positiveness. The case $c < b$ is analogous. Hence $b = b\alpha$ for all reals b, and α is the identity.

For complex numbers there is the nonidentical automorphism $z \to \bar{z}$. There are infinitely many others, but it can be shown that this and the identity are the only continuous automorphisms of the complex field.

Lemma 3-4.9. A collineation κ that leaves invariant three points A, B, and C on a line induces an automorphism of the coordinate field. If this field is the real field, κ preserves all points on AB.

Proof. Let D be collinear with A, B, and C; $(AB|CD) = r$, $(AB|DE) = s$, $(AB|C, D\kappa) = r'$, and $(AB|D\kappa, E\kappa) = s'$. Obviously, putting $r = 0$ and 1 yields $0' = 0$ and $1' = 1$. Then (cf. Exercise 3 of Section 3–3) $(AB|CE) = rs$ and $(AB|C, E\kappa) = r's' = (rs)'$. If $s = r^{-1}$, then $(rs)' = 1 = r'(r^{-1})'$ and $(r^{-1})' = (r')^{-1}$. Moreover $(-1)' = -1$ because, by Theorem 3–3.6, the point for which $r = -1$ can be obtained by constructions that are invariant under κ. Consequently $(-s)' = (-1)'s' = -s'$. In Exercise 1 of Section 3–3, we obtained $(AC|BD) = 1 - r$ from $(AB|CD) = r$. This implies

$$(AC|B, D\kappa) = (1 - r)' = 1 - r'.$$

Then $(r - s)' = r'(1 - s/r)' = r'[1 - (s/r)'] = r' - s'$. If $s = -t$, this yields $(r + t)' = r' + t'$, which together with $(rs)' = r's'$ proves that $r \to r'$ is an automorphism of the coordinate field. By Lemma 3–4.8, it is the identity if the coordinates are real.

Theorem 3–4.10. Every collineation of $P^2(R)$ is a projectivity.

Proof (Fig. 3–22). Let A, B, C, and D be four points, no three of which are collinear. First suppose that they are invariant under a collineation. Then the six sides of the quadrangle $ABCD$ are invariant lines and their intersections are invariant points. Each side contains three invariant points, and by Lemma 3–4.9 is pointwise invariant. Every line of $P^2(R)$ intersects the sides in at least three distinct invariant points and thus each line of $P^2(R)$ is pointwise invariant, so that the collineation is the identity.

Now suppose that the quadrangle $ABCD$ is mapped onto another quadrangle $A'B'C'D'$ by two collineations κ and λ. Then $\kappa\lambda^{-1}$ takes $ABCD$ into itself and consequently $\kappa\lambda^{-1} = I$. This implies that $\kappa = \lambda$ is the only collineation mapping $ABCD$ on $A'B'C'D'$. By Theorem 3–2.5, κ is a projectivity.

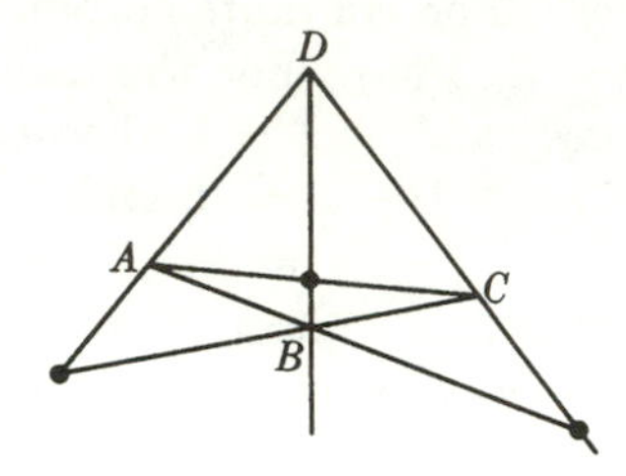

Figure 3–22

For a projective plane P^2 over an arbitrary field K, in view of Lemma 3–4.9, Theorem 3–4.10 changes into the following theorem.

Fundamental Theorem of Projective Geometry. Every collineation of $P^2(K)$ has the form

$$[x_0, x_1, x_2] \to [x_0\alpha, x_1\alpha, x_2\alpha]L,$$

where L is a linear transformation and α an automorphism of K.

A mapping of this kind is called a *semilinear transformation.*

One can generalize the Fundamental Theorem for higher dimensions. If, as the Fundamental Theorem shows, collineations and projectivities are essentially the same, we can now state clearly what constitutes the content of projective geometry. Collineations are the transformations which map straight lines on straight lines and preserve incidence. Projective geometry is the geometry whose group is the group of all collineations. Therefore, linear projective geometry is the geometry which deals with incidence relations between projective subspaces, with the consequences of these relations, and with nothing else. Projective geometry has fewer properties than affine geometry, just as affine geometry has fewer properties than euclidean geometry. The more comprehensive the group of a geometry is, the poorer is the geometry in properties and theorems. The projective group is the most comprehensive among those studied so far: $\mathcal{P}^n \supset \mathcal{A}^n \supset \mathcal{M}^n$.

Exercises

1. Given that π and π' are two one-dimensional projectivities in $P^1(R)$ such that there is a point P with $P = P\pi = P\pi'$ but no further invariant points, prove that $\pi\pi' = \pi'\pi$.

2. (a) Find canonical forms for one-dimensional projectivities in $P^1(R)$, in accordance with Theorem 1–9.11.
   (b) Show that a one-dimensional projectivity in $P^1(R)$ with a single invariant point cannot be involutory.
3. Prove: A projectivity of $P^1(R)$ which interchanges two distinct points is necessarily involutory.
4. Prove that in $P^1(R)$ every hyperbolic involutory projectivity is opposite (that is, has a negative determinant).
5. A projectivity α of $P^1(R)$ has the single invariant point $[1, -1]$. Moreover $[0, 1]\alpha = [3, 1]$. Find α. Is α uniquely determined?
6. Prove in $P^2(R)$:
   (a) A minor perspectivity induces on every line through the center, except the axis, a projectivity with a single invariant point.
   (b) A major perspectivity induces on every line through its center a projectivity with two distinct invariant points.
7. Show that the group mentioned in Theorem 3–4.5 is abelian.
8. Prove in $P^2(R)$: A two-dimensional projectivity which leaves a line pointwise invariant is a minor perspectivity if there are no other invariant points, and is a major perspectivity otherwise.
9. (a) Show that shear transformations are perspectivities.
   (b) Are these perspectivities necessarily major or minor?
10. In the augmented E^2:
    (a) Are line reflections perspectivities?
    (b) Are dilatations from 0 perspectivities? In case they are, find center and axis.
11. (a) Which rotations in E^2 are nonidentical perspectivities and which are not?
    (b) Find center and axis for those that are, and determine whether they are major or minor.
12. Find the trace of $[1, 2, 3]$ under the group of perspectivities whose center is $[1, 0, 0]$ and whose axis is $\langle 0, 1, 1\rangle$.
13. The figure that describes Desargues' theorem consists of 10 lines and 10 points. How many of the points can be taken as perspectivity centers for the theorem so that the same figure is obtained?
14. Find the three-dimensional dual to Desargues' theorem.
15. Find the plane dual of Pappus' theorem. How is it related to Pappus' theorem itself?
16. Classify the following statements as euclidean, affine, or projective.
    (a) $ABCD$ is a quadrangle,
    (b) its sides are equal,
    (c) opposite sides are parallel,
    (d) its angles are right,
    (e) its diagonals bisect each other,
    (f) the point F is inside $ABCD$,
    (g) the sides are tangents to a conic.

3–5 CONICS IN $P^2(R)$

Ellipses, parabolas, and hyperbolas. In the affine plane we found three affinely different types of conics, namely ellipses, hyperbolas, and parabolas. Every two conics of the same type can be transformed into each other by an affinity, and if we augment the affine plane so as to get a model of P^2, each such affinity corresponds to a projectivity. Are the three types distinct also from a projective point of view? The standard equations of an ellipse, a hyperbola, and a parabola, written in homogeneous coordinates, are

$$x_1^2 + x_2^2 - x_0^2 = 0, \tag{1}$$

$$x_1x_2 - x_0^2 = 0, \tag{2}$$

$$x_2^2 - x_0x_1 = 0, \tag{3}$$

respectively. After a first look at these equations one sees that Eq. (2) goes into Eq. (3) by the transformation

$$x_0 \to x_2, \qquad x_1 \to x_1, \qquad x_2 \to x_0.$$

The matrix of this transformation is nonsingular, since its determinant is -1, and thus this is a projectivity. Hence from a projective point of view hyperbolas and parabolas cannot be distinguished from each other. The same is true for hyperbolas and ellipses. For, by a transformation

$$x_0 \to x_1, \qquad x_1 \to x_0 + x_2, \qquad x_2 \to x_0 - x_2,$$

Eq. (2) goes into Eq. (1). Again the transformation is a projectivity.

The geometrical difference between the three affinely distinct types of conics lies in their relation to the improper line of the augmented $A^2(R)$. Intersecting each of the standard conics with the line $x_0 = 0$ yields the following results.

Equation (1) results in $x_1^2 + x_2^2 = 0$; that is, $x_0 = x_1 = x_2 = 0$, with no real point.

Equation (2) yields $x_1x_2 = 0$, which has two solutions, $x_1 = 0, x_2 \neq 0$, and $x_1 \neq 0, x_2 = 0$, the points $[0, 1, 0]$ and $[0, 0, 1]$.

Equation (3) yields $x_2^2 = 0$, and consequently the single point $[0, 1, 0]$. Thus we have the following.

Theorem 3–5.1. In $P^2(R)$ any one of the set of all ellipses (and circles), hyperbolas, and parabolas can be mapped into any other by a projectivity. The affine distinction of the three types of conics is based on the number of improper points on the conic [regarded as a conic in the augmented $A^2(R)$], none on the ellipse, one on the parabola, and two on the hyperbola.

Indeed, affinities preserve the improper line of the augmented plane, and since they also preserve incidence, the number of points that a conic has in common with the improper line is an affine invariant. Geometrically it can be seen that the improper point of a parabola lies on its axis, while the improper points of a hyperbola lie on its asymptotes (Fig. 3–23).

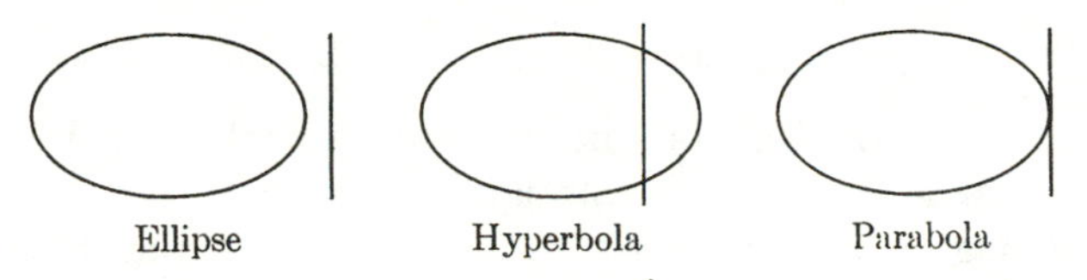

Figure 3–23

The general quadratic function. Besides the equations of the conics treated here, there are also other quadratic functions which we have ignored so far. In the following we shall attempt to develop a classification of these functions in the projective plane. The most general quadratic equation in homogeneous coordinates is

$$F = \sum_j \sum_k b_{jk} x_j x_k = 0,$$

with j and k ranging through 0, 1, 2, and with $b_{jk} = b_{kj}$. Explicitly

$$F = b_{00}x_0^2 + b_{11}x_1^2 + b_{22}x_2^2 + 2b_{01}x_0x_1 + 2b_{12}x_1x_2 + 2b_{02}x_0x_2.$$

The expression F is called a *homogeneous quadratic form* in the variables x_0, x_1, and x_2.

Theorem 3–5.2. Let $F = \sum_j \sum_k b_{jk}x_jx_k$, with j, $k = 0, 1, 2$ and $b_{jk} = b_{kj}$, not all $b_{jk} = 0$. By a linear transformation, F can be brought into the canonical form $c_0x_0^2 + c_1x_1^2 + c_2x_2^2$, where each of c_0, c_1, and c_2 has one of the values 0, 1, or -1.

Proof. First consider the case that at least one of b_{00}, b_{11}, and b_{22} is nonzero. Suppose that $b_{00} \neq 0$, otherwise renumber. (Renumbering is a linear transformation.) Then

$$\begin{aligned} F &= b_{00}x_0^2 + 2b_{01}x_0x_1 + 2b_{02}x_0x_2 + \cdots \\ &= b_{00}(x_0 + b_{01}x_1/b_{00} + b_{02}x_2/b_{00})^2 + \cdots, \end{aligned}$$

where $\cdots$ stands for terms containing only the variables x_1 and x_2. Now put

$$y_0 = x_0 + b_{01}x_1/b_{00} + b_{02}x_2/b_{00}, \qquad y_1 = x_1, \qquad y_2 = x_2.$$

This is a linear transformation which results in $F = y_0^2 + \cdots$ Then the procedure may be repeated for the remaining variables.

In the case where all the $b_{jj} = 0$, certainly one $b_{jk} \neq 0$. Suppose that $b_{01} \neq 0$, otherwise renumber. Then put

$$x_0 = y_0 - y_1, \qquad x_1 = y_0 + y_1, \qquad x_2 = y_2,$$

and obtain

$$F = 2b_{01}x_0x_1 + 2b_{02}x_0x_2 + 2b_{12}x_1x_2 = 2b_{01}y_0^2 - 2b_{01}y_1^2 + \cdots$$

The dots mean terms which contain no squares. Thus this case has been reduced to the first case by a linear transformation.

After obtaining $F = d_0z_0^2 + d_1z_1^2 + d_2z_2^2$, a final linear transformation $z_j' = \sqrt{|d_j|}z_j$, with $j = 0, 1, 2$, brings about the canonical form. This completes the proof.

If one or more of the c_j in the canonical form vanish, we say that $F = 0$ represents a *degenerate conic.*

Theorem 3–5.3. A homogeneous quadratic form F, defined as in Theorem 3–5.2, yields a nondegenerate conic $F = 0$ if and only if $\det(b_{jk}) \neq 0$. The quadratic forms of degenerate and nondegenerate conics are projectively distinct.

Proof. The quadratic form can be written in matrix notation as $F = \mathbf{x}B\mathbf{x}'$, where $\mathbf{x}$ is the row vector (x_0, x_1, x_2) and $B = (b_{jk})$. A linear transformation with matrix A takes $\mathbf{x}$ into $\mathbf{x}A$, and the equation becomes $\mathbf{x}AB(\mathbf{x}A)' = \mathbf{x}ABA'\mathbf{x}' = 0$. This is a new homogeneous quadratic form with matrix ABA'. If the resulting conic is $c_0x_0^2 + c_1x_1^2 + c_2x_2^2 = 0$, then

$$ABA' = \begin{bmatrix} c_0 & 0 & 0 \\ 0 & c_1 & 0 \\ 0 & 0 & c_2 \end{bmatrix}.$$

The determinant of the right-hand side is zero or nonzero according to whether the resulting conic is degenerate or nondegenerate. Since $\det A = \det A' \neq 0$, $\det B$ is zero for a degenerate conic and nonzero for a nondegenerate conic, regardless of the choice of A. This completes the proof.

For the nondegenerate conics there are two possibilities. If $c_0 = c_1 = c_2$, the conic cannot contain any points with real coordinates. $F = 0$ is then called an *imaginary conic.* In every other case the conic contains real points and is called a *real conic.* The properties of being an imaginary or a real conic are projectively invariant because real points become real points under real projectivities.

Projectivities preserving a conic. In Section 2–6, the affinities which preserve a given conic were determined. The same question will now be treated for the projective plane and for imaginary and real nondegenerate conics.

Theorem 3-5.4. The group of two-dimensional projectivities with real coefficients which preserve the imaginary conic

$$x_0^2 + x_1^2 + x_2^2 = 0 \tag{4}$$

is transitive on the points of the conic. The group is isomorphic to the orthogonal group $\mathcal{O}_+^3$.

Proof. Let $y_j = \sum_k a_{jk}x_k$, with j and k ranging through 0, 1, 2, and

$$\det (a_{jk}) \neq 0.$$

Then the invariance of the conic is expressed by

$$y_0^2 + y_1^2 + y_2^2 = \sum_j \left(\sum_k a_{jk}x_k\right)^2 = 0, \tag{5}$$

where x_0, x_1, and x_2 are connected by Eq. (4). The values $x_0 = 1$, $x_1 = i$, and $x_2 = 0$, satisfy Eq. (4) and, when substituted in Eq. (5), they yield

$$(a_{00} + a_{01}i)^2 + (a_{10} + a_{11}i)^2 + (a_{20} + a_{21}i)^2$$
$$= a_{00}^2 - a_{01}^2 + a_{10}^2 - a_{11}^2 + a_{20}^2 - a_{21}^2 + 2i(a_{00}a_{01} + a_{10}a_{11} + a_{20}a_{21}) = 0.$$

On separating the imaginary and real parts we get

$$a_{00}^2 + a_{10}^2 + a_{20}^2 = a_{01}^2 + a_{11}^2 + a_{21}^2$$

and

$$a_{00}a_{01} + a_{10}a_{11} + a_{20}a_{21} = 0.$$

Substitution of $x_0 = 1$, $x_1 = 0$, and $x_2 = i$ yields in the same fashion

$$a_{00}^2 + a_{10}^2 + a_{20}^2 = a_{02}^2 + a_{12}^2 + a_{22}^2$$

and

$$a_{00}a_{02} + a_{10}a_{12} + a_{20}a_{22} = 0.$$

Since we are dealing with projectivities, the matrix (a_{jk}) can be replaced by (ra_{jk}), where r is an arbitrary nonzero real number. We choose r so that

$$r^2(a_{00}^2 + a_{10}^2 + a_{20}^2) = r^2(a_{01}^2 + a_{11}^2 + a_{21}^2) = r^2(a_{02}^2 + a_{12}^2 + a_{22}^2) = 1.$$

This is always possible because $a_{00}^2 + a_{10}^2 + a_{20}^2$ is positive, and if r satisfies the equation, so also does $-r$, which is distinct from r. Then the matrices (ra_{jk}) are orthogonal because they fulfill the conditions of Theorem 2-7.5. These two matrices represent the same projectivity. Exactly one of them has a positive determinant and thus belongs to $\mathcal{O}_+^3$. On the other hand, every such pair of orthogonal 3×3 matrices represents a projectivity which preserves Eq. (4).

Now let $[\mathbf{x}] = [x_0, x_1, x_2]$ be an arbitrary point on the conic of Eq. (4). Its coordinates are complex numbers. A projectivity π with real coefficients a_{jk} taking $[0, 1, i]$ into $[\mathbf{x}]$ has to satisfy $a_{1k} + ia_{2k} = rx_k$, with $k = 0, 1, 2$. The reals a_{1k} and a_{2k} are thus uniquely determined, and if $[\mathbf{x}]$ is on the conic of Eq. (4), the matrix (a_{jk}) can be made orthogonal (cf. Exercise 12). If $[\mathbf{y}] = [y_0, y_1, y_2]$ is another arbitrary point on the conic, let a projectivity π' map $[0, 1, i]$ on $[\mathbf{y}]$. Then $\pi^{-1}\pi'$ takes $[\mathbf{x}]$ into $[\mathbf{y}]$. This completes the proof.

Now we come to the real conics. Since, from a projective point of view, there is only one type of nondegenerate real conic, we may arbitrarily choose one such conic and find the projectivities which leave it invariant. Our choice will be the conic of Eq. (3). By Theorem 2–6.2, all affinities preserving this parabola form a two-parameter group. All these affinities leave invariant the improper point $[0, 1, 0]$ of Eq. (3), whereas under projectivities this point will lose its privileged position. Thus the group of projectivities preserving Eq. (3) might well depend on more than two parameters.

Theorem 3–5.5. There is a three-parameter group $\mathcal{G}$ of real two-dimensional projectivities, isomorphic to the group $\mathcal{H}_+$ of direct hyperbolic isometries, which leaves the conic of Eq. (3) invariant.

Proof. Let a direct hyperbolic isometry be given, $z' = (az + b)/(cz + d)$, with a, b, c, and d real, and $ad - bc > 0$. Put $z = x + iy$ and $z' = x' + iy'$. Then

$$z' = [a(x + iy) + b]/[c(x + iy) + d].$$

Separating real and imaginary parts, we obtain

$$x' = \frac{(ad + bc)x + ac(x^2 + y^2) + bd}{2c\,dx + c^2(x^2 + y^2) + d^2}, \tag{6}$$

$$y' = \frac{(ad - bc)y}{2c\,dx + c^2(x^2 + y^2) + d^2}. \tag{7}$$

Moreover,

$$x'^2 + y'^2 = z'\overline{z'} = \frac{[a(x + iy) + b][a(x - iy) + b]}{[c(x + iy) + d][c(x - iy) + d]},$$

$$x'^2 + y'^2 = \frac{2abx + a^2(x^2 + y^2) + b^2}{2c\,dx + c^2(x^2 + y^2) + d^2}. \tag{8}$$

We introduce new coordinates $\xi = x^2 + y^2$, $\eta = x$, and $\xi' = x'^2 + y'^2$, $\eta' = x'$. The geometrical significance of this coordinate change will be discussed later. With the new coordinates, $\eta^2 - \xi = -y^2$ and $\eta'^2 - \xi' = -y'^2$. It follows from Eq. (7) that $y = 0$ implies $y' = 0$, and consequently $\eta^2 - \xi = 0$ brings about $\eta'^2 - \xi' = 0$; that is, the relation $\eta^2 - \xi = 0$ is preserved under the transformation $\xi \to \xi'$, $\eta \to \eta'$. Now we use homogeneous coordinates, $\xi = x_1/x_0$, $\eta = x_2/x_0$, and $\xi' = x_1'/x_0'$, $\eta' = x_2'/x_0'$, and then Eqs. (6) and (8) become

$$\begin{aligned} sx_0' &= d^2x_0 + c^2x_1 + 2c\,dx_2, \\ sx_1' &= b^2x_0 + a^2x_1 + 2abx_2, \\ sx_2' &= b\,dx_0 + acx_1 + (ad + bc)x_2, \end{aligned} \tag{9}$$

with $ad - bc > 0$ and $s \neq 0$. This is a linear transformation because its determinant is $(ad - bc)^3 > 0$, and it preserves the conic $x_2^2 - x_0x_1 = 0$. Three parameters are involved if the transformation is to be considered as a projectivity in which the ratio $x_0' : x_1' : x_2'$ matters, rather than the value of the coordinates. When we compare Eq. (9) with the affinities of Eq. (3), preserving the parabola on p. 97, we see that those constitute the special case of Eq. (9) in which $c = 0$ and $d = 1$.

We have established a one-to-one correspondence between the transformations from $\mathcal{H}_+$ and the projectivities of Eq. (9). Each direct hyperbolic isometry produces a transformation of the type of Eq. (9), and each such linear transformation gives rise to a transformation from $\mathcal{H}_+$, up to a common nonzero factor r. However, replacing a, b, c, and d by ra, rb, rc, and rd does not change the bilinear transformation. In order to show that this one-to-one correspondence is an isomorphism, we have to verify that the multiplication of the transformations follows the same rules. Indeed, it is easy to see that in both cases a transformation having the coefficients a, b, c, and d, followed by that having the coefficients a', b', c', and d' is the transformation with coefficients a'', b'', c'', and d'', where

$$\begin{bmatrix} a & b \\ c & d \end{bmatrix}\begin{bmatrix} a' & b' \\ c' & d' \end{bmatrix} = \begin{bmatrix} a'' & b'' \\ c'' & d'' \end{bmatrix},$$

with $ad - bc > 0$, $a'd' - b'c' > 0$, and $a''d'' - b''c'' > 0$. This completes the proof.

What was the meaning of the transformation $\xi = x^2 + y^2$, $\eta = x$? All the points on the real axis $y = 0$ are mapped on the curve $\xi = x^2 = \eta^2$, which is the conic $x_2^2 - x_0x_1 = 0$. The proper points of the hyperbolic plane in Poincaré's model, that is, the points $y > 0$, are mapped on points with $\xi > x^2 = \eta^2$, or $x_2^2 - x_0x_1 < 0$. The set of these points will be temporarily defined as the *interior* of the conic in Eq. (3). (Another, more general, definition of the interior of a conic will be given at the end of this section.) Hence the hyperbolic plane has been mapped into the interior of the conic and even onto it, because every point $[1, \xi, \eta]$ with $\xi > \eta^2$ (which cannot be improper) corresponds to a point $\eta + i\sqrt{\xi - \eta^2}$ in the upper half-plane. The transformations of Eq. (9) preserve the conic, just as the direct hyperbolic isometries preserve its preimage, the real axis.

Are there any projectivities besides Eq. (9) which leave the conic of Eq. (3) invariant? The answer is the following theorem.

Theorem 3-5.6. All real two-dimensional projectivities that leave the conic of Eq. (3) invariant form a group $\mathcal{G}'$ which has the group $\mathcal{G}$ as subgroup of index 2. The remaining elements of $\mathcal{G}'$ form the coset $\mathcal{G}G$, where G is the transformation $x_0 \to -x_0$, $x_1 \to -x_1$, and $x_2 \to x_2$. The group $\mathcal{G}'$ is transitive on the points of Eq. (3).

First we prove a lemma.

Lemma 3-5.7. If A, B, C, and A', B', C' are points on the conic of Eq. (3), then there is exactly one real two-dimensional projectivity taking A into A', B into B', C into C', and preserving the conic.

We say that $\mathcal{P}^2$ is *triply transitive* on the points of the conic.

Proof of the lemma. Among the projectivities of Eq. (9) there is at least one transformation taking [1, 0, 0] into A and [0, 1, 0] into B. Call this projectivity T_1, and let $CT_1^{-1} = C_1$. Also let T_2 be a transformation (Eq. 9) mapping [1, 0, 0] on A', [0, 1, 0] on B', and let $C'T_2^{-1} = C_1'$.

Now let us study transformations (Eq. 9) which preserve the conic in Eq. (3) and leave [1, 0, 0] and [0, 1, 0] invariant. For such transformations we obtain $b = c = 0$. Then Eq. (9) becomes

$$x_0' = d^2x_0, \qquad x_1' = a^2x_1, \qquad x_2' = a\,dx_2.$$

As a projectivity, this can be written in the form

$$sx_0' = rx_0, \qquad sx_1' = x_1/r, \qquad sx_2' = x_2, \tag{10}$$

with some $s \neq 0$, and every transformation (Eq. 10) with positive r can be obtained in this way.

The transformation of Eq. (10) can, by suitable choice of r, take any point of the conic in Eq. (3) except [1, 0, 0] and [0, 1, 0] into any other point of the conic except these two. For instance, let $[x_0, x_1, x_2]$ be one of its points and $[x_0', x_1', x_2']$ the desired image point. We have $x_0x_1 = x_2^2$ and $x_0'x_1' = x_2'^2$. Now $x_2 \neq 0$ and $x_2' \neq 0$ because we have excluded the points [0, 1, 0] and [1, 0, 0]. For the same reason, x_0 and x_0' have to be nonzero. Set $s = x_2/x_2'$ and $r = sx_0'/x_0$. Then $sx_0' = rx_0$ and $sx_1' = x_1/r$. However, this procedure might involve negative values of r. In that case the transformations of Eq. (10) still are projectivities, but not necessarily of the type in Eq. (9), because $ad - bc$ might have become negative. Let T_3 be the transformation of Eq. (10) which takes C_1 into C_1'. Then the projectivity $T_1^{-1}T_3T_2$ maps A, B, C, on A', B', C', respectively, as required.

We have to prove that this is the only projectivity mapping the three points on the three image points. It is known from elementary geometry that through every point of a nondegenerate conic there is a single line that does not intersect the conic in any other point. This is the *tangent* to the conic in this point. The tangents to the conic of Eq. (3) in A and B meet in a point, say D. No three of the four points A, B, C, D can be collinear because a line has at most two points in common with a nondegenerate conic, and a tangent, by definition, meets the conic in only one point. Since incidences are preserved under projectivities, D will be mapped by $T_1^{-1}T_3T_2$ on a point D', which is the common point of the tangents to the conic in A' and B'. Hence four points, no three of which are collinear, have been mapped on four other points that fulfill the same condition. By Theorem 3–2.5, there is only one projectivity doing this, and hence $T_1^{-1}T_3T_2$ is unique. Thus the lemma is proved.

Proof of the theorem. In every mapping of the conic of Eq. (3) onto itself, some given three of its points are mapped onto three other given points of the conic, and the only projectivity doing so, as described in the proof of the lemma, is composed of projectivities of the type in Eq. (9) and possibly also others of the type in Eq. (10). But Eq. (10) is a special case of Eq. (9) if $r > 0$, and a transformation (Eq. 10) with negative r is obtained from Eq. (10) with positive r by right or left multiplication with G. Hence every projectivity preserving Eq. (3) is of the type $T_1^{-1}T_3T_2$, which is in $\mathcal{G}$ if $T_3 \in \mathcal{G}$, or of the type $T_1^{-1}T_4GT_2$, where T_1, T_4 and $T_2 \in \mathcal{G}$. Now it can easily be checked that the effect of multiplying an element (Eq. 9) of $\mathcal{G}$ with G is the same as multiplying the matrix

$$\begin{bmatrix} a & b \\ c & d \end{bmatrix} \quad \text{by the matrix} \quad \begin{bmatrix} -1 & 0 \\ 0 & 1 \end{bmatrix}.$$

The product is a matrix with negative determinant. The group $\mathcal{G}$ is isomorphic to the multiplicative group $\mathcal{G}_1$ of all nonsingular 2×2 matrices with positive determinant. The resulting group $\mathcal{G}'$ is isomorphic to the multiplicative group $\mathcal{G}_1'$ of all nonsingular 2×2 matrices. It can be shown without difficulty that $\mathcal{G}_1$ is of index 2 in $\mathcal{G}_1'$, and hence also $\mathcal{G}$ is of index 2 in $\mathcal{G}'$.

The transitivity of $\mathcal{G}'$ follows from the lemma.

Corollary 1. The group $\mathcal{G}'$ is isomorphic to the group $\mathcal{H}$ of hyperbolic isometries.

Proof. $\mathcal{G} \cong \mathcal{H}_+$ and $\mathcal{H}$ is developed from $\mathcal{H}_+$ exactly as $\mathcal{G}'$ is from $\mathcal{G}$.

Corollary 2. The group of real two-dimensional projectivities leaving invariant an arbitrary given conic is a conjugate of $\mathcal{G}'$ and is therefore isomorphic to $\mathcal{H}$.

Proof. The conic can be transformed into the special conic of Eq. (3) by a projectivity, say M. All elements of $\mathcal{G}'$, and only these, preserve Eq. (3), which in turn is taken by M^{-1} into the given conic. Hence $M\mathcal{G}'M^{-1}$ is the group of projectivities preserving the conic.

Corollary 3. $\mathcal{G}'$ is transitive on the interior points of the conic of Eq. (3).

Proof (Fig. 3-24). A point [y] is defined to be in the interior of the conic of Eq. (3) by the inequality $y_2^2 - y_0y_1 < 0$. Every line through such a point intersects the conic of Eq. (3) in two distinct points, as will be proved in Exercise 17. Let P and Q be distinct interior points and R be the intersection of the tangents at U_1 and U_2, where U_1 and U_2 are the points in which PQ meets the conic. R cannot lie on the conic. Let PR intersect the conic in S and let QR intersect it in T. The element of $\mathcal{G}'$ that maps U_1 and U_2 on themselves and S on T also maps P on Q because R is preserved.

It may surprise the reader that what is obviously a concept of orientation has been used in Theorem 3-5.6. The transformations from $\mathcal{G}$ are direct, while those from $\mathcal{G}G$ are opposite, and we have seen that there is no orientation in the projective plane. The reason for the sudden appearance of orientability is that the conic as a *one*-parameter projective subset is indeed orientable. A point which moves continuously along a conic does so either in one sense or in the opposite sense, and these two are clearly distinguishable.

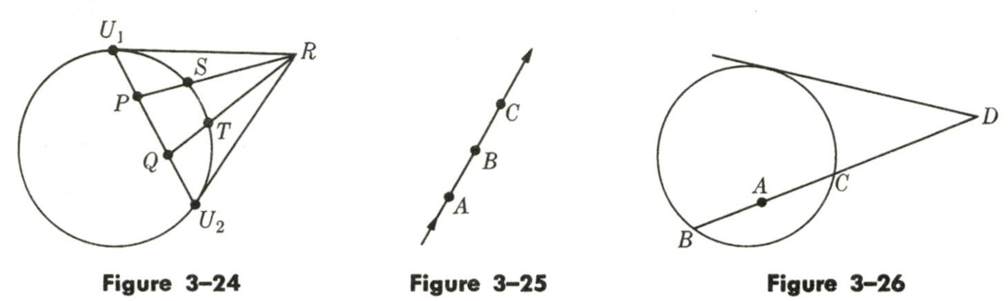

Figure 3-24 **Figure 3-25** **Figure 3-26**

Interior of a conic. This brings us to a related question. We saw that in the projective plane no "betweenness" of collinear points can be defined. In fact (Fig. 3-25), if A, B, and C are distinct collinear points, each of the three lies between the other two. On the other hand, for four distinct collinear points it can be determined whether two of them separate the other two and are separated by them at the same time (cf. Exercise 2 of Section 3-2). In close connection with this separation of point couples, there is the problem of whether the *interior* of a conic can be defined in a projectively meaningful way. This is important, because we would like to replace the temporarily adopted special definition by one that is generally valid and geometrically motivated. Certainly it would not do to say (Fig. 3-26) that a point A lies in the interior of a conic because on every secant through A, A lies between the conic points B and C. The only meaningful (that is, projectively invariant) statement could be that on every secant an exterior point D and an interior point A are separated by B and C. However, this only distinguishes between exterior and interior without defining either. We therefore introduce the following.

Definition. A point is *exterior* to a nondegenerate conic if it does not lie on the conic and if there exists a (real) tangent to the conic through it. Any point which is neither on the conic nor exterior is *interior* to the conic.

The equivalence of the two definitions of the interior will be discussed in Exercise 16.

Theorem 3–5.8. Projectivities map the interior of a nondegenerate conic onto the interior of the nondegenerate image conic.

Proof. The existence or nonexistence of a tangent to the conic through a point is a projective property because only incidence and collinearity are involved. The mapping is injective because the inverse of the projectivity maps every point inside the image conic on some point inside the preimage conic.

Exercises

1. Find the projectivities that take the ellipse $b^2x_1^2 + a^2x_2^2 = a^2b^2x_0^2$ into
   (a) the parabola $x_2^2 = 2px_0x_1, \quad p > 0$
   (b) the hyperbola $b^2x_1^2 - a^2x_2^2 = a^2b^2x_0^2$.
2. Perform coordinate transformations in order to find the class of each of the following conics by reducing them to canonical form.
   (a) $5x_0^2 + 5x_1^2 + 5x_2^2 + 4x_0x_1 + 4x_1x_2 + 4x_0x_2 = 0$
   (b) $3x_0^2 - 5x_1^2 - 3x_2^2 - 4x_0x_1 - 4x_1x_2 + 4x_0x_2 = 0$
   (c) $x_0x_1 + 3x_0x_2 = 0$
   (d) $x_0x_1 - x_1x_2 + x_0x_2 = 0$
   (e) $2x_0^2 + 4x_1^2 + x_2^2 - 4x_0x_1 - 2x_0x_2 - 4x_1x_2 = 0$
3. Prove: A conic in whose equation all the square terms, but none of the other terms, vanish, is nondegenerate.
4. The number of nonzero coefficients c_j in the canonical form of $F = \mathbf{x}B\mathbf{x}'$ is called the *rank* of the matrix B. Show that, if $F = 0$ represents two distinct lines, the rank of B is 2, and if $F = 0$ represents one line, the rank is 1.
5. What, if any, real points do the degenerate conics contain if they are given by the following canonical forms?

   (a) $x_0^2 + x_1^2 = 0$ (b) $x_0^2 - x_1^2 = 0$ (c) $x_0^2 = 0$

6. The plane dual of a conic is called a *line conic.*
   (a) What is it geometrically?
   In particular, what are
   (b) $u_0^2 - u_2^2 = 0$? (c) $u_0^2 + u_1^2 = 0$?
   (d) $u_0^2 = 0$? (e) $u_0^2 - u_1^2 - u_2^2 = 0$?
7. Prove that in $P^2(C)$ every real line intersects every conic in one point or in two points. (C is the complex field.)
8. Prove:
   (a) If a line in $P^2(C)$ passes through two conjugate points $[x_0, x_1, x_2]$ and $[\bar{x}_0, \bar{x}_1, \bar{x}_2]$, it is real.
   (b) If a line in $P^2(C)$ passes through two real points, it is real.

9. Show that in $P^2(C)$ every conic intersects every line.

10. Prove that two homogeneous quadratic forms with complex coefficients are equivalent under linear transformations with complex coefficients if and only if their matrices have the same rank.

11. Find the two points of $P^2(C)$ common to all circles about the origin $[1, 0, 0]$. These points are called the *circle points*.

12. Show that in the proof of Theorem 3–5.4 the matrix (a_{jk}) can be determined so that it is orthogonal.

13. Find all the projectivities that map the circle $x_0^2 - x_1^2 - x_2^2 = 0$ onto itself.

14. (a) Which conic contains the points $P_1 = [1, 0, 1]$, $P_2 = [1, 1, 0]$, $P_3 = [1, -1, 0]$, $Q_1 = [5, 3, 4]$, $Q_2 = [1, 0, -1]$, and $Q_3 = [13, 5, 12]$?
    (b) Find a projectivity that takes the conic into itself and P_j into Q_j, with $j = 1, 2, 3$.
    (c) Show that this is the only such projectivity.

15. (a) Find a projectivity that takes the conic $x_1^2 + x_2^2 = x_0^2$ into itself and changes the orientation on the conic.
    (b) What does this projectivity do to the interior of the conic?

16. Using the general definition of the interior, prove that a point $[\mathbf{y}]$ lies in the interior of the conic of Eq. (3) if and only if $y_2^2 - y_0y_1 < 0$.

17. Prove: In $P^2(R)$ every line through an interior point of a conic intersects the conic in two distinct points.

18. (a) Show that the definition of the exterior and the interior as given for conics is applicable also to triangles. (Remember that a tangent to a conic was defined as a line having a single point on the conic.)
    (b) A *convex* curve in E^2 is defined as a closed curve containing in its interior the whole segment AB whenever A and B lie on the curve. Use a curve that is nonconvex according to this definition as an example in which the projective definition of interior and exterior is not reasonable.

3–6 POLARITIES

Correlations and polarities. We continue our discussion of transformations in the projective plane over the real or complex field. All the transformations that we dealt with were mappings of points onto points. However, the duality principle of the projective plane suggests a new possibility. We do not consider lines as sets of points, but rather look upon the projective plane as consisting of two classes of objects, points and lines, between which a relation is defined, that of incidence. If this concept of the projective plane is adopted, then it is natural to attempt an extension of the definition of a linear transformation. We will have to discuss mappings of points on lines and lines on points, after having treated mappings of points on points and lines on lines at length in the earlier sections of this chapter. Theoretically these new linear transformations will be mappings of the one-subspaces of a V^3 onto the one-subspaces of the dual space $(V^*)^3$, and vice versa. These so-called *correlations* have the same form as projectivities, the only difference being that here the preimages are given in point coordinates and the images in line coordinates, and vice versa. Hence, under a correlation with nonsingular matrix B, a line $\langle \mathbf{u} \rangle$ is mapped on a point $[\mathbf{u}(B^{-1})']$, while a point $[\mathbf{x}]$ is mapped on a line $\langle \mathbf{x}B \rangle$, in accordance with Theorem 3–2.3. Since the equation of a line $\mathbf{x}\mathbf{u}' = 0$ then goes into $(\mathbf{x}B)[\mathbf{u}(B^{-1})']' = \mathbf{x}\mathbf{u}' = 0$, collinearity and incidences are preserved under correlations. Thus, if a point Q is incident with a line q and if α is a correlation, then the line $Q\alpha$ is incident with the point $q\alpha$. We will have to pay special attention to involutory correlations, which are called *polarities*. If α is a polarity and if $Q\alpha = q$, then $q\alpha = Q\alpha^2 = Q$. Thus, if a polarity maps a point Q on a line q, then the same polarity takes q back into Q. The point Q is called the *pole* of the line q, q being the *polar* of Q.

If again α is a polarity and if the point Q is incident with the line p, then $Q\alpha$ is incident with $p\alpha$. We say that Q and $p\alpha$ are *conjugate points* and that p and $Q\alpha$ are *conjugate lines* under α. If $Q\alpha$ is incident with Q, Q is a *self-conjugate* point and $Q\alpha$ a *self-conjugate line.*

The choice of the name "conjugate" is misleading because in most of the mathematical uses of this word a one-to-one correspondence is implied (for instance, the conjugate complex number), while here a point is conjugate to all the points on its polar.

Polarities and conics. Now polarities will be introduced in a special way, namely with respect to a conic. Let a homogeneous quadratic form be given, and denoted $F(\mathbf{x}, \mathbf{x}) = \mathbf{x}B\mathbf{x}'$, with $B = (b_{jk})$ a nonsingular 3×3 matrix. As explained on page 160, $F(\mathbf{x}, \mathbf{x}) = 0$ represents a nondegenerate conic. Similarly, we define $F(\mathbf{y}, \mathbf{x}) = \mathbf{y}B\mathbf{x}'$. In this case F is called a *homogeneous bilinear form,* of which the homogeneous quadratic form is a special case. Let $[\mathbf{y}]$ be a fixed point and $[\mathbf{x}]$ variable. Then $F(\mathbf{y}, \mathbf{x}) = 0$ is the equation of a line, namely

$$x_0(b_{00}y_0 + b_{10}y_1 + b_{20}y_2) + x_1(b_{01}y_0 + b_{11}y_1 + b_{21}y_2) + x_2(b_{02}y_0 + b_{12}y_1 + b_{22}y_2) = 0.$$

This yields a correlation $[\mathbf{y}] \rightarrow \langle \mathbf{y}B \rangle$. Is this correlation a polarity?

Theorem 3–6.1. Let B be a nonsingular 3×3 matrix. Then $[\mathbf{y}] \rightarrow \langle \mathbf{y}B \rangle$ is a polarity if and only if B is symmetric, that is, $B = B'$.

Proof. As we saw, $[\mathbf{y}] \rightarrow \langle \mathbf{y}B \rangle$ is a correlation. Its inverse is $\langle \mathbf{y} \rangle \rightarrow [\mathbf{y}B^{-1}]$. However, by Theorem 3–2.3, the image of $\langle \mathbf{y} \rangle$ under B has to be $[\mathbf{y}B'^{-1}]$. Thus, if the correlation is to be involutory, that is, equal to its inverse, then necessarily $[\mathbf{y}B^{-1}] = [\mathbf{y}B'^{-1}]$ for all $[\mathbf{y}]$. This happens if and only if $B = rB'$ for some

$r \neq 0$. Repeated, this yields $B = rB' = r(rB')' = r^2B$, and hence $r^2 = 1$ and $r = \pm 1$. Now $B = -B'$ would imply

$$\det B = \det(-B') = (-1)^3 \det B' = -\det B$$

and finally $\det B = 0$, a contradiction. Hence $B = B'$. Conversely, the correlation is a polarity if B is symmetric.

Corollary. Every nondegenerate conic determines a polarity.

Proof. The conic has a nonsingular symmetric matrix, by Theorem 3–5.3.

Theorem 3–6.2. Under a polarity with respect to a conic, a point is self-conjugate if and only if it lies on the conic. In this case its polar is tangent to the conic.

Proof. A point $[\mathbf{y}]$ is self-conjugate if and only if $F(\mathbf{y}, \mathbf{y}) = 0$. But this is equivalent to its lying on the conic whose equation is $F(\mathbf{x}, \mathbf{x}) = 0$. Suppose that a second point $[\mathbf{z}]$ lies on the polar of $[\mathbf{y}]$ as well as on the conic. Being a point on the conic, $[\mathbf{z}]$ is self-conjugate, $F(\mathbf{z}, \mathbf{z}) = 0$. Since $[\mathbf{z}]$ lies on the polar of $[\mathbf{y}]$, $[\mathbf{y}]$ lies on the polar of $[\mathbf{z}]$. Thus both points lie on both polars, which is possible for distinct points only when the polars coincide. But this again is impossible unless $[\mathbf{y}] = [\mathbf{z}]$. Hence the polar has only one point in common with the conic, and is a tangent.

It can be checked easily that the well-known equations of tangents to conics, as they are studied in elementary analytic geometry, are indeed equations of polars with the poles as the contact points.

A geometric construction is possible for finding the polar q of a point Q and the pole Q of a line q in the case of a real nondegenerate conic.

Case 1 (Fig. 3–27). The point Q is outside the conic. Then there are two tangents to the conic passing through Q. The join q of the two contact points R and S is the polar of Q. Conversely, if q is given, intersecting the conic in two points R and S, construct Q as the intersection of the tangents at R and S.

Proof. R and S are self-conjugate points. Q lies on the polars of R and S, hence its polar q is the join of R and S.

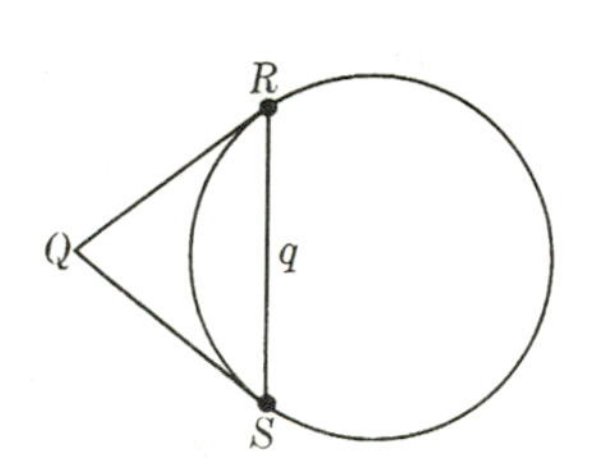

Figure 3–27

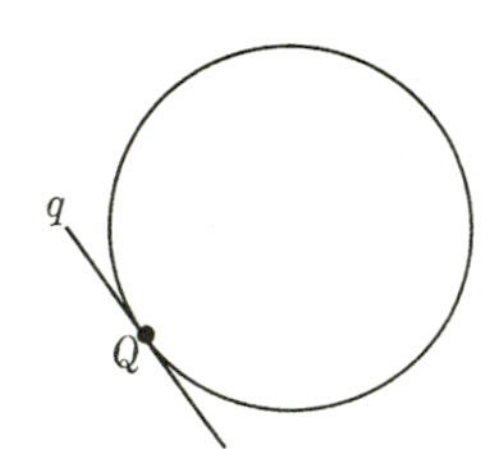

Figure 3–28

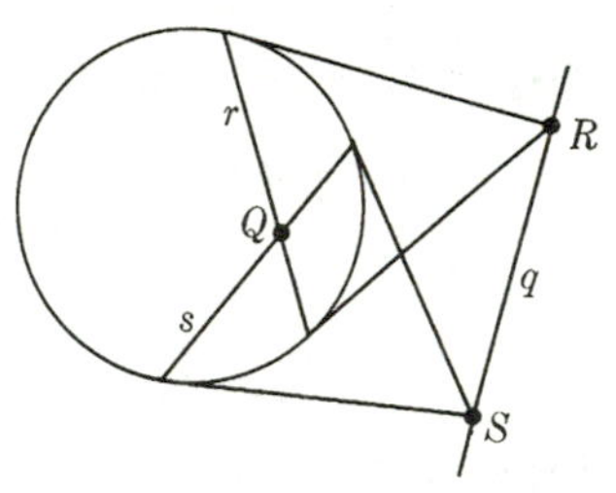

Figure 3–29

Case 2 (Fig. 3–28). Q is on the conic and q is the tangent at Q.

Proof. Theorem 3–6.2.

Case 3 (Fig. 3–29). Q is inside the conic. Draw two lines r and s through Q and find their poles R and S as described in Case 1. The join $RS = q$. Conversely, if a line q is given outside the conic, choose two points R and S on it and find their polars r and s, as in Case 1. The intersection $Q = r \times s$ is the pole of q.

Proof. R and S lying on q is equivalent to their polars r and s intersecting in the pole Q of q.

The constructions were, of course, independent of the choice of the coordinate system. However, the algebraic treatment depended heavily on the coordinate system and can be considered as having a geometric meaning only if it can be shown to yield the same results after an arbitrary coordinate change. This, indeed, is the case. Under a coordinate transformation with matrix A, the form F goes into a form with matrix ABA', as was developed in the proof of Theorem 3–5.3. Now all we have to prove is that this new matrix is again nonsingular and symmetric. We have $\det ABA' = (\det A)^2 \det B \neq 0$ because both A and B are nonsingular. Moreover, $(ABA')' = AB'A' = ABA'$ in view of B being symmetric.

Line conics. Our next concern will be with the dualization of the concepts developed so far. A nondegenerate conic was given in point coordinates by an equation $\mathbf{x}B\mathbf{x}' = 0$, B nonsingular and symmetric. For a fixed point $[\mathbf{y}]$ on the conic, $\mathbf{y}B\mathbf{x}' = 0$ is a tangent, and its equation can be written as $\mathbf{u}\mathbf{x}' = 0$, where $\langle\mathbf{u}\rangle$ is a tangent with $\langle\mathbf{u}\rangle = \langle\mathbf{y}B\rangle$. Now if $[\mathbf{y}]$ runs through all the points of the conic, we will always have $\mathbf{y}B\mathbf{y}' = 0$ or

$$\mathbf{y}BB^{-1}B\mathbf{y}' = (\mathbf{y}B)B^{-1}(\mathbf{y}B)' = \mathbf{u}B^{-1}\mathbf{u}' = 0.$$

The equation $\mathbf{u}B^{-1}\mathbf{u}' = 0$ represents the tangents $\langle\mathbf{u}\rangle$ of the original conic. Just as the conic was composed of points $[\mathbf{x}]$, our new *line conic* consists of tangent lines which envelop the point conic. Its matrix B^{-1} is nonsingular, and it is also symmetric because $(B^{-1})' = (B')^{-1} = B^{-1}$ (cf. Exercise 5 of Section 2–2). Thus we have

Theorem 3–6.3. If $\mathbf{x}B\mathbf{x}' = 0$ is a nondegenerate conic, then, in line coordinates, $\mathbf{u}B^{-1}\mathbf{u}' = 0$ is the corresponding line conic consisting of the tangents to the point conic.

From these considerations it should be clear that polarities can be explained in terms of line coordinates exactly as was done with point coordinates.

Thus polarities and nondegenerate conics are associated with each other to such an extent that a treatment or a classification of conics is frequently performed by discussing polarities instead. This is not surprising, since both are represented by the same symmetric matrices. The following theorem illustrates this attitude.

Theorem 3–6.4. A projectivity π preserves a nondegenerate conic C if and only if π commutes with the polarity associated with the conic.

Proof. Suppose that π preserves C. If $[\mathbf{x}]$ is on C, then also $[\mathbf{x}]\pi$ is on C. Let β be the polarity belonging to C. Then $[\mathbf{x}]\pi\beta$ is the tangent to C at $[\mathbf{x}]\pi$, by Theorem 3–6.2. In view of Theorem 3–2.3, the projectivity π^{-1} takes this tangent into the line

$$[\mathbf{x}]\pi\beta(\pi^{-1})'^{-1} = [\mathbf{x}]\pi\beta\pi',$$

which has a single common point with $C\pi^{-1} = C$ and is therefore the tangent to C at $[\mathbf{x}]\pi\pi^{-1} = [\mathbf{x}]$. However, the tangent to C at $[\mathbf{x}]$ is given by $[\mathbf{x}]\beta$. Hence $[\mathbf{x}]\pi\beta\pi' = [\mathbf{x}]\beta$ for all points $[\mathbf{x}]$ on C, that is, for at least four points, no three of which are collinear. By Theorem 3–2.5,

$$\pi\beta\pi' = \beta \qquad \text{and} \qquad \pi\beta = \beta\pi'^{-1}.$$

Conversely, assume $C\pi = C' \neq C$. Then $[\mathbf{x}]\pi$ is on C' and $[\mathbf{x}]\pi\beta$ is not tangent to C' for all possible choices of $[\mathbf{x}]$. For, if all the $[\mathbf{x}]\pi\beta$ were tangents to C', β would be the polarity of C', which would imply $C = C'$. Furthermore, $[\mathbf{x}]\pi\beta\pi'$ is not tangent to C for all $[\mathbf{x}]$ on C. Hence $\pi\beta\pi' \neq \beta$ and $\pi\beta \neq \beta\pi'^{-1}$.

The correlations do not form a group because the identity is not a correlation. The product of two correlations obviously is a projectivity which takes points into points and lines into lines. The following is obvious.

Theorem 3–6.5. The projectivities and the correlations of P^2 form a group of which $\mathcal{P}^2$ is a subgroup of index 2, with all the correlations forming a coset.

How can the polarity concept be extended to P^3? The conic will have to be replaced by a quadric and the point-line duality by the point-plane duality of P^3. The same techniques apply. A quadric has the equation $F(\mathbf{x}, \mathbf{x}) = \mathbf{x}B\mathbf{x}'$, with B a symmetric 4×4 matrix and $\mathbf{x}$ a vector from a V^4. The polar $F(\mathbf{y}, \mathbf{x}) = 0$ of a point $[\mathbf{y}]$ is now a plane, the *polar plane*. The main difference between P^2 and P^3 concerning polarity lies in the fact that not every involutory correlation in P^3 is a polarity because $B = r^2B$ (cf. the proof of Theorem 3–6.1) does not necessarily imply $r = 1$. A discussion of the other case will be left for the exercises.

EXERCISES

1. Carry out the proof of Theorem 3–6.5 in every detail.
2. Discuss the case of involutory correlations of P^3 which are not polarities, the so-called *null systems*. If the matrix of a null system is $B = (b_{jk})$, prove:
   (a) $B = -B'$ (b) $b_{jj} = 0$ for $j = 0, 1, 2, 3$
   (c) Every point is self-conjugate. (d) Every plane is self-conjugate.
3. Prove that a correlation that maps each of the vertices of some triangle on its opposite side is a polarity.
4. Prove: A line which is incident with two or more self-conjugate points cannot be self-conjugate.
5. Find the polarities which map $[1, 0, 0]$ on $\langle 1, 0, 0\rangle$, $[0, 1, 0]$ on $\langle 0, 1, 0\rangle$, and $[0, 0, 1]$ on $\langle 0, 0, 1\rangle$.
6. If the conic $4x_0^2 - x_1^2 + x_2^2 - 2x_0x_2 = 0$ is given, find with respect to it
   (a) the tangent at $[1, 2, 0]$
   (b) the join of the contact points of the tangents through $[-1, 1, 1]$
   (c) the pole of $\langle -1, 1, 1\rangle$.
7. Find the polar of the center of an ellipse in the affine plane.
8. In E^2, the point $[p/2, 0]$ is called the *focus* of the parabola $y^2 = 2px$.
   (a) Find the polar of the focus, the so-called *directrix*.
   (b) Show that every point of the parabola has equal distance from the focus and from the directrix.
9. Let C be a nondegenerate conic, P be a point not on C, and p be the polar of P with respect to C. If a line l through P intersects C in the points Q_1 and Q_2, and if $l \times p = R$, prove that $(Q_1Q_2|PR) = -1$.
10. (a) What happens to lines under three-dimensional polarities?
    (b) What is the image of the x_1-axis under the polarity which belongs to the quadric $x_0^2 - x_1^2 - x_2^2 - x_3^2 = 0$?
11. Find the tangent plane
    (a) to $4x_0^2 - x_1^2 - 4x_2^2 - x_3^2 = 0$ at $|1, 2, 0, 0]$
    (b) to $x_1x_2 - x_3x_0 = 0$ at $[1, 0, 0, 0]$
    (c) to $x_1x_3 - 2x_2x_0 + x_2^2 = 0$ at $[1, 1, -1, -3]$.
12. Prove: The plane $3x - 2y - z + 6 = 0$ in A^3 is tangent to the quadric $z = xy$. Find the contact point.
13. (a) Use polarity for finding in E^2 the contact points of those tangents to the circle $x^2 + y^2 = 1$ which pass through $[1, -1]$.
    (b) Use the same procedure for finding in $P^2(C)$ the contact points of the tangents to the conic $c^2x_0^2 - x_1^2 - x_2^2 = 0$ which pass through $[1, 0, 0]$. Explain geometrically why we find that this is impossible in the real projective plane. (Assume $c \neq 0$.)

14. Verify Theorem 3-6.4 for the projectivities that preserve the conic $x_2^2 = x_0x_1$.
15. Show that the projectivity with matrix

$$\begin{bmatrix} 3 & -2 & 1 \\ 1 & -2 & -1 \\ 2 & -2 & 2 \end{bmatrix}$$

takes the conic $x_0^2 - x_1^2 - x_2^2 = 0$ into itself
(a) directly (b) by using Theorem 3-6.4.

16. Use Theorem 3-6.4 to prove the second statement of Theorem 3-5.4.

3-7 THE CAYLEY-KLEIN PROJECTIVE MODELS OF NONEUCLIDEAN GEOMETRIES

Points and lines in the hyperbolic plane. In Section 1-9 we discussed Poincaré's model of the hyperbolic plane. The British mathematician Arthur Cayley (1821-1895) laid the foundations for another model of the hyperbolic plane which was later developed by Felix Klein. In contrast to Poincaré's model, which lies in the euclidean plane, the *Cayley-Klein model* lies in the projective plane. As we will see, this model serves the hyperbolic plane and another plane geometry, the *elliptic plane*. These two geometries are called the plane *noneuclidean* geometries.

It should be clear that in these models we are dealing with projective geometry. Just as we obtained euclidean geometry by imposing a metric on affine geometry, we will modify the projective plane, and later also projective three-space, by the two kinds of noneuclidean metrics.

For the purpose of introducing the hyperbolic plane, let a nondegenerate real conic in P^2 be given, and be called the *absolute conic* C. The polarity belonging to the absolute conic is the *absolute polarity*. For the sake of simplicity we will assume C to be the circle C_h: $x_0^2 - x_1^2 - x_2^2 = 0$. The points of the model are all the real points inside C_h (Fig. 3-30). The "hyperbolic lines" are those segments of projective lines which lie inside C_h. Hyperbolic lines which intersect each other on C_h, like k and l, are called parallel. Thus again through a given point Q there exist exactly two parallels j and k to a given hyperbolic line l which does not pass through Q. Pairs of hyperbolic lines are, therefore, classified as concurrent (having a common hyperbolic point), parallel, or *ultra-*

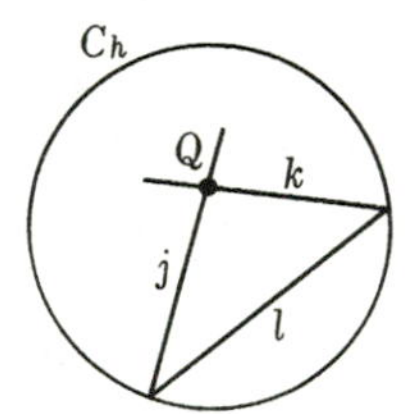

Figure 3-30

parallel, which means that the projective lines carrying them intersect outside C_h. Consequently no duality exists between points and lines of the model because through every two points there is a line, but not every two lines have a point in common.

The projectivities which carry C_h into itself and preserve orientation form a group $\mathcal{C}$ which, by Theorem 3–5.6 and its corollaries, is isomorphic to $\mathcal{H}_+$. By Theorem 3–5.8, the projectivities of $\mathcal{C}$ map the interior of C_h onto itself. Thus all the points of the model are mapped on all the points of the model and, since the projectivities are collineations, projective lines on projective lines. Consequently, the segments of projective lines interior to C_h go into segments of the same kind, or in short, hyperbolic lines into hyperbolic lines. Incidence, and hence also parallelism, is preserved. All this explains the name "model of the hyperbolic plane."

In order to justify fully the assertion that the Cayley-Klein model describes the same geometry as Poincaré's model, we would need an explicit exhibition of an incidence-preserving bijective mapping between the points and the lines of one model and those of the other model. This has actually been outlined in the proof of Theorem 3–5.5, where the upper half-plane of the complex number plane was mapped onto the interior of the conic $x_2^2 - x_0x_1 = 0$. It would be easy, though cumbersome, to establish the full details of the incidence-preserving bijectivity. However, we will postpone this matter until Section 4–7, where a different proof will be given.

The distance between two points P and Q is defined as follows (Fig. 3–31). Let the projective line PQ have the two points M and N on C_h. Then the hyperbolic distance is

$$d(P, Q) = |c \log_e (PQ|MN)|. \quad (1)$$

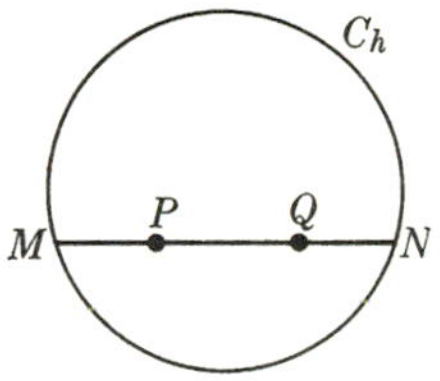

Figure 3–31

The hyperbolic distance, as defined in Section 1–9, did not use a specified basis of the logarithm. Thus a proportionality factor was involved which we call c. For the hyperbolic plane we fix $c = \frac{1}{2}$. By Exercise 2 of Section 3–3, the cross ratio $(PQ|MN)$ is always positive because P, Q and M, N do not separate each other. Hence $\log (PQ|MN)$ is always real and so is $d(P, Q)$.

For given points P, M, and N and a given value $b > 0$, there is a unique Q such that $c \log_e (PQ|MN) = b$, by the corollary to Theorem 3–3.2. Since there is also a unique Q' for $c \log_e (PQ'|MN) = -b$, there are two points, Q and Q', on the line MPN, whose distance from P is $|b|$, one on each of the half-lines PM and PN. When Q approaches M or N, $d(P, Q)$ tends toward ∞. Thus, in this model of the hyperbolic plane, C_h plays the role of a "line at infinity" just as the improper line does in the augmented affine plane.

All the mappings from $\mathcal{C}$ leave the cross ratio and hence $d(P, Q)$ invariant. The group $\mathcal{C}$ is therefore a group of isometries in our model.

A useful formula can be found for the distance $d([\mathbf{x}], [\mathbf{y}])$. We will develop it for a general absolute conic C, rather than for the special case C_h, in view of later applications. Let C have the equation $F(\mathbf{x}, \mathbf{x}) = 0$, with F a homogeneous quadratic form of rank 3. The join of $[\mathbf{x}]$ and $[\mathbf{y}]$ contains the points $[\mathbf{x} - r\mathbf{y}]$, for all r. The points of intersection of the join with C have

$$F(\mathbf{x} - r\mathbf{y}, \mathbf{x} - r\mathbf{y}) = 0,$$

or

$$r^2F(\mathbf{y}, \mathbf{y}) - 2rF(\mathbf{x}, \mathbf{y}) + F(\mathbf{x}, \mathbf{x}) = 0.$$

This yields

$$r_{1,2} = \frac{F(\mathbf{x}, \mathbf{y}) \pm \sqrt{F(\mathbf{x}, \mathbf{y})^2 - F(\mathbf{x}, \mathbf{x})F(\mathbf{y}, \mathbf{y})}}{F(\mathbf{y}, \mathbf{y})}.$$

The cross ratio

$$\begin{aligned}\rho = (\mathbf{x}, \mathbf{y} \mid \mathbf{x} - r_1\mathbf{y}, \mathbf{x} - r_2\mathbf{y}) &= \frac{r_1}{r_2} \\ &= \frac{F(\mathbf{x}, \mathbf{y}) + \sqrt{F(\mathbf{x}, \mathbf{y})^2 - F(\mathbf{x}, \mathbf{x})F(\mathbf{y}, \mathbf{y})}}{F(\mathbf{x}, \mathbf{y}) - \sqrt{F(\mathbf{x}, \mathbf{y})^2 - F(\mathbf{x}, \mathbf{x})F(\mathbf{y}, \mathbf{y})}},\end{aligned}$$

and $d([\mathbf{x}], [\mathbf{y}]) = |c \log_e \rho|$. We borrow from calculus the formula

$$\log_e a = 2i \arccos \frac{a + 1}{2\sqrt{a}}$$

and get $d([\mathbf{x}], [\mathbf{y}]) = \pm 2ic \arccos (\rho + 1)/2\sqrt{\rho}$, where $\pm$ means that the nonnegative value for this real number has to be chosen. After substituting the value of ρ and performing a few computations, we obtain

$$d([\mathbf{x}], [\mathbf{y}]) = \pm 2ic \arccos \frac{F(\mathbf{x}, \mathbf{y})}{\sqrt{F(\mathbf{x}, \mathbf{x})F(\mathbf{y}, \mathbf{y})}}. \tag{2}$$

For instance, in the case $C = C_h$, this yields

$$d([\mathbf{x}], [\mathbf{y}]) = \pm i \arccos \frac{x_0y_0 - x_1y_1 - x_2y_2}{\sqrt{(x_0^2 - x_1^2 - x_2^2)(y_0^2 - y_1^2 - y_2^2)}}. \tag{3}$$

In Eq. (3), "i arccos" may be written as "arccosh." Then it becomes clear that, in order to obtain a real value for the distance, the fraction after i arccos has to be ≥ 1. This can always be achieved, because this expression is a nonzero real number. If it is negative, it can be made positive by choosing the coordinate values $[-x_0, -x_1, -x_2]$ instead of $[x_0, x_1, x_2]$.

Does the hyperbolic distance satisfy the requirements Ds 1 through 3 of page 100? It obviously satisfies the first two, namely $d([\mathbf{x}], [\mathbf{x}]) = 0$ and $d([\mathbf{x}], [\mathbf{y}]) = d([\mathbf{y}], [\mathbf{x}]) > 0$ if $[\mathbf{x}] \neq [\mathbf{y}]$, but the verification of Ds 3, the triangle inequality, is more difficult.

Let P, Q, and R be three points of the hyperbolic plane. By virtue of Corollary 3 to Theorem 3–5.6, we may choose $P = [1, 0, 0]$, and by rotation about this point we may obtain Q on $\langle 0, 0, 1\rangle$. Since none of P, Q and R lie on C_h, we can normalize their coordinates so that each coordinate triple satisfies the relation $x_0^2 - x_1^2 - x_2^2 = 1$. Then we can write

$$Q = [f, \sqrt{f^2 - 1}, 0], \qquad R = [b, c, \sqrt{b^2 - c^2 - 1}],$$

with $f > 0$, $f^2 > 1$, $b > 0$, and $b^2 - 1 \geqq c^2$. From Eq. (3) we obtain

$$d(P, Q) = \pm i \arccos f = \operatorname{arccosh} f \text{ (the positive value)}$$

and similarly

$$d(P, R) = \operatorname{arccosh} b,$$

$$d(Q, R) = \operatorname{arccosh}\,(bf + c\sqrt{f^2 - 1}) \text{ (the positive value)}.$$

Then, by the addition formula,

$$\cosh\,(\alpha + \beta) = \cosh \alpha \cosh \beta + \sinh \alpha \sinh \beta,$$

we have

$$\cosh\,[d(P, Q) + d(P, R)] = bf + \sqrt{(b^2 - 1)(f^2 - 1)},$$

in view of $\cosh^2 \alpha - \sinh^2 \alpha = 1$. As a consequence of $b^2 - 1 \geqq c^2$,

$$bf + \sqrt{(b^2 - 1)(f^2 - 1)} \geqq bf + c\sqrt{f^2 - 1}$$

and hence

$$\cosh\,[d(P, Q) + d(P, R)] \geqq \cosh d(Q, R).$$

For positive arguments the function cosh is monotonically increasing, and thus

$$d(P, Q) + d(P, R) \geqq d(Q, R),$$

the triangle inequality for hyperbolic distances. The equality sign appears if and only if $b^2 - 1 = c^2$, that is, if R lies on PQ. This completes the proof that hyperbolic distance, as defined here, satisfies all the distance requirements.

Angles in the hyperbolic plane. We add another definition, that of perpendicular hyperbolic lines. Two lines p and q are called (hyperbolically) *perpendicular* if their projective carriers are conjugate to each other with respect to the polarity determined by C_h. Obviously, p is perpendicular to q if and only if q is perpendicular to p. Through every point P on a line l (Fig. 3–32) there is exactly one line perpendicular to l because, if L is the pole of l, then PL is this line. Since the mappings from $\mathfrak{C}$ preserve incidence, they preserve polarity and hence also perpendicularity. Two lines PL and QL, perpendicular to the same line l, need not be parallel, in contrast to the situation in euclidean geometry.

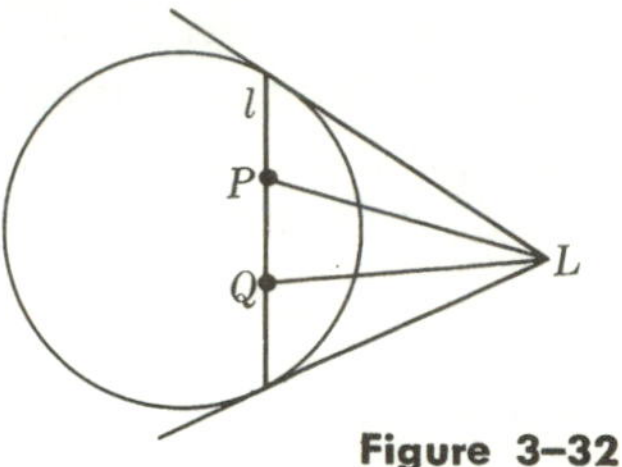

Figure 3-32

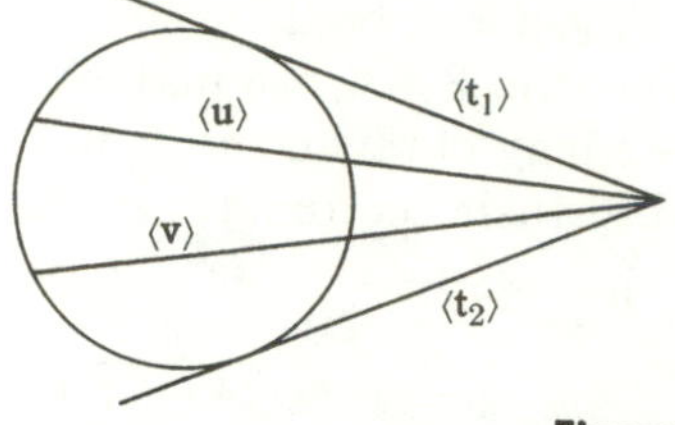

Figure 3-33

This treatment of perpendicularity raises the question of how to define angles in the hyperbolic plane. Angles in the euclidean plane were defined as rotations. Thus a study of hyperbolic rotations is indicated. As elaborated in Section 1-9, all the rotations about any proper hyperbolic point P, defined as the elements of $\mathcal{H}_+$ which have P as their only invariant point, form a group. These rotation groups for different points P are conjugate to each other and hence isomorphic. Let us follow an analogous procedure in $\mathcal{C}$ and discuss the group $\mathcal{R}$ of hyperbolic rotations about the center of C_h, the point $[1, 0, 0]$. All the elements of $\mathcal{R}$ are projectivities that leave $[1, 0, 0]$ invariant, and since they also preserve incidences, the polar of $[1, 0, 0]$ is an invariant line. This polar is the improper line $x_0 = 0$. Hence all the elements of $\mathcal{R}$ are affinities. These affinities are linear, since they preserve $[1, 0, 0]$. They preserve C_h and are, therefore, rotations or line reflections. However, we permitted only transformations which preserve orientation, and consequently they have to be rotations. We have proved the following.

Theorem 3-7.1. The group of hyperbolic rotations about a hyperbolic point is isomorphic to the group of euclidean rotations about a given point.

This theorem shows that the hyperbolic angle measure can be introduced in the same way as the angle measure in the euclidean plane, and that the hyperbolic angle is invariant under direct hyperbolic isometries. We shall attempt to define angle measure dually, corresponding to the distance between points as defined in Eq. (1). In order to define the distance between two points, we used their join and its points of intersection with C_h. The dual procedure would be (Fig. 3-33) to draw the two tangents $\langle \mathbf{t}_1 \rangle$ and $\langle \mathbf{t}_2 \rangle$ to C_h through the vertex of the angle $\sphericalangle \langle \mathbf{u} \rangle \langle \mathbf{v} \rangle$ and then to consider the value of $|\log (\mathbf{uv}|\mathbf{t}_1\mathbf{t}_2)|$. This procedure, however, seems not to be applicable in our model. Namely, $\langle \mathbf{u} \rangle$ and $\langle \mathbf{v} \rangle$ have to intersect inside C_h, and then no tangent to C_h passes through the point $\langle \mathbf{u} \rangle \times \langle \mathbf{v} \rangle$. In order to overcome this difficulty we could use complex coordinates.

However, an easier procedure is obtained by dualizing Eq. (2). The point conic C now is replaced by the dual line conic δC, whose equation is $E(\mathbf{u}, \mathbf{u}) = 0$, and Eq. (2) becomes

$$\sphericalangle \langle \mathbf{u} \rangle \langle \mathbf{v} \rangle = \pm 2ic \arccos \frac{E(\mathbf{u}, \mathbf{v})}{\sqrt{E(\mathbf{u}, \mathbf{u}) E(\mathbf{v}, \mathbf{v})}} \cdot \tag{4}$$

By Theorem 3-6.3, the matrix of E is the inverse of that of F.

We compute an angle whose vertex is [1, 0, 0] in the hyperbolic model with $\delta C = \delta C_h$. By Corollary 3 to Theorem 3–5.6, we may make $\langle \mathbf{u} \rangle = \langle 0, 0, 1 \rangle$ one of the sides of the angle and $\langle \mathbf{v} \rangle = \langle 0, m, -1 \rangle$, which in an orthonormal system would mean that $\langle \mathbf{u} \rangle$ is the x-axis and $\langle \mathbf{v} \rangle$ the line through the origin with slope m. Indeed, assume $m = \tan \alpha$. Now substitute these values in Eq. (4), observe that the matrix for C_h is equal to its inverse, and get

$$\sphericalangle \langle \mathbf{u} \rangle \langle \mathbf{v} \rangle = \pm 2ic \arccos \frac{1}{\sqrt{m^2 + 1}}$$

$$= \pm 2ic \arccos \frac{1}{\sqrt{\tan^2 \alpha + 1}} = \pm 2ic\alpha.$$

Consequently we obtain the euclidean angle between $\langle \mathbf{u} \rangle$ and $\langle \mathbf{v} \rangle$ if we make $c = 1/2i$. We will keep this value fixed for hyperbolic angle measure, so that Eq. (4) becomes

$$\sphericalangle \langle \mathbf{u} \rangle \langle \mathbf{v} \rangle = \arccos \frac{E(\mathbf{u}, \mathbf{v})}{\sqrt{E(\mathbf{u}, \mathbf{u}) E(\mathbf{v}, \mathbf{v})}} \cdot \tag{5}$$

In the case of the arccos function the absolute value symbol is not needed.

In the exercises, in particular, it will be shown that the hyperbolic perpendicularity, as defined above by means of polarity, corresponds to euclidean perpendicularity.

To summarize, we have the following theorem.

Theorem 3–7.2. In the Cayley-Klein model of the hyperbolic plane, let an angle be measured by

$$\sphericalangle \langle \mathbf{u} \rangle \langle \mathbf{v} \rangle = \left| \frac{1}{2i} \log_e (\mathbf{u}\mathbf{v}|\mathbf{t}_1\mathbf{t}_2) \right|,$$

where $\langle \mathbf{t}_1 \rangle$ and $\langle \mathbf{t}_2 \rangle$ are the two tangents to C_h through $\langle \mathbf{u} \rangle \times \langle \mathbf{v} \rangle$. Then every angle whose vertex is [1, 0, 0] has equal euclidean and hyperbolic values.

The angle measure defined in Eq. (5) is actually the absolute value of an angle measure. In other words, the rotation that takes $\langle \mathbf{u} \rangle$ into $\langle \mathbf{v} \rangle$ is identified with its inverse. It is standard procedure sometimes to do so with angles (in analogy with distances), but it cannot be done when dealing with rotations.

Theorem 3–7.3. In the hyperbolic plane the angle sum of every triangle is less than π.

First we prove a lemma.

Lemma 3–7.4. If a hyperbolic line l intersects a diameter d of C_h in a point other than the center of C_h, then the acute euclidean angle $\sphericalangle ld$ is greater than the acute hyperbolic angle. If the euclidean angle is right, l and d are hyperbolically perpendicular.

Proof. For disposing of the second case, a glance at Fig. 3–34 is sufficient. In the first case we use Corollary 3 to Theorem 3–5.6 and perform a hyperbolic isometry such that d becomes the line $x_2 = 0$ (Fig. 3–35). Let $d \times l = [1, f, 0]$ with $0 < |f| < 1$, and let the equation of l in an orthonormal system be $x_2 = (x_1 - fx_0)m$, with $m = \tan \alpha$ and, of course, $\alpha \neq \pi/2$. We have for d, $\langle \mathbf{d} \rangle = \langle 0, 0, 1 \rangle$ and for l, $\langle \mathbf{l} \rangle = \langle -fm, m, -1 \rangle$. By Eq. (5), the hyperbolic angle

$$\beta = \arccos (-f^2m^2 + m^2 + 1)^{-1/2}$$

$$= \arccos \frac{1}{\sqrt{m^2(1 - f^2) + 1}},$$

which yields $\tan \beta = m\sqrt{1 - f^2}$. Now either $\alpha = \sphericalangle l\, d$ or $\alpha = \pi - \sphericalangle l\, d$. In the first case $m = \tan \sphericalangle l\, d > 0$, and then $\tan \beta < \tan \sphericalangle l\, d$ and $\beta < \sphericalangle l\, d$. In the second case $m < 0$,

$$0 > \tan \beta = m\sqrt{1 - f^2} > m = \tan \alpha.$$

Then β is obtuse and $\beta > \alpha = \pi - \sphericalangle l\, d$. Consequently the acute hyperbolic angle $\pi - \beta < \sphericalangle l\, d$, as claimed.

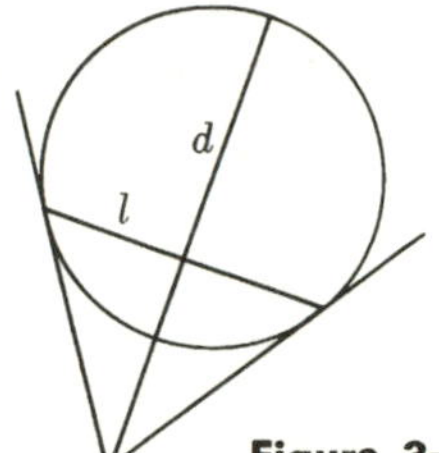

Figure 3–34

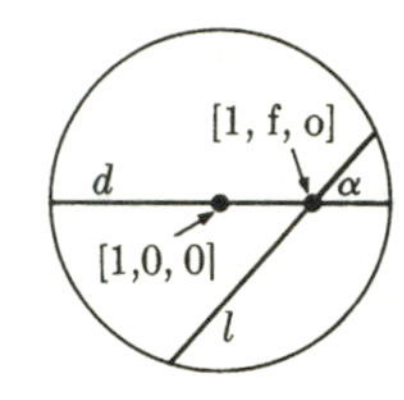

Figure 3–35

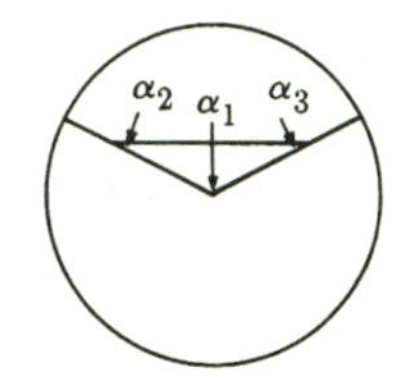

Figure 3–36

Proof of Theorem 3–7.3 (Fig. 3–36). We perform a hyperbolic isometry which takes one of the vertices of the triangle into $[1, 0, 0]$. This is possible in view of Corollary 3 to Theorem 3–5.6. If there is a right or obtuse angle in the triangle, we choose its vertex as the one that goes into $[1, 0, 0]$. If the angle at this vertex has the euclidean measure α_1 and the remaining euclidean angles are α_2 and α_3, then $\alpha_1 + \alpha_2 + \alpha_3 = \pi$. Let the corresponding hyperbolic values be β_1, β_2, and β_3. Then $\alpha_1 = \beta_1$ by Theorem 3–7.2. By virtue of the lemma, $\beta_2 < \alpha_2$ and $\beta_3 < \alpha_3$. Consequently, $\beta_1 + \beta_2 + \beta_3 < \pi$.

The elliptic plane. The absolute conic of the Cayley-Klein model was assumed to be real. We chose the circle C_h as representative and this did not restrict the generality because every real nondegenerate conic in P^2 can be obtained from C_h by a projectivity which also maps the interior of the conic onto the interior of C_h. Quite another case arises when the absolute conic is nondegenerate but imaginary. This class of conics may be represented by the imaginary conic C_e: $x_0^2 + x_1^2 + x_2^2 = 0$. The geometry obtained from the model with C_e as the absolute conic is called the *elliptic geometry*. It is the second case of a plane noneuclidean geometry represented by a Cayley-Klein model.

As with the hyperbolic geometry, the name "elliptic" is slightly confusing (no ellipse is involved here!) and is retained mainly for historical reasons.

There may be some doubt in the reader's mind concerning the use of a conic without a single real point as the absolute conic of a geometry with real points. However, we do not actually use the conic itself but rather the quadratic form and the polarity which belong to it. This becomes abundantly clear when we scrutinize the formulas of Eqs. (2) and (4) which replace the formula of Eq. (1). The absolute polarity, that is, the polarity associated with the absolute conic, is a perfectly good transformation of the real points and lines of P^2, even when the absolute conic has no real points. The polarity in this case is distinguished merely by the fact that it has no (real) self-conjugate points. Moreover, the points of the geometry are still all real points of P^2, and the absolute conic or the absolute polarity serve only to introduce a metric (that is, a distance function) and an angle measure for P^2.

The points of P^2 which have to be considered as belonging to the Cayley-Klein model of elliptic plane geometry are all the real points, because C_e does not have any real points to form a boundary between the interior and the exterior of C_e, as was the case for the hyperbolic plane. In fact, since no real tangents to C_e are possible, all points of the plane are interior. The isometry group of the elliptic plane is isomorphic to $\mathcal{O}^3_+$, by Theorem 3–5.4. This group is isomorphic to the group of all direct isometries preserving the unit sphere in E^3. Accordingly, by Theorem 2–8.3, we should expect the elliptic isometries to be related to rotations about lines. This will now be clarified by computing the elliptic distance between two points, utilizing Eq. (2).

By an elliptic isometry we move the first point to $Q_1 = [1, 0, 0]$ and the second point to some point $Q_2 = [1, b, 0]$ on the projective line $x_2 = 0$. Then Eq. (2) yields

$$d(Q_1, Q_2) = \pm 2ic \arccos \frac{1}{\sqrt{1 + b^2}} = \pm 2ic \arctan b.$$

If we again choose the proportionality factor as $c = 1/2i$, we obtain $d(Q_1, Q_2) = \pm \arctan b$. If $b = 0$, this distance is 0. When Q_2 moves away from Q_1 in either direction, $d(Q_1, Q_2)$ grows, until for $b \to \infty$ or $Q_2 \to [0, 1, 0]$ it tends toward its maximum value $\pi/2$. In the other direction it also reaches this maximum of $\pi/2$ when $b \to -\infty$ or $Q_2 \to [0, 1, 0]$.

Theorem 3–7.5. In the elliptic plane, straight lines are closed and their length is $2\pi ic = \pi$.

Angles can again be defined as in the hyperbolic case. It can be shown (see Exercise 9) that an elliptic angle with vertex at $[1, 0, 0]$ coincides with the euclidean and hyperbolic angle value. The constant c again is fixed as $1/2i$.

As a consequence of Theorem 3-7.5, there exists an ambiguity in the determination of the distance between two points. In fact, two points determine two segments between them, and if one of them has length d, then the other has length $\pi - d$. The same applies, of course, to angles between two lines. The ambiguity of choice between the acute and the obtuse angle in such cases in analytic geometry is a familiar phenomenon. Analytically, this means that in the elliptic case the two interpretations $[\mathbf{x}] = [x_0, x_1, x_2]$ and

$$[\mathbf{x}] = [-x_0, -x_1, -x_2]$$

indeed do yield two different valid results for the elliptic distance when Eq. (2) is used. The same is true for angles.

The finite length of elliptic lines shows that there is a perfect duality between elliptic distances and elliptic angle measures.

A point has to move along a line through a distance of π until it returns to its initial position.	A line has to move about a point through an angle of π until it returns to its initial position.

Since the direct elliptic isometries form a group isomorphic to a group of motions of the unit sphere (or any sphere), the bundle model is perhaps the best intuitive visualization of the elliptic plane. In it the straight lines of E^3 through a fixed point P are the elliptic points. The planes through P are the elliptic lines. Incidence of a line through P and a plane through P means incidence of an elliptic point and an elliptic line. This model could be simplified by considering intersections of the euclidean lines and planes with a fixed sphere about P. Then pairs of antipodal points correspond to elliptic points and great circles to elliptic lines. Thus a strong relation is suggested between the elliptic plane and geometry on the sphere.

Every direct euclidean isometry preserving the sphere is a rotation about a diameter of the sphere. Every opposite isometry preserving the sphere is the product of such a direct isometry and a point reflection in the sphere center. This point reflection produces in the elliptic plane nothing but the identity transformation. Thus direct and also opposite isometries preserving the sphere both induce the same transformations of the elliptic plane as the rotations about a diameter of the sphere. This diameter corresponds to a unique point of the elliptic plane, and hence we have the following.

Theorem 3-7.6. Every nonidentical isometry of the elliptic plane is a rotation about one of its points.

It can be shown that the sphere model is conformal, that is, euclidean angles on the sphere coincide in their value with that of elliptic angles. The angle sum of a spherical triangle always exceeds π. In the exercises it will be shown that this is true for triangles in the elliptic plane.

The euclidean plane as limiting case. We have encountered a few facts suggesting that the euclidean metric somehow occupies an intermediate position between the two noneuclidean metrics. The angle sum of a triangle is $<\pi$ in the hyperbolic plane, π in the euclidean plane, and $>\pi$ in the elliptic plane. Through a given point there exist two parallels to every line in the hyperbolic plane, one parallel in the euclidean plane, and no parallels in the elliptic plane (because every two lines meet). The hyperbolic Cayley-Klein model contains the interior of a conic, the euclidean plane contains all of P^2 except one line, the elliptic plane covers the whole of P^2. How does euclidean distance compare to hyperbolic and elliptic distances, and is there an absolute conic that determines euclidean metric in P^2 in the same way as the absolute conics of the hyperbolic and elliptic planes?

Theorem 3–7.7. Let the degenerate conic C_E: $x_0^2 = 0$ be chosen as the absolute conic. Then the distance formula of Eq. (2) yields euclidean distance if the proportionality factor c is suitably determined.

In the proof it will be shown that the degenerate conic C_E is in a sense a type intermediate to C_h and C_e.

Proof. Consider the conic $F(\mathbf{x}, \mathbf{x}) = x_0^2 + \epsilon x_1^2 + \epsilon x_2^2 = 0$. For $\epsilon > 0$ it is nondegenerate and imaginary, like C_e. For $\epsilon = 0$ it is C_E. For $\epsilon < 0$ it is nondegenerate and real, like C_h. Let $[\mathbf{p}] = [p_0, p_1, p_2]$ and $[\mathbf{q}] = [q_0, q_1, q_2]$ be two points. By Eq. (2) we have

$$d = d([\mathbf{p}], [\mathbf{q}]) = \pm 2ic \arccos \frac{p_0q_0 + \epsilon p_1q_1 + \epsilon p_2q_2}{\sqrt{(p_0^2 + \epsilon p_1^2 + \epsilon p_2^2)(q_0^2 + \epsilon q_1^2 + \epsilon q_2^2)}}.$$

When we substitute $\arccos \alpha = \arcsin \sqrt{1 - \alpha^2}$, after a short computation this becomes

$$d = \pm 2ic \arcsin \sqrt{\epsilon} \sqrt{\frac{\epsilon(p_1q_2 - p_2q_1)^2 + (p_2q_0 - p_0q_2)^2 + (p_0q_1 - p_1q_0)^2}{(p_0^2 + \epsilon p_1^2 + \epsilon p_2^2)(q_0^2 + \epsilon q_1^2 + \epsilon q_2^2)}}.$$

Now let ϵ tend toward zero and determine c so that $2ic\sqrt{\epsilon} = 1$. Since

$$\lim_{\alpha \to 0} \frac{\arcsin \alpha}{\alpha} = 1,$$

we may remove the arcsin and obtain

$$\lim_{\epsilon \to 0} d = \sqrt{\frac{(p_0q_1 - p_1q_0)^2 + (p_0q_2 - p_2q_0)^2}{p_0^2q_0^2}}$$

$$= \sqrt{\left(\frac{q_1}{q_0} - \frac{p_1}{p_0}\right)^2 + \left(\frac{q_2}{q_0} - \frac{p_2}{p_0}\right)^2},$$

which is the euclidean distance between $[\mathbf{p}]$ and $[\mathbf{q}]$.

Is there a conic that can be considered as the absolute conic of the euclidean plane? Since in Theorem 3-7.7 the conic C_E appeared, we will check to see whether it indeed plays this role. In fact, in the neighborhood of the absolute conic, distances always tend toward infinity, and in the euclidean plane this happens when one approaches the improper line $x_0 = 0$ which coincides with C_E.

However, the assumption of C_E being the absolute conic of E^2 leads to complications. First, there is no polarity associated with C_E. This follows from the fact that C_E is degenerate, and it appears intuitively clear when we try to construct polars. Furthermore, the group of projectivities preserving $x_0 = 0$ is $\mathcal{A}^2$, which certainly is not the group of E^2.

Let us therefore return to the conic $x_0^2 + \epsilon(x_1^2 + x_2^2) = 0$, determine its equation in line coordinates, and then see what happens when ϵ tends toward zero.

By Theorem 3-6.3, the line conic is $\epsilon u_0^2 + u_1^2 + u_2^2 = 0$. When $\epsilon = 0$, we have $u_1^2 + u_2^2 = 0$, or $u_2 = \pm i u_1$. These lines have coordinates $\langle u_0, 1, \pm i\rangle$ and their equations are $u_0 x_0 + x_1 \pm i x_2 = 0$. This set of lines must be invariant under the transformations for which C_E is to be the absolute conic. Their points of intersection with $x_0^2 = 0$ will also have to be preserved or interchanged. They are $[0, 1, i]$ and $[0, 1, -i]$. Let us find out which projectivities preserve or interchange $[0, 1, i]$ and $[0, 1, -i]$.

First we assume that each of them is preserved. If $[\mathbf{x}] \to [\mathbf{x}A] = [\mathbf{x}']$, with $A = (a_{jk})$, then substitution of the invariant points yields $a_{10} = a_{20} = 0$, $a_{12} = -a_{21}$ and $a_{11} = a_{22}$, such that

$$
\begin{aligned}
rx_0' &= a_{00}x_0,\\
rx_1' &= a_{01}x_0 + a_{11}x_1 - a_{12}x_2,\\
rx_2' &= a_{02}x_0 + a_{12}x_1 + a_{11}x_2,
\end{aligned}
$$

with $r \neq 0$. Moreover, $a_{00} \neq 0$ in view of det $A \neq 0$. In nonhomogeneous coordinates this becomes

$$
x \to ax + by + e, \qquad y \to -bx + ay + f,
$$

a direct similitude, and every direct similitude can be thus obtained. In the second case, where $[0, 1, \pm i]$ are interchanged, the same method yields all the opposite plane similitudes.

Thus instead of obtaining the group $\mathfrak{M}^2$ of isometries, we found $\mathcal{S}^2$ to be the group of projectivities preserving C_E and the corresponding conic in line coordinates. This illustrates the difficulties arising from the degenerate character of the euclidean metric. Similitudes preserve angles and distance ratios, but not the distances themselves. We lost the invariance of distances when, in the proof of Theorem 3-7.7, we had to make c infinite. This procedure introduces an indeterminate factor. On the other hand, it can be proved (cf. Exercise 18) that the angle can be properly determined with respect to the absolute conic.

We summarize in the following theorem.

Theorem 3-7.8. The degenerate conic C_E: $x_0^2 = 0$, together with the corresponding imaginary line conic $u_1^2 + u_2^2 = 0$, serves as absolute conic of the plane similarity geometry whose group is $\mathcal{S}^2$. Equivalently, this is the geometry whose transformations preserve or interchange the two imaginary circle points, $[0, 1, i]$ and $[0, 1, -i]$.

Exercises

1. Prove that two perpendiculars to the same hyperbolic line are ultraparallel.
2. If a hyperbolic line is intersected by two other lines which are parallel to each other, are the corresponding angles equal?
3. Show that perpendicular hyperbolic lines indeed enclose a hyperbolic angle of $\pi/2$.
4. Show that no two parallel or concurrent hyperbolic lines can have a common perpendicular, while any two ultraparallel lines have a common perpendicular.
5. The distance between ultraparallel hyperbolic lines is measured along their common perpendicular. Compute the hyperbolic distance in the Cayley-Klein model between the lines $\langle 0, 0, 1\rangle$ and $\langle 1, 0, 2\rangle$.
6. Assuming that the absolute conic is C_h, compute the sum of hyperbolic angles in the triangle whose vertices are $[1, 0, 0]$, $[2, -1, 0]$, and $[2, 0, 1]$.
7. (a) Find the midpoints of the elliptic segment with endpoints $[1, 0, 0]$ and $[1, 1, 0]$.
   (b) Why are there two midpoints? Why does this fact conform with the duality between segments and angles in the elliptic plane?
8. Since the elliptic line $\langle 0, 1, 0\rangle$ is of finite length, there must be a midpoint between $[1, 0, 0]$ and itself. Compute this point on $\langle 0, 1, 0\rangle$.
9. Show that an elliptic angle whose vertex is $[1, 0, 0]$ has the same angular measure as the euclidean angle.
10. Prove: In the elliptic plane the angle sum of every triangle exceeds π.
11. (a) Use the hyperbolic definition of perpendicularity to introduce perpendicularity in the elliptic plane.
    (b) Find the elliptic perpendicular to $\langle 0, u_1, u_2\rangle$ at any of its points, to $\langle 1, u_1, u_2\rangle$ at any of its points, and compare the results when drawn in an orthonormal coordinate system.
12. Dualize the perpendicularity of elliptic lines by formally carrying out the dual to the construction of perpendicular lines with respect to the absolute conic C_e. What is the elliptic distance between the resulting "perpendicular points?" Why are there no "perpendicular points" in the hyperbolic plane?
13. Prove the triangle inequality for elliptic distances.

14. Use the sphere model to determine what is a line reflection in the elliptic plane.
15. Show that the following mapping in the bundle model represents a polarity of the elliptic plane: Every line of the bundle is mapped on the plane perpendicular to it through the bundle center, and vice versa.
16. (a) Assuming that the angles in the bundle model are equal to corresponding elliptic angles and distances, show that in the polarity of Exercise 15 a distance d is mapped on an angle $\pi - d$.
    (b) Under this polarity, what is the dual statement to Exercise 10?
    (c) What is the dual statement to the triangle inequality?
17. Between the points $[1, 0, 0]$ and $[2, 1, 1]$ in the Cayley-Klein models, find
    (a) the hyperbolic distance
    (b) the elliptic distance.
    (c) Compare with the euclidean distance.
18. (a) Show that the euclidean angle measure coincides with that obtained by considering the euclidean metric as a limiting case of the noneuclidean metrics.
    (b) In particular, show that the euclidean angle measure is proportional to $|\log(\mathbf{uv}|\mathbf{jk})|$, where $\langle\mathbf{u}\rangle$ and $\langle\mathbf{v}\rangle$ are the sides of the angle and $\langle\mathbf{j}\rangle$ and $\langle\mathbf{k}\rangle$ the joins of the angle vertex with the circle points $[0, 1, \pm i]$.

3-8 QUATERNIONS AND ELLIPTIC THREE-SPACE

The quaternion algebra. Noneuclidean spaces of higher dimensions are known and have been thoroughly investigated. However, we will restrict ourselves to elliptic three-space. The treatment could be carried out by using the same methods as in the plane. However, since the computations become much simpler if we use quaternions, we will introduce them before approaching the elliptic space.

Let an n-dimensional real vector space W be closed under a multiplication of the vectors (not to be confused with the inner product!) which is such that W satisfies the following postulates for all vectors $\mathbf{a}$, $\mathbf{b}$, and $\mathbf{c}$, in W, and all reals α.

Al 1. W is an n-dimensional real vector space.
Al 2. $\mathbf{a}(\alpha\mathbf{b}) = (\alpha\mathbf{a})\mathbf{b} = \alpha(\mathbf{ab})$.
Al 3. $\mathbf{a}(\mathbf{b} + \mathbf{c}) = \mathbf{ab} + \mathbf{ac}$, $(\mathbf{b} + \mathbf{c})\mathbf{a} = \mathbf{ba} + \mathbf{ca}$.

W is called an *algebra of rank n over the reals.* Sometimes additional postulates hold.

Al 4. $(\mathbf{ab})\mathbf{c} = \mathbf{a}(\mathbf{bc})$.
Al 5. $\mathbf{ab} = \mathbf{ba}$.
Al 6. W contains a unit element $\mathbf{e}$, such that $\mathbf{ea} = \mathbf{ae} = \mathbf{a}$ for all $\mathbf{a}$ in W. The equations $\mathbf{ax} = \mathbf{b}$ and $\mathbf{ya} = \mathbf{b}$, with $\mathbf{a} \neq \mathbf{0}$, have unique solutions $\mathbf{x}$ and $\mathbf{y}$ in W.

When Al 1, 2, 3, and 4 hold, W is an *associative algebra.* When Al 1, 2, 3, and 5 hold, W is a *commutative algebra,* and when Al 1, 2, 3, and 6 hold, W is a *division algebra.* An associative division algebra is a skew field, and an associative and commutative division algebra is a field. It should be recalled that in every vector space (as defined on page 58) $\mathbf{a}\alpha = \alpha\mathbf{a}$.

As an example we may mention the complex numbers, which form an associative and commutative division algebra of rank 2 over the reals.

We introduce an algebra Q of rank 4 over the reals with a basis $\mathbf{1}$, $\mathbf{i}$, $\mathbf{j}$, $\mathbf{k}$ and the multiplication table

	1	i	j	k
1	1	i	j	k
i	i	−1	k	−j
j	j	−k	−1	i
k	k	j	−i	−1

For easier memorization, note that

$$\mathbf{ij} = \mathbf{k},\ \mathbf{jk} = \mathbf{i},\ \mathbf{ki} = \mathbf{j};$$

$$\mathbf{ji} = -\mathbf{k},\ \mathbf{kj} = -\mathbf{i},\ \mathbf{ik} = -\mathbf{j};$$

$$\mathbf{i}^2 = \mathbf{j}^2 = \mathbf{k}^2 = -\mathbf{1}.$$

The elements of Q are called *quaternions.* It follows that $\mathbf{1}$ is the unit element of Q. Q is noncommutative because $\mathbf{ij} = -\mathbf{ji} \neq \mathbf{ji}$. It can easily be checked using the multiplication table that the multiplication of the basis elements is associative, and consequently the multiplication in Q is associative. This can be verified by repeated use of Al 2 and 3.

Instead of writing correctly $q_0\mathbf{1}$, with real q_0, we will adopt the easier, less rigid, notation $q_0\mathbf{1} = q_0$. (Algebraically this procedure reflects the existence of an isomorphic mapping of the reals into Q, with the quaternions $q_0\mathbf{1} + 0\mathbf{i} + 0\mathbf{j} + 0\mathbf{k}$ as images.)

We designate q_0 as the *real part* and $q_1\mathbf{i} + q_2\mathbf{j} + q_3\mathbf{k}$ as the *imaginary part* of the quaternion $\mathbf{q} = q_0 + q_1\mathbf{i} + q_2\mathbf{j} + q_3\mathbf{k}$. Its *conjugate* is $\overline{\mathbf{q}} = q_0 - q_1\mathbf{i} - q_2\mathbf{j} - q_3\mathbf{k}$. A quaternion with zero real part is called *pure.* It is easy to verify that for quaternions $\mathbf{p}$ and $\mathbf{q}$,

$$(\overline{\mathbf{pq}}) = \overline{\mathbf{q}}\,\overline{\mathbf{p}} \quad \text{and} \quad \overline{\mathbf{p}} + \overline{\mathbf{q}} = \overline{\mathbf{p} + \mathbf{q}}.$$

The product $\mathbf{q}\overline{\mathbf{q}} = q_0^2 + q_1^2 + q_2^2 + q_3^2$ is called the *norm* of $\mathbf{q}$, denoted $N(\mathbf{q})$. Now $N(\mathbf{q}) = 0$ if and only if $\mathbf{q} = \mathbf{0}$. Hence for every $\mathbf{q} \neq \mathbf{0}$, $\mathbf{q}[\overline{\mathbf{q}}/N(\mathbf{q})] = \mathbf{1}$; that is, every nonzero quaternion has an inverse. We have therefore

Theorem 3–8.1. The quaternions form an associative, noncommutative division algebra of rank 4 over the reals.

Theorem 3–8.2. For all quaternions $\mathbf{p}$ and $\mathbf{q}$,

$$N(\mathbf{p})N(\mathbf{q}) = N(\mathbf{pq}).$$

Proof.

$$N(\mathbf{pq}) = (\mathbf{pq})(\overline{\mathbf{pq}}) = \mathbf{pq}\,\overline{\mathbf{q}}\,\overline{\mathbf{p}} = \mathbf{p}N(\mathbf{q})\overline{\mathbf{p}} = \mathbf{p}\overline{\mathbf{p}}N(\mathbf{q}) = N(\mathbf{p})N(\mathbf{q}).$$

Corollary. The quaternion algebra has no divisors of zero, that is, $\mathbf{pq} = \mathbf{0}$ implies $\mathbf{p} = \mathbf{0}$ or $\mathbf{q} = \mathbf{0}$.

Proof. $\mathbf{pq} = \mathbf{0}$ implies $N(\mathbf{pq}) = 0$. Hence $N(\mathbf{p})N(\mathbf{q}) = 0$. Consequently $N(\mathbf{p}) = 0$ or $N(\mathbf{q}) = 0$, and finally $\mathbf{p} = \mathbf{0}$ or $\mathbf{q} = \mathbf{0}$.

Quaternion representation of elliptic three-space. In P^3, the point $[\mathbf{x}] = [x_0, x_1, x_2, x_3]$ can be represented by the nonzero quaternion $\mathbf{x} = x_0 + x_1\mathbf{i} + x_2\mathbf{j} + x_3\mathbf{k}$ and all its nonzero scalar multiples. If we choose as the representative of a point $[\mathbf{x}]$ the one quaternion $\mathbf{x}$ whose norm is 1 and whose first nonzero component is positive, we may omit the brackets in the following. In other words, we shall use *normalized* quaternions for representing points.

We consider the mapping $R_\mathbf{a}: \mathbf{x} \to \mathbf{xa}$, with a fixed nonzero quaternion $\mathbf{a}$. A similar mapping is $L_\mathbf{a}: \mathbf{x} \to \mathbf{ax}$.

Theorem 3–8.3. The mappings $R_\mathbf{a}$ and $L_\mathbf{a}$ for all quaternions $\mathbf{a} \neq \mathbf{0}$ are projectivities which, respectively, form groups $\mathcal{E}_R$ and $\mathcal{E}_L$. These groups are transitive on the points of P^3; that is, for two given points $\mathbf{x}$ and $\mathbf{y}$ in P^3 there exists exactly one element $R_\mathbf{a} \in \mathcal{E}_R$ and one element $L_\mathbf{a} \in \mathcal{E}_L$ such that $\mathbf{x}R_\mathbf{a} = \mathbf{x}L_\mathbf{b} = \mathbf{y}$. All the quaternions are assumed to be normalized.

Proof. That the mappings are one-to-one follows readily from elementary properties of the quaternions. Hence they are transformations. To show that they are projectivities we verify their linearity by employing Al 3. As the determinant of the projectivities $R_\mathbf{a}$ and $L_\mathbf{a}$ we find $(N(\mathbf{a}))^2$ (cf. Exercise 14). The closure of the groups follows from $\mathbf{x}R_\mathbf{a}R_\mathbf{b} = \mathbf{xab} = \mathbf{x}R_{\mathbf{ab}}$ and $\mathbf{x}L_\mathbf{a}L_\mathbf{b} = \mathbf{bax} = \mathbf{x}L_{\mathbf{ba}}$. The identity is $R_1 = L_1$ and the inverses are $R_\mathbf{a}^{-1} = R_{\mathbf{a}^{-1}}$ and $L_\mathbf{a}^{-1} = L_{\mathbf{a}^{-1}}$. The transitivity is a consequence of $\mathbf{x}(\mathbf{x}^{-1}\mathbf{y}) = \mathbf{x}R_{\mathbf{x}^{-1}\mathbf{y}} = \mathbf{y}$ and the corresponding procedure for $\mathcal{E}_L$.

Theorem 3–8.4. The transformations from $\mathcal{E}_R$ and $\mathcal{E}_L$ leave invariant the imaginary quadric J,

$$x_0^2 + x_1^2 + x_2^2 + x_3^2 = 0.$$

Proof. It can be shown that Theorem 3–6.4 holds also in the three-dimensional case. This makes it possible to avoid dealing with imaginary points, which do not have an easy quaternion representation. Thus, if $\pi \in \mathcal{E}_R$ and if β is the polarity belonging to J, we have to prove $\pi\beta = \beta\pi'^{-1}$. But, since the matrix of β is I_4, this becomes $\pi\pi' = I$. Indeed, it is easy to show that $\pi = R_{\mathbf{a}}$ implies $\pi' = R_{\bar{\mathbf{a}}}$.

The imaginary quadric J is the three-dimensional analog to the imaginary absolute conic C_e of the elliptic plane. Thus we call the projectivities of P^3 which leave J invariant *three-dimensional elliptic transformations*, and the corresponding geometry *elliptic three-space*. In the following we will study the elliptic transformations from $\mathcal{E}_R$ and $\mathcal{E}_L$ and then investigate whether there exist elliptic transformations other than these. Except when we deal with the points on J we will restrict ourselves to real points, and assume that all the quaternions representing them are normalized.

If we define the elliptic distance of two points $\mathbf{x}$ and $\mathbf{y}$ as

$$d(\mathbf{x}, \mathbf{y}) = |(1/2i) \log_e (\mathbf{x}, \mathbf{y}|M, N)|,$$

where M and N are the points where the join of $\mathbf{x}$ and $\mathbf{y}$ intersects J, then the elliptic transformations preserve elliptic distance because, as projectivities, they preserve cross ratios and leave J invariant. In the plane $x_3 = 0$, this new elliptic distance coincides with the distance as defined for the elliptic plane because $x_3 = 0$ intersects J in C_e. Hence we have the following.

Theorem 3–8.5. The three-dimensional elliptic transformations are elliptic isometries.

Theorem 3–8.6. $\mathcal{E}_R$ and $\mathcal{E}_L$ have only the identity in common.

Proof. Suppose that T belongs to both, and $T = R_{\mathbf{a}} = L_{\mathbf{b}}$. Then for all $\mathbf{x}$, $\mathbf{xa} = c\mathbf{bx}$, with c a nonzero scalar. The case $\mathbf{x} = 1$ yields $\mathbf{a} = c\mathbf{b}$, or if $\mathbf{a} = a_0 + a_1\mathbf{i} + a_2\mathbf{j} + a_3\mathbf{k}$ and $\mathbf{b} = b_0 + b_1\mathbf{i} + b_2\mathbf{j} + b_3\mathbf{k}$, then $a_s = cb_s$, with $s = 0, 1, 2, 3$, for some real $c \neq 0$. If we successively substitute $\mathbf{i}$ and $\mathbf{j}$ for $\mathbf{x}$, we eventually obtain $a_r = cb_r = -cb_r$, with $r = 1, 2, 3$, and consequently $a_1 = a_2 = a_3 = b_1 = b_2 = b_3 = 0$. Hence $\mathbf{a} = c\mathbf{b}$, a scalar $\neq 0$. This completes the proof.

Now consider products of elements of $\mathcal{E}_R$ and $\mathcal{E}_L$. For instance, $R_{\mathbf{a}}L_{\mathbf{b}}$ acts on $\mathbf{q}$ as $\mathbf{q}R_{\mathbf{a}}L_{\mathbf{b}} = \mathbf{bqa} = \mathbf{q}L_{\mathbf{b}}R_{\mathbf{a}}$. Thus every element of $\mathcal{E}_R$ commutes with every element of $\mathcal{E}_L$, and every product of a finite number of factors from $\mathcal{E}_R$ and $\mathcal{E}_L$ can be written in the form $L_{\mathbf{b}}R_{\mathbf{a}} = R_{\mathbf{a}}L_{\mathbf{b}}$. All these products form a group which we will call $\mathcal{E}^3_+$.*

Theorem 3–8.7. $\mathcal{E}^3_+$ is the group of all direct three-dimensional elliptic transformations.

Proof. Let B be some direct elliptic transformation and let B: $1 \to \mathbf{q}$. Then $BR_{\mathbf{q}}^{-1}$ preserves 1. In order to find all direct elliptic transformations it suffices, therefore, to determine all those which leave 1 invariant and then to multiply

---

* In the language of group theory $\mathcal{E}^3_+$ is the direct product of $\mathcal{E}_R$ and $\mathcal{E}_L$.

them with elements of $\mathcal{E}_R$. In every projectivity preserving 1 and the absolute quadric J, the polar plane of 1 with respect to J will also be preserved. This polar plane is $x_0 = 0$. Now, if $x_0^2 + x_1^2 + x_2^2 + x_3^2 = 0$ and $x_0 = 0$ are preserved, then $x_1^2 + x_2^2 + x_3^2 = 0$ will also be left unchanged. Therefore, in the plane $x_0 = 0$, the conic C_e is invariant. But C_e is preserved by exactly those elliptic transformations which map the plane $x_0 = 0$ onto itself. By Theorem 3–5.4, the group of these transformations is isomorphic to $\mathcal{O}_+^3$. Now, by the corollary to Theorem 2–8.6, the group $\mathcal{O}_+^3$ can be generated by all the rotations about any two lines through the invariant point of $\mathcal{O}_+^3$. For the direct elliptic transformations in $x_0 = 0$, this means that their group is generated by the rotations about two distinct points. Consider those about the point [1, 0, 0] in the plane $x_0 = 0$; that is, the point $[0, 1, 0, 0] = \mathbf{i}$ in elliptic three-space. Since this point is invariant, we obtain transformations with a matrix

$$A = \begin{bmatrix} 1 & 0 & 0 & 0 \\ 0 & 1 & 0 & 0 \\ 0 & 0 & \cos\phi & -\sin\phi \\ 0 & 0 & \sin\phi & \cos\phi \end{bmatrix}.$$

We construct a transformation

$$T\colon \mathbf{x} \to (a + b\mathbf{i})^{-1}\mathbf{x}(a + b\mathbf{i}), \qquad a^2 + b^2 = 1.$$

Obviously, $T \in \mathcal{E}_+^3$ and $1T = 1$, $\mathbf{i}T = \mathbf{i}$. For $\mathbf{j}$ and $\mathbf{k}$ we obtain

$$\mathbf{j}T = (a^2 - b^2)\mathbf{j} - 2ab\mathbf{k},$$
$$\mathbf{k}T = 2ab\mathbf{j} + (a^2 - b^2)\mathbf{k}.$$

If we now substitute $a = \cos(\phi/2)$ and $b = \sin(\phi/2)$, we see that T has the matrix A. In view of the condition $a^2 + b^2 = 1$, every transformation of the type T can be written as a transformation with matrix A. This procedure can be repeated for the second invariant point, $\mathbf{j}$, just as was done for $\mathbf{i}$, and again all the direct transformations obtained will be elements of $\mathcal{E}_+^3$. This proves the theorem.

Corollary. The group $\mathcal{E}^3$ of all three-dimensional elliptic transformations consists of $\mathcal{E}_+^3$ and one coset $\mathcal{E}_+^3\, T$, where T is any opposite elliptic transformation, as for instance $T\colon \mathbf{x} \to \bar{\mathbf{x}}$.

Proof. T is an elliptic transformation because the absolute polarity commutes with it. Moreover, if $\mathbf{x} = x_0 + x_1\mathbf{i} + x_2\mathbf{j} + x_3\mathbf{k}$, then

$$\bar{\mathbf{x}} = x_0 - x_1\mathbf{i} - x_2\mathbf{j} - x_3\mathbf{k} \qquad \text{and} \qquad \det T = -1 < 0.$$

Now the projectivities of P^3 are divided into direct and opposite projectivities, according to whether their determinants are positive or negative. In short, there is an orientation in P^3 (and any other P^{2n+1}), in contrast to P^2 (or any other P^{2n}). No 4×4 matrix with positive determinant represents the same projectivity as such a matrix with negative determinant. Multiplication of all the entries by -1 changes the sign of a 3×3 determinant, but not of a 4×4 determinant. Composition with T takes every element of $\mathcal{E}^3_+$ into an element with negative determinant, and every such element can be written as an element of $\mathcal{E}^3_+$ multiplied with T.

Clifford parallels. There is an analogy between $\mathcal{E}_R$ and $\mathcal{E}_L$ and the group $\mathcal{T}$ in euclidean space. Both are transitive on the points of the space, and thus the operations $R_{\mathbf{a}}$ and $L_{\mathbf{a}}$ remind us of drawing a vector from some point. Considerable interest is attached to the one-parameter subgroups of $\mathcal{T}$ and their traces. One-parameter subgroups of $\mathcal{E}_R$ and $\mathcal{E}_L$ will be discussed in the following theorem.

Theorem 3–8.8. Let $\mathbf{p}$ be a fixed nonzero pure quaternion. The set of quaternions $\mathbf{x} = a + \alpha\mathbf{p} \neq 0$ for all real a and α form a multiplicative group. The $R_{\mathbf{x}}$ and the $L_{\mathbf{x}}$ with these $\mathbf{x}$ form one-parameter subgroups $\mathcal{C}_R(\mathbf{p})$ and $\mathcal{C}_L(\mathbf{p})$ of $\mathcal{E}_R$ and $\mathcal{E}_L$, respectively.

Proof. Checking closure, we obtain

$$(a + \alpha\mathbf{p})(b + \beta\mathbf{p}) = c + \gamma\mathbf{p},$$

with $c = ab - \alpha\beta N(\mathbf{p})$ and $\gamma = a\beta + b\alpha$. For the inverse we have

$$(a + \alpha\mathbf{p})^{-1} = \frac{a - \alpha\mathbf{p}}{N(a + \alpha\mathbf{p})}\cdot$$

The subgroups $\mathcal{C}_R(\mathbf{p})$ and $\mathcal{C}_L(\mathbf{p})$ are one-parameter groups because the normalized quaternion $a + \alpha\mathbf{p}$ depends on one parameter only.

Theorem 3–8.9. (i) The trace of every point under every $\mathcal{C}_R(\mathbf{p})$ is a projective straight line.

(ii) For a fixed $\mathbf{p}$, any two of these lines either coincide or have no common points.

(iii) No two of these lines, if distinct, lie in one plane.

(iv) For every projective line in P^3 there is a unique $\mathbf{p}$ such that the line is a trace under $\mathcal{C}_R(\mathbf{p})$.

(v) The above four statements hold when $\mathcal{C}_R$ is replaced by $\mathcal{C}_L$.

Proof. (i) All the points $\mathbf{y} = a + \alpha\mathbf{p} \neq 0$ for all a and α lie on a projective line through 1 and $\mathbf{p}$, and all the points on this line can be represented in this form. The points $\mathbf{x}(a + \alpha\mathbf{p}) = \mathbf{y}L_{\mathbf{x}}$, with $x \neq 0$, also form a line because the projectivity $L_{\mathbf{x}}$ is a collineation, and these are exactly the points of the trace of $\mathbf{x}$ under $\mathcal{C}_R(\mathbf{p})$.

(ii) Let $\mathbf{x}$ and $\mathbf{y}$ be distinct points. Then the lines consist of the points $\mathbf{x}\mathcal{C}_R(\mathbf{p})$ and $\mathbf{y}\mathcal{C}_R(\mathbf{p})$, respectively. These are cosets with respect to the same subgroup and are therefore either identical or disjoint.

(iii) If two distinct lines were to lie in a plane, they would intersect, a contradiction to (ii).
(iv) Let $\mathbf{x}$ and $\mathbf{y}$ be two distinct points on a line. Then $\mathbf{x}^{-1}\mathbf{y} = q_0 + q_1\mathbf{i} + q_2\mathbf{j} + q_3\mathbf{k} = q_0 + \alpha\mathbf{p}$, say, and $\mathbf{x}(q_0 + \alpha\mathbf{p}) = \mathbf{y}$. Hence $\mathbf{y}$ lies on the trace of $\mathbf{x}$ under $\mathcal{C}_R(\mathbf{p})$. Can a line be a trace under two different groups $\mathcal{C}_R(\mathbf{p})$ and $\mathcal{C}_R(\mathbf{q})$? If so, then for some $\mathbf{x} \neq 0$, $\mathbf{x}(a + \alpha\mathbf{p}) = \mathbf{x}(b + \beta\mathbf{q})$. Consequently, $a + \alpha\mathbf{p} = b + \beta\mathbf{q}$ are equal normalized quaternions and hence $\mathbf{p} = \mathbf{q}$.
(v) Obvious.

Definition. Two lines are *right Clifford parallels* if they are traces under the same $\mathcal{C}_R(\mathbf{p})$. They are *left Clifford parallels* if they are traces under the same $\mathcal{C}_L(\mathbf{p})$.

This "parallelism" was introduced by the British mathematician William K. Clifford (1845–1879).

As we saw in Theorem 3–8.9, right Clifford parallels and left Clifford parallels have some of the properties which one would expect from parallels. Right (left) parallels do not intersect, there is exactly one right (left) parallel to a given line through every point outside this line, and right (left) Clifford parallelism is an equivalence relation. On the other hand, Clifford parallels do not lie in the same plane. We will develop some further properties.

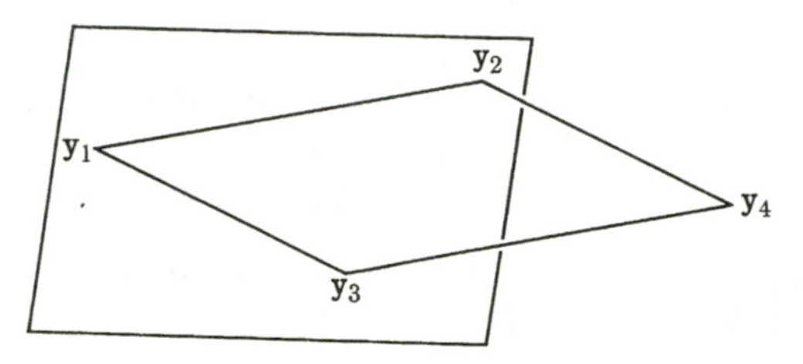

Figure 3–37

Let $\mathbf{y}_1 = \mathbf{x}$, $\mathbf{y}_2 = \mathbf{x}(a + \alpha\mathbf{p})$ and $\mathbf{y}_3 = (b + \beta\mathbf{q})\mathbf{x}$ (Fig. 3–37) be vertices of a triangle, where a, b, α, and β are reals, and $\mathbf{p}$ and $\mathbf{q}$ are pure quaternions. Now let $\mathbf{y}_4 = (b + \beta\mathbf{q})\mathbf{x}(a + \alpha\mathbf{p})$. Then the join of $\mathbf{y}_1$ and $\mathbf{y}_2$ is a right Clifford parallel of the join of $\mathbf{y}_3$ and $\mathbf{y}_4$, while the join of $\mathbf{y}_1$ and $\mathbf{y}_3$ is a left Clifford parallel of the join of $\mathbf{y}_2$ and $\mathbf{y}_4$. The figure with the vertices $\mathbf{y}_1, \mathbf{y}_2, \mathbf{y}_3, \mathbf{y}_4$ is a *Clifford parallelogram* which is, however, not a plane figure, in view of Theorem 3–8.9. Every transformation from $\mathcal{E}^3_+$ preserves elliptic distance. Since $\mathbf{y}_2 = \mathbf{y}_1 R_{a+\alpha\mathbf{p}}$ and $\mathbf{y}_4 = \mathbf{y}_3 R_{a+\alpha\mathbf{p}}$, we have $d(\mathbf{y}_2, \mathbf{y}_4) = d(\mathbf{y}_1, \mathbf{y}_3)$, and similarly, $\mathbf{y}_3 = \mathbf{y}_1 L_{b+\beta\mathbf{q}}$ and $\mathbf{y}_4 = \mathbf{y}_2 L_{b+\beta\mathbf{q}}$ which implies $d(\mathbf{y}_3, \mathbf{y}_4) = d(\mathbf{y}_1, \mathbf{y}_2)$. Thus we have the following:

Theorem 3–8.10. Opposite sides of Clifford parallelograms have equal lengths. Clifford parallelograms do not lie in a plane.

Now let $\mathbf{x}$, $\mathbf{p}$, and $\mathbf{q}$ be defined as before. We consider the point quadruple $\mathbf{x}, \mathbf{qx}, \mathbf{xp}, \mathbf{qxp}$. It is possible that these all lie on a line, and if so then all the points

$\mathbf{x}\mathcal{C}_R(\mathbf{p})$, $\mathbf{x}\mathcal{C}_L(\mathbf{q})$, and $\mathbf{x}\mathcal{C}_R(\mathbf{p})\mathcal{C}_L(\mathbf{q})$ also lie on this line. What happens in the other case, when $\mathbf{x}$, $\mathbf{xp}$, and $\mathbf{qx}$ form a triangle? In other terms, what kind of point set (hopefully: surface) is generated by all the Clifford parallelograms?

We choose the points $\mathbf{x}$, $\mathbf{xp}$, $\mathbf{qx}$, and $\mathbf{qxp}$ as reference points of a coordinate system in P^3. This involves a projectivity which maps $\mathbf{x}$ on 1, $\mathbf{qx}$ on $\mathbf{i}$, $\mathbf{xp}$ on $\mathbf{j}$, and $\mathbf{qxp}$ on $\mathbf{k}$. Then the point $(b + \beta\mathbf{q})\mathbf{x}(a + \alpha\mathbf{p})$ goes into $(b + \beta\mathbf{i})(a + \alpha\mathbf{j}) = ab + a\beta\mathbf{i} + \alpha b\mathbf{j} + \alpha\beta\mathbf{k} = [x_0, x_1, x_2, x_3]$, say.

Consequently we obtain the relation $x_0x_3 - x_1x_2 = 0$. That is, the point $(b + \beta\mathbf{q})\mathbf{x}(a + \alpha\mathbf{p})$ ranges over a so-called *Clifford surface*, which is projectively equivalent to a quadric whose equation in nonhomogeneous coordinates would be $z = xy$, a hyperbolic paraboloid. The hyperbolic paraboloid is well known to be a "ruled surface," that is, to be generated entirely by straight lines. These straight lines, in our case, are the sides of the Clifford parallelograms.

Theorem 3–8.11. If l_1 and l_2 are two distinct concurrent projective lines, then all the right Clifford parallels to l_1 and all the left Clifford parallels to l_2 generate a Clifford surface, which is projectively equivalent to a hyperbolic paraboloid.

If one wants to visualize Clifford parallels, parallelograms, and surfaces, one has to keep in mind that in elliptic three-space every plane is an elliptic plane. This can be verified by transforming the plane into $x_0 = 0$ which is intersected by J in C_e. As a consequence, according to Theorem 3–7.5, every straight line in elliptic three-space is of finite length.

Exercises

1. Let $\mathbf{p} = 1 - \mathbf{i} + \mathbf{j} + 2\mathbf{k}$, and $\mathbf{q} = 2\mathbf{i} - \mathbf{k}$. Find
   (a) $\mathbf{pq}$ (b) $\mathbf{p\bar{q}}$ (c) $\mathbf{pq}^{-1}$ (d) $\mathbf{p}^2$.
2. Show that, if $\mathbf{p}$ and $\mathbf{q}$ are two pure quaternions, $\mathbf{pq} = \mathbf{p} \times \mathbf{q} - (\mathbf{p}, \mathbf{q})$, where $\mathbf{p} \times \mathbf{q}$ is the cross product, as used in calculus and physics, and $(\mathbf{p}, \mathbf{q})$ is the inner product of $\mathbf{p}$ and $\mathbf{q}$, considered as three-dimensional vectors.
3. (a) Show that ± 1, $\pm\mathbf{i}$, $\pm\mathbf{j}$, $\pm\mathbf{k}$ form a multiplicative group.
   (b) Is the group abelian?
   (c) Find its subgroups.
4. What is the center of the multiplicative group of nonzero quaternions?
5. Find three isomorphic mappings of the field of complex numbers into the quaternion algebra.
6. If each of two integers m and n is the sum of four squares of integers, use quaternions to prove that mn is also the sum of four squares of integers.
7. In $P^3(C)$ find the points that are common to all the spheres about $[1, 0, 0, 0]$. Show that the circle points (Exercise 11 of Section 3–5) of the plane $x_0 = 0$ are among them.

8. (a) Show that the transformations $\mathbf{x} \to \mathbf{px}$ and $\mathbf{x} \to \mathbf{xq}$ do not have invariant points unless they are the identity.
   (b) Is the same true for $\mathbf{x} \to \mathbf{pxq}$?
9. Let $\mathbf{q} = \frac{1}{3}(-1 + 2\mathbf{i} - 2\mathbf{k})$, and T: $\mathbf{x} \to \mathbf{qx}$.
   (a) What are the invariant points of T, if any?
   (b) Is T involutory?
10. Let $\mathbf{x} = \mathbf{i}$, $\mathbf{p} = \mathbf{i} - \mathbf{k}$, and $\mathbf{q} = \mathbf{j} + \mathbf{k}$. Verify that the Clifford parallelogram with vertices $\mathbf{x}$, $\mathbf{px}$, $\mathbf{xq}$, and $\mathbf{pxq}$ is not coplanar.
11. Show that in a Clifford parallelogram the sum of two angles adjacent to one side is π.
12. Explain why $\mathbf{x}$, $\mathbf{px}$, $\mathbf{qx}$, and $\mathbf{pqx}$ do not yield a Clifford parallelogram.
13. Is a Clifford surface projectively equivalent to an ellipsoid? Why?
14. Show that the determinant of the projectivity $R_{\mathbf{a}}$ is $(N(\mathbf{a}))^2$.

3-9 OTHER GEOMETRIES

In this section we intend to summarize the results obtained so far and take a brief look at possible further developments without going into detailed proofs. Instead of proofs, we shall give, at the end of the section, a list of books for suggested reading.

We have obtained an outline of important results in the two plane noneuclidean geometries and some facts about elliptic three-space. Of course, these geometries have been developed in full detail, analytically as well as synthetically, and the theories that can be found in the literature are almost as complete as the corresponding treatment of euclidean geometry. For instance, a noneuclidean trigonometry is available. Also, noneuclidean coordinate systems have been introduced and discussed. Noneuclidean geometries in higher dimensions exist and have interesting properties. Thus our treatment so far cannot be considered as more than a sample.

Most of the projective geometry in our exposition was done over the real field. Sometimes we used complex coordinates in order to reach new conclusions in real projective and noneuclidean geometry. But we never treated projective geometry over the field of complex numbers in its own right. This can be done, and exhaustive studies of complex projective geometry appear in the literature.

Since the complex numbers form an associative and commutative division algebra of rank 2 over the reals, it is reasonable to inquire about projective geometries over algebras of higher rank over the reals. The next step would be a projective geometry over the quaternions, whose rank over the reals is 4. Now we used quaternions for the treatment of a projective model of elliptic three-space. However, this is quite a different subject. The quaternions were only a technical device for studying real projective geometry, just as in Chapter 1 the complex numbers were used for dealing with the euclidean plane over the field of real numbers. Projective geometry over the quaternions (i.e., where each of the coordinates is a quaternion) has also been studied. One further step is the so-called *octave geometry*.

The *octaves* or *Cayley numbers* form the *Cayley algebra* of rank 8 over the reals. We start from the quaternions and introduce a new symbol **e**. The octaves are all possible expressions $\mathbf{p} + \mathbf{q}\mathbf{e}$, where **p** and **q** are quaternions, with the conventions $\mathbf{p} + 0\mathbf{e} = \mathbf{p}$, and $0 + \mathbf{q}\mathbf{e} = \mathbf{q}\mathbf{e}$. When we define addition by

$$(\mathbf{p} + \mathbf{q}\mathbf{e}) + (\mathbf{r} + \mathbf{s}\mathbf{e}) = (\mathbf{p} + \mathbf{r}) + (\mathbf{q} + \mathbf{s})\mathbf{e}$$

and scalar multiplication with α by $(\mathbf{p} + \mathbf{q}\mathbf{e})\alpha = \mathbf{p}\alpha + (\mathbf{q}\alpha)\mathbf{e}$, we see that we obtain a V^8 with the basis $\{1, \mathbf{i}, \mathbf{j}, \mathbf{k}, \mathbf{e}, \mathbf{ie}, \mathbf{je}, \mathbf{ke}\}$. We define multiplication of octaves by the following multiplication table of the basis elements:

	1	i	j	k	e	ie	je	ke
1	1	i	j	k	e	ie	je	ke
i	i	−1	k	−j	ie	−e	−ke	je
j	j	−k	−1	i	je	ke	−e	−ie
k	k	j	−i	−1	ke	−je	ie	−e
e	e	−ie	−je	−ke	−1	i	j	k
ie	ie	e	−ke	je	−i	−1	−k	j
je	je	ke	e	−ie	−j	k	−1	−i
ke	ke	−je	ie	e	−k	−j	i	−1

The Cayley algebra is noncommutative and nonassociative, as the examples

$$\mathbf{ij} = \mathbf{k} \neq -\mathbf{k} = \mathbf{ji} \qquad \text{and} \qquad (\mathbf{ij})\mathbf{e} = \mathbf{ke} \neq -\mathbf{ke} = \mathbf{i}(\mathbf{je})$$

show. Again we define a *conjugate* of an octave, $\overline{(\mathbf{p} + \mathbf{q}\mathbf{e})} = \bar{\mathbf{p}} - \mathbf{q}\mathbf{e}$, and the *norm* of an octave $\mathbf{o} = \mathbf{p} + \mathbf{q}\mathbf{e}$, $N(\mathbf{o}) = \mathbf{o}\bar{\mathbf{o}}$. It can be shown that $\mathbf{o}\bar{\mathbf{o}} = \bar{\mathbf{o}}\mathbf{o} = N(\mathbf{p}) + N(\mathbf{q})$. As in the cases of complex numbers and quaternions, $N(\mathbf{o}) = 0$ if and only if $\mathbf{o} = \mathbf{0}$, because the norm is the sum of the squares of the components of the octave. The rule $N(\mathbf{o}_1)N(\mathbf{o}_2) = N(\mathbf{o}_1\mathbf{o}_2)$ can be proved to hold as it does for quaternions, and consequently the Cayley algebra is a division algebra.

An algebra with a basis $\mathbf{e}_1, \ldots, \mathbf{e}_n$ and a norm

$$N(\alpha_1\mathbf{e}_1 + \cdots + \alpha_n\mathbf{e}_n) = \alpha_1^2 + \cdots + \alpha_n^2$$

is called a *normed algebra*, provided $N(\mathbf{p})N(\mathbf{q}) = N(\mathbf{pq})$ for all elements **p** and **q** of the algebra. The following is an important result concerning normed algebras.

The Hurwitz-Albert Theorem. A normed algebra of finite rank over the reals, with an identity element, is either the field of reals, the field of complex numbers, the quaternion algebra, or the Cayley algebra.

This theorem combines results found in 1898 by the German mathematician A. Hurwitz, with work done in 1947 by the American mathematician A. A. Albert.

The following shows how length has been introduced over a field other than R. If $\mathbf{p} = (p_1, \ldots, p_n)$ and $\mathbf{q} = (q_1, \ldots, q_n)$ are complex vectors, the inner product is defined by $(\mathbf{p}, \mathbf{q}) = \sum_k p_k \bar{q}_k$, and $\mathbf{p}^2 = |\mathbf{p}|^2 = \sum_k p_k \bar{p}_k = \sum_k |p_k|^2$ is real. Then almost all requirements of VS4 are met. The only difference is that the symmetry $(\mathbf{p}, \mathbf{q}) = (\mathbf{q}, \mathbf{p})$ of the inner product has to be replaced by $(\mathbf{p}, \mathbf{q}) = (\overline{\mathbf{q}, \mathbf{p}})$. A vector space satisfying this modified version of VS 4 is called *unitary*.

The significance of the Hurwitz-Albert theorem for geometry is obvious. When we look for algebras over which to construct a geometry that permits introduction of length by means of inner products and norms, only the four algebras mentioned in the Hurwitz-Albert theorem appear as possibilities. We have encountered the geometries over the reals and over the complex numbers. The remaining cases, quaternion geometry and octave geometry, have been explored (the latter rather recently, especially by the Dutch mathematician H. Freudenthal), and their groups of transformations have been studied. All this is outside the scope of this book, and the reader is referred to the literature listed at the end of this section.

Another theorem which sheds light on the role of the four algebras mentioned above is the following.

The generalized Frobenius Theorem. The field of reals and the field of complex numbers are the only associative and commutative division algebras of finite rank over the reals. The skew field of quaternions is the only associative and noncommutative division algebra of finite rank over the reals. The Cayley algebra is the only alternative, nonassociative division algebra of finite rank over the reals.

Here *alternative* means the following special case of associativity: $a^2b = a(ab)$, $ba^2 = (ba)a$ for all a and b in the algebra.

The first two parts of the theorem are the work of the German G. Frobenius (1849–1917). The third result was found in 1950 by two American mathematicians, R. H. Bruck and E. Kleinfeld.

Thus it appears that every time we increase the rank of the algebra used for constructing a geometry, we have to sacrifice a desirable property. When we exchange the reals for the complex numbers we sacrifice the order, because, in contrast to the reals, we cannot always decide which of two complex numbers is greater than the other. When we move on to the quaternions, we lose commutativity, and finally, the octaves are not even associative, but satisfy only the rather weak alternative laws instead. However, it is important to note that reasonable geometries can be constructed over these quite general algebraic structures, and in the next chapter we shall meet such geometries.

In Section 2–10 we discussed affine planes over finite fields. These can be made into projective planes by adjoining an improper line. Theorem 2–10.2 would then assume the more symmetric form:

"For every prime p there is a projective plane whose points are the $p^2 + p + 1$ ordered triples $[x_0, x_1, x_2]$ with distinct ratios $x_0 : x_1 : x_2$. All the x_0, x_1, and x_2 are elements of $GF(p)$, and at least one of the elements of each triple has to be nonzero. This plane has $p^2 + p + 1$ distinct lines. Through each point pass exactly $p + 1$ lines, and on each line lie exactly $p + 1$ points."

As we saw in Section 2–10, there are Galois fields $GF(p^n)$ for all prime powers p^n, and projective planes can be constructed over them. Beyond this, finite projective planes are possible over more general algebras. The generality can be carried so far as to define a *rudimentary projective plane* of order n as a set of points and lines merely satisfying the three assumptions, "Every two distinct $\begin{Bmatrix}\text{lines}\\ \text{points}\end{Bmatrix}$ are incident with exactly one $\begin{Bmatrix}\text{point}\\ \text{line}\end{Bmatrix}$," and "Every $\begin{Bmatrix}\text{point}\\ \text{line}\end{Bmatrix}$ is incident with exactly $n + 1$ $\begin{Bmatrix}\text{lines}\\ \text{points}\end{Bmatrix}$." The problem which then arises is to determine the numbers n for which at least one such plane exists. No projective plane has yet been found for n not a prime power. On the other hand, it has not been proved that such planes are impossible, except in certain cases which were determined in 1949 by the two Americans R. H. Bruck and H. J. Ryser.

The Bruck-Ryser Theorem. If n yields a remainder of 1 or 2 when divided by 4, there cannot be a rudimentary projective plane of order n, unless n can be expressed as a sum of two squares of integers.

Thus there can be no plane for $n = 6$ or $n = 14$, but nothing is known, for instance, for $n = 10$ because 10 is not a prime power (which would yield a plane over a Galois field) and because, in view of $10 = 3^2 + 1^2$, the Bruck-Ryser theorem does not apply.

Another line of study is the group-theoretic investigation of the transformation groups of geometries over arbitrary fields. Here purely algebraic considerations occupy the foreground rather than the geometrical facts that originated the groups. Fruitful relations between such groups and other algebraic theories have been found.

Suggestions for further reading

General

E. Artin, *Geometric Algebra*, New York: Interscience, 1957. This work covers various geometries over arbitrary fields, from an advanced algebraic standpoint.

G. de B. Robinson, *The Foundations of Geometry*, Toronto: University of Toronto Press, 1940. This is a short easy-to-read introduction to the subject, covering the axiomatic aspect as well.

Projective Geometry

H. S. M. Coxeter, *The Real Projective Plane*, Cambridge: Cambridge University Press, 1955. A synthetic introduction to plane projective geometry over the reals.

R. A. Rosenbaum, *Introduction to Projective Geometry and Modern Algebra*, Reading, Mass.: Addison-Wesley, 1963. An introductory work on the subject.

A. Seidenberg, *Lectures in Projective Geometry*, Princeton: Van Nostrand, 1962. This work is a brief, although more advanced, treatment.

O. Veblen and J. W. Young, *Projective Geometry*, 2 vols., Boston: Ginn & Co., 1910 and 1918. This is a very detailed treatise which covers real and complex projective geometry.

Noneuclidean geometry

H. Busemann and P. J. Kelly, *Projective Geometry and Projective Metrics*, New York: Academic Press, 1953. A moderately advanced book which treats the various metrics for projective geometry.

H. S. M. Coxeter, *Non-Euclidean Geometry*, Toronto: University of Toronto Press, 1957. A concise, comprehensive treatment, moderately advanced.

F. Klein, *Vorlesungen über nicht-euklidische Geometrie*, Berlin: Springer, 1928. (Reprint, New York: Chelsea, 1959.) An easy-to-read classic.

H. Schwerdtfeger, *Geometry of Complex Numbers*, Toronto: University of Toronto Press, 1962. Chapter 3 gives a moderately advanced account of two-dimensional noneuclidean geometries.

D. M. Y. Sommerville, *The Elements of non-Euclidean Geometry*, London: Bell, 1914. (Reprint, New York: Dover, 1958.) A very elementary work which gives details of synthetic geometry.

Geometrical groups

J. Dieudonné, *La géométrie des groupes classiques*, 2nd ed., Berlin: Springer, 1963. An advanced work.

H. Freudenthal, *Oktaven, Ausnahmegruppen und Oktavengeometrie*, Utrecht: Utrecht University, 1960. An advanced introduction to octave geometry and its groups.

CHAPTER 4

Axiomatic Plane Geometry

Geometric axioms. Whereas so far we have constructed our geometries in an analytic way over given fields, we will now make a new start and construct geometries from first geometrical principles, without any algebraic assumptions. In this we will restrict ourselves to two-dimensional geometries, not because higher-dimensional geometry is uninteresting, but rather because this book is not intended to cover more than an outline of the important methods. The undefined objects of our geometries are *points* and *lines*. Between them there is an undefined relation called *incidence*. The unproved statements from which we start are called *axioms* or, synonymously, *postulates*.

Systems of axioms have to be *consistent;* that is, they should contain no contradictions. The standard method of proving the consistency of a geometry is to introduce coordinates and then to show that the set of coordinate values constitutes an algebraic structure as, for instance, a field. Every contradiction of the geometry would then have to appear as a contradiction in the algebraic structure. Since mathematicians believe more in their understanding of algebraic structures than in their geometric intuition, this procedure provides the best possible proof for the consistency of the axiom system.

Another requirement for axiom systems is that they be *categorical.* A system is categorical if the geometry described by the axioms is unique up to a one-to-one correspondence of its basic elements, with preservation of the relations between them. If the basic elements are points and lines, and if their only basic relation is incidence, then the categorical property of an axiom system means that, for every two geometries described by it, there are one-to-one correspondences between the points and the lines of one geometry and the points and the lines, respectively, of the second geometry, such that incidence is preserved. In short, a categorical axiom system describes essentially only one geometry.

As an example we will discuss an axiom system for a geometry which was mentioned in Section 3–9, namely the projective plane over $GF(2)$. A model for this plane is obtained by augmenting $A^2(2)$ of Section 2–10 in the usual way. The resulting figure is Fig. 4–1 (the "circle" in the figure is one of the lines of the geometry!). We propose the following axioms for this geometry.

I. For every two points there is exactly one line incident with both.
II. For every two lines there is exactly one point incident with both.
III. There exist four points no three of which are incident with a line.
IV. Every line is incident with exactly three points.

It is easy to verify that the augmented $A^2(2)$ satisfies the four axioms. $\Phi = GF(2)$ is a field over which a vector space, $V^2(\Phi)$, and hence a projective plane, $P^2(\Phi)$, can be defined without any contradiction. Hence the axiom system is consistent.

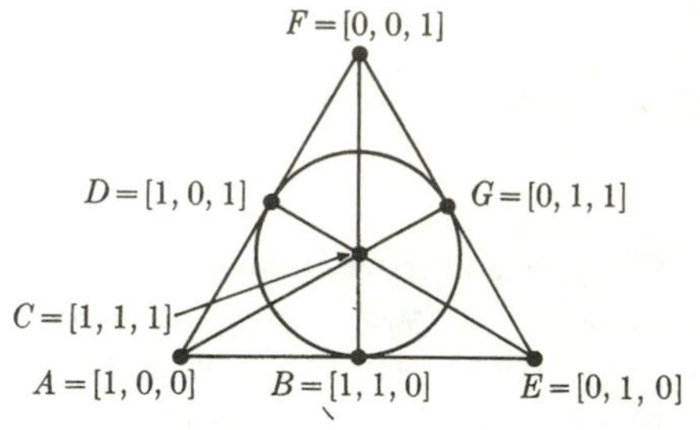

Figure 4-1

Now, to outline a proof that the system is categorical, let us construct an arbitrary geometry based on points and lines, which are related by incidence, and on the Axioms I through IV. By Axiom III there are four points, no three of which lie on a line. We call these points A, B, C, and D and couple them with points of $P^2(\Phi)$ by a one-to-one correspondence, as in Fig. 4-1. By I, the lines AB, AC, AD, BC, BD, and CD exist and can be shown to be distinct. By II, the intersections of these lines exist. $AB \times CD = E$ is distinct from A, B, C, and D, and so is $AD \times BC = F$. The points E and F are distinct because otherwise $AB = AD$, contradicting III. Thus the line EF exists and is distinct from AB, AC, AD, BC, BD, and CD, because otherwise distinct points would have to coincide. So far, the lines AB, AD, BC, and CD have a third point each, namely, E, F, F, and E, respectively. By IV, there has to be exactly one more point on each of AC, BD, and EF. Now, let $AC \times EF = G$. Again G cannot coincide with any of the other six points. For instance, $B = G$ would imply $B = E$. The line BG must meet AF in one of its three points A, D, or F. $BG \times AF = A$ would result in $A = B$, and $BG \times AF = F$ in $F = G$. Thus B, D, and G lie on a line. We have obtained exactly 7 points and 7 lines between which precisely the incidence relations of $P^2(\Phi)$ hold. Since the construction in no way depended on the particular choice of the geometry, we have proved that the system is categorical.

It should be emphasized that the points and lines of the example indeed are undefined elements. They do not have to bear any resemblance to points and lines as defined and used in Chapters 2 and 3.

As an example, we replace "point" by "color," "line" by "ball," and let the relation be "ball has color," so that the resulting "geometry" deals with statements about colored balls. Axioms I through IV now read:

I. Every two colors appear on exactly one ball.
II. Every two balls have exactly one color in common.
III. There are four colors such that no three of them appear on one ball.
IV. Each ball has three colors.

Then, in view of the categorical property of the axiom system, there are exactly 7 balls and 7 colors. If the colors are A, B, C, D, E, F, and G, then the 7 balls contain colors as follows: ABE, ADF, ACG, BCF, BDG, CDE, EFG (meaning that one ball has the colors A, B, and E, another the colors A, D, and F, etc.), or in any other arrangement obtained from this list by permuting the seven colors.

Thus, in the words of H. G. Forder, "Our Geometry is an abstract Geometry. The reasoning could be followed by a disembodied spirit who had no idea of a physical point; just as a man blind from birth could understand the Electromagnetic Theory of Light."*

Our plan for this chapter calls for the construction of a rudimentary projective plane, based on incidence axioms alone. Gradually more axioms will be added and the corresponding algebraic coordinate structures will be studied, with special emphasis on the collineations whose existence is assured by the axioms. The additional axioms will be of the type of incidence properties, such as those of Desargues and Pappus. Eventually the coordinates will constitute a field. Then new axioms, of order and continuity, will be added in order to make the coordinate field isomorphic to the field of reals. Certain modifications will be made for dealing with affine rather than projective planes. This brings us back to the starting point of Chapter 2, and were we to be completely logical, this axiomatic treatment should appear before Chapters 2 and 3, rather than after them.

As an alternative to the analytic introduction of euclidean geometry in Section 2–7, an axiomatic treatment is provided by means of congruence axioms. Then it will be shown that hyperbolic geometry in the plane satisfies all axioms of the euclidean plane except one. The equivalence of the two models of the hyperbolic plane, Poincaré's model of Section 1–9 and the Cayley-Klein model of Section 3–7, will be proved. Finally, the position of the two noneuclidean geometries with respect to other plane geometries will be discussed on the grounds of the axiomatic treatment.

4–1 RUDIMENTARY PROJECTIVE AND AFFINE PLANES AND THEIR COORDINATIZATION

Rudimentary planes. A *rudimentary projective plane* consists of points and lines between which a relation called incidence is defined. The following axioms hold.

PP 1. For every two points there is exactly one line incident with both.
PP 2. For every two lines there is exactly one point incident with both.
PP 3. There exist four points no three of which are incident with a line.

---

* *The Foundations of Euclidean Geometry*, p. 43. See reference at end of chapter.

In the following we will again use the various terms that replace the word "incident" in a less rigid style, such as for instance "lie on," "belong to," "pass through," collinear, and concurrent. The word "rudimentary" will be omitted from now on.

Theorem 4–1.1. There exist four lines no three of which are concurrent.

Proof. If A, B, C, and D are the points mentioned in PP 3, then AB, BC, CD, and DA are lines with the desired property.

This theorem is the dual statement to PP 3. This proves that in the plane defined by PP 1 through 3, the duality principle holds, since PP 1 and PP 2 are dual to each other.

Theorem 4–1.2. In a projective plane there are at least three points on every line and at least three lines through every point.

Proof. If A, B, C, and D are such as mentioned in the last proof, then every line not through A is intersected by AB, AC, and AD in three distinct points. Every line through A, except AB, is intersected by BC and BD in two distinct points different from A, and thus contains at least three points. Finally, AB contains the three distinct points A, B, and $AB \times CD$. The second part of the theorem is dual to the first.

Thus the smallest projective plane has three points on each line, as shown in Fig. 4–1. On the other hand, the projective planes of Chapter 3 also satisfy PP 1 through 3, and thus there are finite as well as infinite projective planes.

In order to introduce coordinates in a projective plane we choose four points as reference points. These points have to be such that no three of them are collinear, and we call them Q_0, Q_1, Q_2, and U, as we did in Section 3–2.

We make our plane a (*rudimentary*) *augmented affine plane* by distinguishing Q_1Q_2 as the improper line. By removing it we obtain an *affine plane*, and again the term "augmented" will be omitted when no ambiguity is to be feared.

From PP 1 through 3 we obtain the well-known properties of parallels, for instance, the equivalence property of parallelism and the fact (*Euclid's parallel axiom*) that through a given proper point P there is exactly one line parallel to a given line not through P (cf. Exercises 3 and 4).

To points on Q_0U we assign symbols $[b, b]$, where b belongs to a set Σ of labels containing at least two elements called 0 and 1. The points Q_0 and U receive the labels $Q_0 = [0, 0]$ and $U = [1, 1]$. The only point of Q_0U in the augmented plane without such a label will be the improper point $Q_0U \times Q_1Q_2$. The remaining proper points are coordinatized by the rule $[x, x]Q_2 \times [y, y]Q_1 = [x, y]$ (Fig. 4–2).

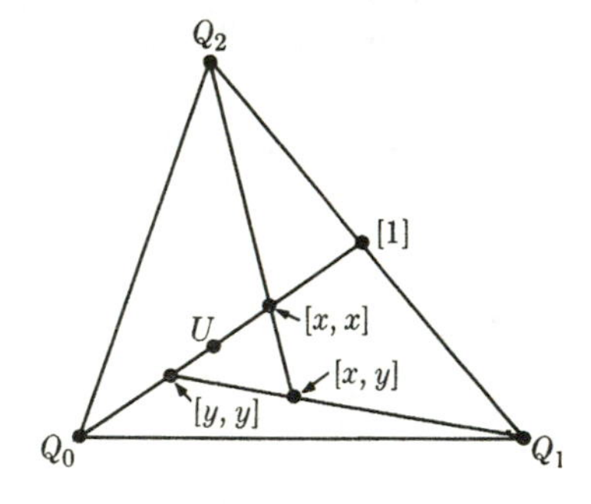

Figure 4–2

For the improper points of the augmented plane we use a special notation, $Q_0[1, m] \times Q_1Q_2 = [m]$. This provides names for all improper points other than Q_2. In particular, $Q_0U \times Q_1Q_2 = [1]$ and $Q_1 = [0]$. We define also $Q_2 = [\infty]$, ∞ being a mere symbol, not an element of Σ.

Coordinatization by ternary rings. We will attempt to introduce an algebraic operation on the elements of Σ. Let $a, b, c \in \Sigma$, and

$$[0, c][b] \times [a, a][\infty] = [a, t],$$

(see Fig. 4-3). Then t is an element of Σ because $[a, t]Q_1 \times Q_0U = [t, t]$, and it is a function of a, b, and c. We use the notation

$$t = a * b * c.$$

This is a *ternary operation*, mapping all ordered triples of elements of Σ into Σ, in symbols of a cartesian product, $\Sigma \times \Sigma \times \Sigma \to \Sigma$, just as ordinary addition or multiplication are *binary operations*, mapping $\Sigma \times \Sigma \to \Sigma$. The set Σ under this ternary operation is called a *ternary ring*, $(\Sigma, *)$.

If a ranges through all elements of Σ, then the points $[x, y]$ on the line $[0, c][b]$ are $[x, x * b * c]$. Hence we may say that the equation of this line is

$$y = x * b * c. \tag{1}$$

Every proper line l, not through Q_2, has an equation of this kind. For $l \times Q_0Q_2$ is always of the form $[0, c]$, and $l \times Q_1Q_2$ of the form $[b]$, with b and c in Σ. The equations of the lines through Q_2 are of the form $x = a$, with $a \in \Sigma$.

We draw Fig. 4-3 again as Fig. 4-4, using an orthonormal coordinate system. This is done purely as an explanatory device and should not mislead the reader into thinking that any right angles have been defined. Then we see that Eq. (1) generalizes the line equation $y = xb + c$ of the real euclidean plane, while $[b]$ corresponds to the point where all the lines of slope b meet.

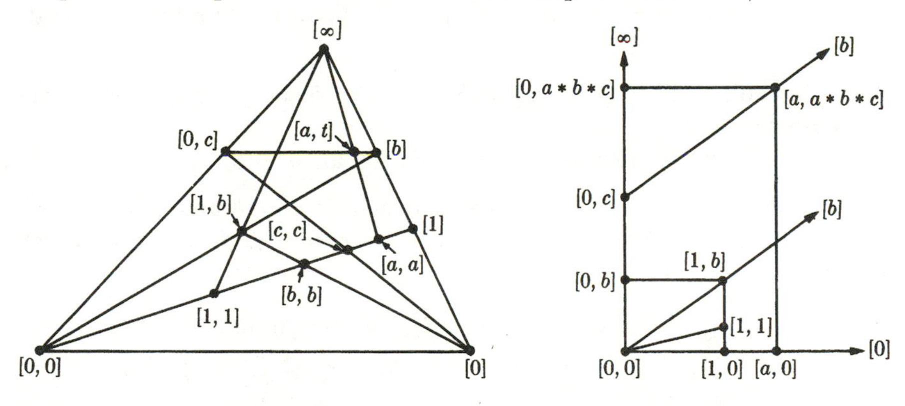

Figure 4-3 **Figure 4-4**

Algebraic structures, like fields, groups, and algebras, are characterized by sets of axioms. The following axioms for ternary rings are easily seen to be satisfied by the ternary ring as defined above in a geometric way.

A ternary ring is a set Σ, containing at least two distinct elements 0 and 1, and closed under a ternary operation $(*)$, such that for all a, b, c, d, in Σ the following hold.

TR 1. $a * 0 * c = c$.

TR 2. $0 * b * c = c$.

TR 3. $1 * b * 0 = b$.

TR 4. $a * 1 * 0 = a$.

TR 5. The equation $a * b_1 * c_1 = a * b_2 * c_2$ has exactly one solution for a in Σ if $b_1 \neq b_2$.

TR 6. $a * b * c = d$ has exactly one solution c in Σ.

TR 7. The simultaneous equations $a_1 * b * c = d_1$ and $a_2 * b * c = d_2$ have a solution for b and c in Σ if $a_1 \neq a_2$.

We introduce an addition and a multiplication by

$$a * 1 * c = a + c, \tag{3}$$

$$a * b * 0 = ab. \tag{4}$$

In the special case in which

$$a * b * c = ab + c, \tag{2}$$

the postulates TR are satisfied by skew fields and fields, which are therefore special ternary rings. When Eq. (2) is satisfied, the ring is called *linear*.

In contrast to Eq. (2), which does not hold in every ternary ring, Eqs. (3) and (4) are nothing but definitions of new binary operations and, as such, hold in every ternary ring. Under addition alone, $(\Sigma, +)$ forms a *loop*, and so does $(\Sigma', \cdot)$, that is, Σ less the zero element under multiplication. (The proof is left for the exercises.) By a loop we mean a set S, closed under a binary operation $(\circ)$, and satisfying the following axioms.

Lp 1. For all a and b in S, the equations $a \circ x = b$ and $y \circ a = b$ have unique solutions x and y in S.

Lp 2. S contains an element e such that for all $a \in S$, $a \circ e = e \circ a = a$.

In a multiplicative loop we use the notations $a \backslash c = b$ and $c/b = a$, when $ab = c$. In an additive loop we write $a \vdash c = b$ and $c \dashv b = a$, when $a + b = c$. Obvious rules are

$$a(a \backslash c) = (c/b)b = a \backslash (ac) = (cb)/b = c,$$

$$a + (a \vdash c) = (c \dashv b) + b = a \vdash (a + c) = (c + b) \dashv b = c,$$

for all a, b, c in the loop.

Groups are special loops. In particular, a loop in which the associative law holds is a group (Exercise 10).

Three-nets and isotopic loops. We will now restrict ourselves to those proper lines of the augmented plane which pass through $[\infty]$ and those in whose equations (Eq. 1) either $b = 0$ or $b = 1$. The equations of these lines are expressible in terms of $(\Sigma, +)$ alone. The equations are of the types $x = t$, $y = t$, $y = x + t$, with t ranging over all elements of Σ. These are three sets of parallels or, in projective language, the three pencils of all lines through the improper points $[0]$, $[\infty]$, and $[1]$. What we obtain is a so-called *three-net*, defined as a set of points and lines in an affine plane such that: (a) each line of the set belongs to exactly one of the three pencils with proper or improper centers, (b) through each proper point, other than the pencil centers, passes exactly one line from each of the three pencils, (c) lines from different pencils intersect in a point of the net. If all three pencil centers are improper (as in our case), the net is a *parallel three-net.*

Let us specialize Fig. 4-4 for the case $b = 1$. Then the construction of $a * 1 * c = a + c$, which remains entirely in the three-net, is described by Fig. 4-5.

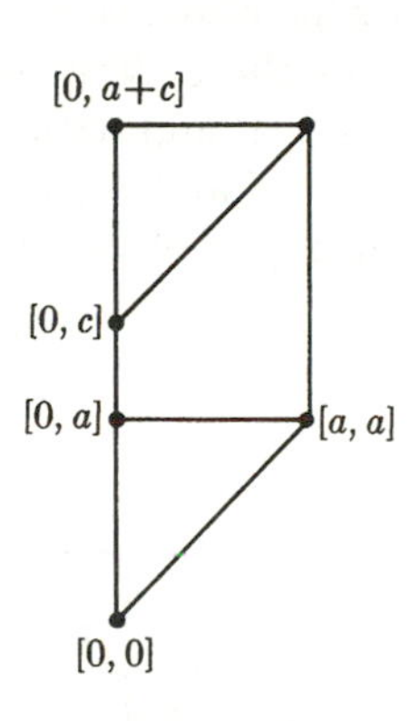

Figure 4-5

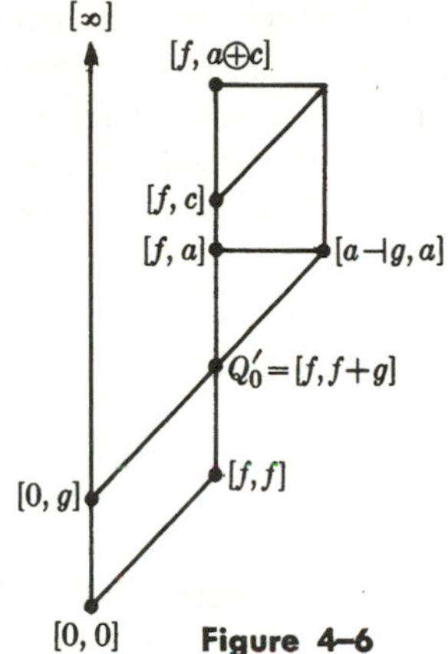

Figure 4-6

In a parallel three-net a coordinate transformation can be introduced. Suppose that the point $Q_0' = [f, f + g]$ is chosen as the new origin, but the three pencil centers $[0]$, $[\infty]$, and $[1]$ are left unchanged (Fig. 4-6). We perform the same construction as in Fig. 4-5, but substitute Q_0' for Q_0, and therefore $Q_0'[0]$ and $Q_0'[\infty]$ for $Q_0[0]$ and $Q_0[\infty]$. Consequently $[f, a]$ and $[f, c]$ replace $[0, a]$ and $[0, c]$, respectively. We call the sum of a and c in the new system $a \oplus c$. The line $Q_0'[1]$ has the equation $y = x + g$ and hence meets $[f, a][0]$, with equation $y = a$, at $[a \dashv g, a]$. The line $[f, c][1]$ has the equation $y = x + (f \vdash c)$ and intersects $[a \dashv g, a][\infty]$ at $[a \dashv g, (a \dashv g) + (f \vdash c)]$. Consequently

$$a \oplus c = (a \dashv g) + (f \vdash c). \tag{5}$$

$(\Sigma, \oplus)$ is a loop. Its closure under the addition (Eq. 5) is obvious. In order to check the validity of Lp 1 for $(\Sigma, \oplus)$, we solve Eq. (5) for a and then for c. Let $a \oplus c = d$. Then

$$a = [d \dashv (f \vdash c)] + g,$$
$$c = f + [(a \dashv g) \vdash d].$$

Both expressions are uniquely determined in view of the loop property of $(\Sigma, +)$. The postulate Lp 2 for $(\Sigma, \oplus)$ holds with $e = f + g$. For, by Eq. (5),

$$(f + g) \oplus x = [(f + g) \dashv g] + (f \vdash x) = f + (f \vdash x) = x,$$
$$x \oplus (f + g) = (x \dashv g) + [f \vdash (f + g)] = (x \dashv g) + g = x.$$

The loop $(\Sigma, \oplus)$ is called *isotopic* to the loop $(\Sigma, +)$, or an *isotope* of $(\Sigma, +)$. Isotopy is an equivalence relation, as will be proved in Exercise 13.

We summarize.

Theorem 4–1.3. An augmented affine plane can be coordinatized by one ternary ring $(\Sigma, *)$ for each choice of reference points Q_0, Q_1, Q_2, and U, no three of which are collinear. The proper lines have equations of the types $y = x * b * c$ and $x = t$, and each equation of these types represents a line. The three-net with pencil centers Q_1, Q_2 and $[1]$ and with Q_0 as origin is then coordinatized by the loop $(\Sigma, +)$. If, for f and g arbitrary in Σ, the origin of the net is moved to $[f, f + g]$, then $(\Sigma, \oplus)$ is the coordinate loop, with the operation $(\oplus)$ defined by Eq. (5).

Another three-net in the affine plane can be constructed with the pencil centers Q_0, Q_1, and Q_2. This net is coordinatized by the loop $(\Sigma', \cdot)$. For, if we redraw Fig. 4–4 with $c = 0$, we obtain Fig. 4–7, which yields equations $y = xb$ for the pencil with $[0, 0]$ as center. The other two pencils of the net are $x = t$ and $y = t$ for all $t \in \Sigma'$. The postulates TR 1 and 2 yield

$$a \cdot 0 = 0 \cdot a = 0 \qquad (6)$$

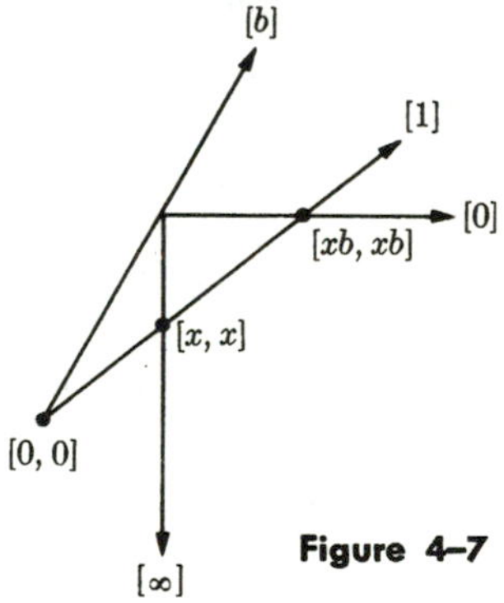

for all a in Σ.

[b]
[1]
[0]
[xb, xb]
[x, x]
[0, 0]
[∞]

Figure 4–7

In the multiplicative loop $(\Sigma', \cdot)$, isotopes can be introduced precisely as in the additive case. The isotopes coordinatize nets with different unit points $[1, 1]$. The proof will be left for the exercises.

We will prove one further theorem about isotopes.

Theorem 4–1.4. If a loop L and all its isotopes are commutative, then L is a group, and all its isotopes are isomorphic to L.

Proof. Let L have the operation $(+)$. For every b in L there is an isotopic loop $(L, \oplus)$ such that

$$x \oplus y = x + (b \vdash y),$$

according to Eq. (5). Substitute $b \vdash y = z$. Then

$$x \oplus (b + z) = x + z.$$

Both $(L, +)$ and $(L, \oplus)$ are commutative. Thus, for all elements a and c of L,

$$\begin{aligned}(a + b) + c &= (a + b) \oplus (b + c) \\ &= (b + c) \oplus (b + a) = (b + c) + a = a + (b + c).\end{aligned}$$

This makes $(L, +)$ associative, that is, a group.

In view of Eq. (5), and using the group properties, we can define every isotope $(L, \circ)$ of this group by

$$x \circ y = (x - g) + (-f + y) = x - (f + g) + y.$$

This implies

$$(x \circ y) - (f + g) = x - (f + g) + y - (f + g),$$

which shows that the mapping $x \to x - (f + g)$ maps $(L, \circ)$ on $(L, +)$ isomorphically. This completes the proof.

So far, the plane has been given and $(\Sigma, *)$ has been developed from the properties of the plane. The converse problem is: If a ternary ring satisfying TR 1 through 7 is given, is there an affine plane coordinatized by it? The answer is contained in the following theorem.

Theorem 4–1.5. For every given ternary ring $(\Sigma, *)$, there is an affine plane which it coordinatizes.

Proof. For each ordered pair (p, q) of elements of Σ we define a point $[p, q]$ of the plane. For each p in Σ an improper point $[p]$ is introduced. Moreover, another improper point $[\infty]$ shall be in the augmented plane. All the improper points are collinear, and so are the points $[\infty]$ and $[t, y]$ for fixed t and for y ranging through all of Σ. Also collinear are the points $[x, x * b * c]$ and $[b]$ for fixed b and c and for x running through the elements of Σ. We have to prove that the points and lines introduced in this way satisfy the axioms PP 1 through 3. This involves the discussion of various cases, and we will go into the details of a few sample cases only.

(i) Let $[p, q]$ and $[r, s]$ be two points, $p \neq r$, $q \neq s$. Their join cannot have an equation of the type $x = t$, because $p \neq r$, nor is it the improper line. Hence its equation has to be of the form $y = x * b * c$. This requires the simultaneous equations

$$q = p * b * c,$$
$$s = r * b * c,$$

to have a unique solution for b and c. By TR 7 this solution exists. Is it unique? Suppose there are two solutions, (b_1, c_1) and (b_2, c_2). Then

$$p * b_1 * c_1 = p * b_2 * c_2,$$
$$r * b_1 * c_1 = r * b_2 * c_2.$$

If $b_1 = b_2$, then by TR 6, $c_1 = c_2$. If $b_1 \neq b_2$, then TR 5 implies that $p = r$, a contradiction. Hence the solution for b and c is unique, and PP 1 is satisfied in this case.

(ii) Let $y = x * b * c$ and $y = x * d * e$ be two distinct lines. If $b = d$, the improper point $[b]$ lies on both lines. A second common point would have to be proper, say $[x, y]$. Then $x * b * c = x * b * e$, and by TR 6 this implies $c = e$, a contradiction to the lines being distinct. Now let $b \neq d$. By virtue of TR 5, the equation $x * b * c = x * d * e$ has a unique solution x. Therefore the two lines intersect in exactly one point, as required by PP 2.

(iii) PP 3 requires the existence of four points no three of which are collinear. The points $[0, 0]$, $[0]$, $[\infty]$, and $[1, 1]$ have this property. For instance, the join of $[0, 0]$ and $[0]$ is the line $y = x * 0 * 0 = 0$, which is incident with neither $[\infty]$ nor $[1, 1]$. A complete proof will be left for the exercises.

Corollary. For every given loop there is a parallel-three-net which it coordinatizes.

EXERCISES

1. Show that the axioms PP 1 through 3 are equivalent to a system in which PP 2 is replaced by the weaker axiom "For every two lines there is at least one point incident with both."

2. (a) Prove that if one line of a projective plane is incident with exactly $n + 1$ points ($n \geqq 2$), then every line of the plane has this property.
   (b) Prove that under the same assumption as in (a), every point is incident with exactly $n + 1$ lines.
   (c) How many points does the plane contain?
   (d) How many lines does the plane contain?

3. Prove: In an affine plane the following statements are true.
   (a) For every two points there is exactly one line incident with both.
   (b) Given a point P and a line l not through P, there exists a unique line through P which does not intersect l.
   (c) There exist three noncollinear points.

4. Take the statements (a), (b), and (c) of Exercise 3 as axioms of an affine plane and prove that:
   (a) There are at least two points on every line
   (b) Two lines either do not intersect or have exactly one common point
   (c) If the lines l_1 and l_2 do not intersect, and if l_2 and a line l_3 do not intersect, then l_1 and l_3 do not intersect.

5. Prove that in every ternary ring the following equations have a unique solution for x.
   (a) $x * b * c = d$, with $b \neq 0$ (b) $a * x * c = d$, with $a \neq 0$.

6. In the projective plane of Fig. 4–1, choose two different reference point quadruples and list the ternary ring operations in each of them. Are these two rings necessarily isomorphic?

7. In the finite affine plane $A^2(5)$, as defined in Section 2–10, choose $Q_0 = [0, 0]$, $U = [1, 1]$, Q_1 as the improper point of $x = 0$, and Q_2 as the improper point of $y = 0$. Find:
   (a) $2 * 3 * 4$, (b) the equation of the line through $[0, 4]$ and $[3]$,
   (c) the intersection of this line with the line $y = 0$,
   (d) its intersection with the line $y = x * 2 * 1$.

8. (a) Is $x \to x + a$, $y \to y + b$ a collineation of the plane for every arbitrary a and b?
   (b) Is it a collineation of the parallel three-net?

9. If $(\Sigma, *)$ is a ternary ring, show that $(\Sigma, +)$ and $(\Sigma', \cdot)$ are loops.

10. Prove that every associative loop is a group.

11. (a) Can a loop have more than one identity element?
    (b) Does every element of a loop have a unique inverse?
    (c) In a loop, does $xa = ya$ imply $x = y$?

12. (a) For a given $b \in \Sigma$, draw a figure to find b_1 such that $b + b_1 = 0$.
    (b) Do the same to find b_2 such that $b_2 + b = 0$.
    (c) What incidence theorem has to hold in the plane if $b_1 = b_2$ for every choice of b in Σ?

13. Prove that isotopy of loops is an equivalence relation.

14. Let the multiplication table of a loop be as follows.

	e	a	b	c	d
e	e	a	b	c	d
a	a	b	c	d	e
b	b	e	d	a	c
c	c	d	a	e	b
d	d	c	e	b	a

   (a) Show that it is not a group.

(b) Find the multiplication table of the isotope with $f = a$, $g = c$, where f and g are defined as in Eq. (5).

(c) Which is the identity element of the isotope?

15. Introduce a new multiplication ($\circ$) resulting from leaving Q_0, Q_1, and Q_2 unchanged but replacing U by the point $[a, b]$, where $a \neq 0 \neq b$.
   (a) Show that $(\Sigma', \circ)$ is a loop isotopic to $(\Sigma', \cdot)$.
   (b) Conversely, show that every loop isotopic to $(\Sigma', \cdot)$ can be created in this fashion by suitable choice of a and b.

16. Work out a complete proof of Theorem 4–1.5.

4–2 VEBLEN-WEDDERBURN PLANES

The minor affine Desargues property. To our axioms PP 1 through 3 we now add another, the minor affine Desargues property, δ_a (Fig. 4–8): If ABC and $A'B'C'$ are two triangles and AA', BB', and CC' are distinct and parallel, and if $AB \parallel A'B'$ and $BC \parallel B'C'$, then $AC \parallel A'C'$.

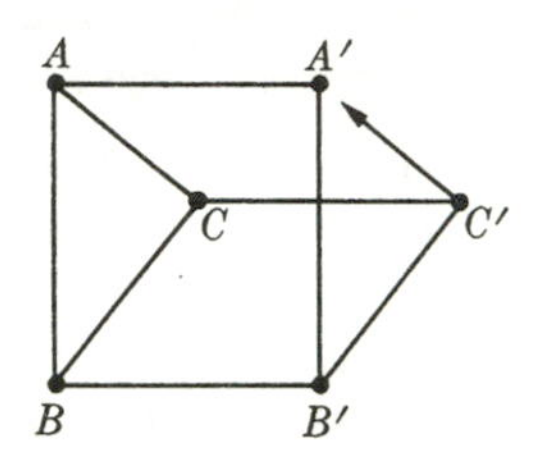

Figure 4–8

Some exceptional cases such as, for instance, B' on BC have to be excluded.

In Section 3–4 we could *prove* the Desargues property in the coordinate model of the projective plane over a field. In the rudimentary plane such a proof is impossible; that is, in some planes the theorem holds while in others it does not. In P^2 over a field the Desargues property is always valid; it is a so-called *desarguesian plane.* In the following we will describe an affine nondesarguesian plane π, as proposed by F. R. Moulton (1902).

Let an orthonormal coordinate system in a euclidean plane be given. The points of this plane are also the points of π. The euclidean straight lines $x =$ const and $y = mx + n$, with $m \leqq 0$, are straight lines of π. However, each line $y = mx + n$ with $m > 0$ is considered as a line of π only in the lower half-plane, $y \leqq 0$. From its point on the x-axis the line of π continues upward, like a refracted light ray, with the slope $m/2$ (Fig. 4–9). Parallel lines of π are those which have equal slopes below the x-axis and equal slopes above it. The plane π can be augmented by an improper line which intersects a proper line in $[m]$ if its lower part has slope m. As we will show, the augmented π satisfies PP 1 through 3. However, as Fig. 4–10 shows, δ_a does not hold in π, because the corresponding sides AC and $A'C'$ meet in D instead of being parallel.

We proceed to prove that the augmented Moulton model, π', is indeed a projective plane.

PP 1 *holds.* Let $[a_1, b_1]$ and $[a_2, b_2]$ be two distinct proper points of π', and assume $a_1 \leqq a_2$. The following cases are trivial: $a_1 = a_2$; $b_1 = b_2$; $a_1 < a_2$ and $b_1 > b_2$; $a_1 < a_2$ and $b_1 < b_2 \leqq 0$; $a_1 < a_2$ and $0 \leqq b_1 < b_2$. For in

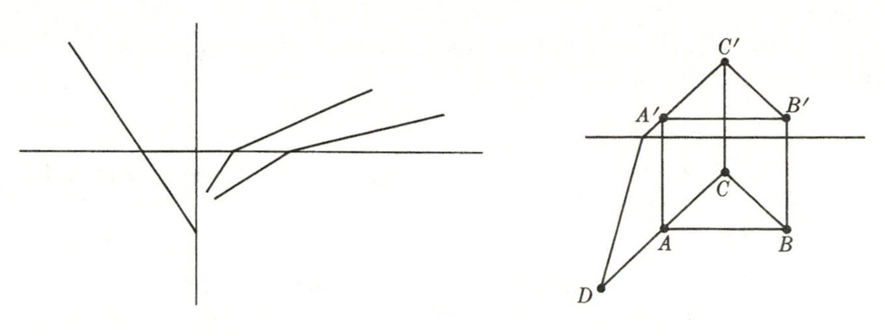

Figure 4-9

Figure 4-10

these cases the points are joined by their euclidean straight line joins. The only remaining case is $a_1 < a_2$, $b_1 < 0 < b_2$. Then the equations

$$b_1 = m(a_1 - a_0) \qquad \text{and} \qquad b_2 = \frac{m}{2}(a_2 - a_0),$$

with a_0 the x-coordinate of the intersection of the join with the x-axis, have a unique solution $m = (2b_2 - b_1)/(a_2 - a_1)$ for the slope. The slope m is positive because the numerator $(b_2 - b_1) + b_2$ and the denominator are both positive.

Now consider improper points. If both points are improper, the join is the improper line. If $[a, b]$ and $[m]$ are the points, then the following cases are trivial: $m \leqq 0$; $m = \infty$; $m > 0$ and $b \leqq 0$. In the remaining case $m > 0$, $b > 0$, the only solution is the broken euclidean line, passing through $[a, b]$ with slope $m/2$ and changing to slope m in the lower half-plane.

PP 2 holds. When the intersecting lines are unbroken euclidean lines or when one of them is improper, the result is obvious. Now assume both lines to be broken. Let their slopes in the lower half-plane be m_1 and m_2, both positive. If $m_1 = m_2$, then their upper as well as their lower parts are parallel, and their only common point in π' is $[m]$. If $m_1 > m_2 > 0$, then let $y = m_1(x - a_1)$ and $y = m_2(x - a_2)$ be the euclidean equations of the lower parts of the two lines. The corresponding upper half-lines are then

$$y = \frac{m_1}{2}(x - a_1) \qquad \text{and} \qquad y = \frac{m_2}{2}(x - a_2).$$

The two euclidean lines carrying the lower half-lines meet at a point whose y-coordinate is

$$\frac{m_1 m_2(a_1 - a_2)}{m_1 - m_2}.$$

According to the sign of this expression we distinguish three cases: (a) $a_1 = a_2$. Then the only intersection of the lines is $[a_1, 0]$. Case (b), $a_1 < a_2$. The intersection lies in the lower half-plane. The upper half-lines do not meet. Case (c), $a_1 > a_2$. The lower half-lines do not meet, but the upper half-lines intersect at exactly one point.

PP 3 *holds*. Obvious.

Lemma 4–2.1 (Fig. 4–11). The property δ_a is equivalent to its following converse: If $AB \parallel A'B'$, $BC \parallel B'C'$, $AC \parallel A'C'$ and $BB' \parallel CC'$, then $AA' \parallel BB'$.

Proof. Let the parallel to BB' through A intersect $A'B'$ in A''. Then δ_a implies $A''C' \parallel AC$, and therefore $A''C' = A'C'$. But $A'C'$ can intersect $A'B'$ only in A', and hence $A' = A''$ and $AA' \parallel BB'$. The proof that the converse implies δ_a will be left for the exercises.

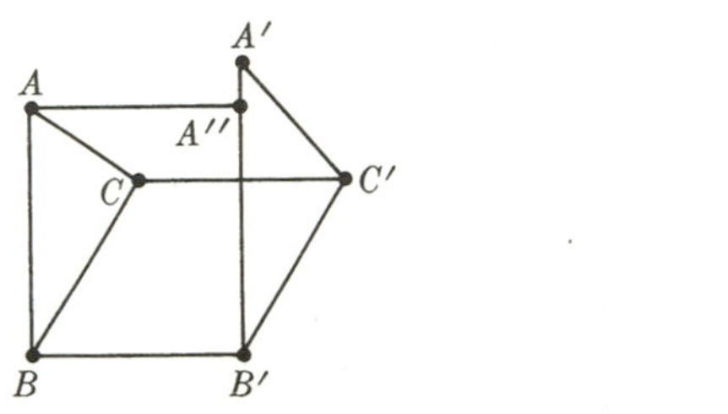

Figure 4–11

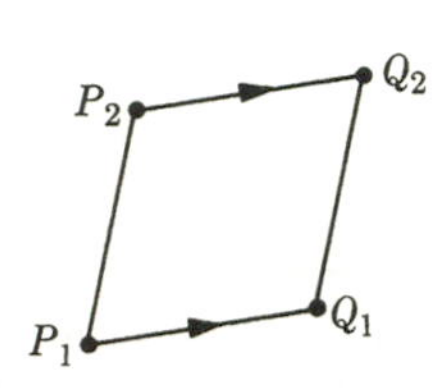

Figure 4–12

Vectors and their group. In Section 2–4, an affine space was derived from a vector space by means of the postulates Af 1 through 4. We will attempt to introduce vectors whose relation to the rudimentary affine plane resembles that of V^2 to A^2. We cannot expect these vectors to be anything but rudimentary analogs of those of V^2.

A *vector* $\overrightarrow{PQ}$ in a rudimentary affine plane will temporarily be defined as an ordered pair of points P and Q. Two vectors $\overrightarrow{P_1Q_1}$ and $\overrightarrow{P_2Q_2}$ are defined to be congruent or, in short, equal if $P_1Q_1 \parallel P_2Q_2$ and $P_1P_2 \parallel Q_1Q_2$ (Fig. 4–12). This definition is formally possible in every affine plane based on PP 1 through 3. However, vector equality will be meaningful only if it is an equivalence relation. Let us, therefore, check whether axioms Ev 1 through 3 of Section 1–3 hold for vector equality. Evidently $\overrightarrow{P_1Q_1} = \overrightarrow{P_2Q_2}$ implies $\overrightarrow{P_2Q_2} = \overrightarrow{P_1Q_1}$, and this settles Ev 1. The requirement $\overrightarrow{PQ} = \overrightarrow{PQ}$, for Ev 2, is trivially satisfied if we extend the definition of parallelism for degenerate cases in a suitable way.

The discussion of Ev 3 is harder. If $\overrightarrow{P_1Q_1} = \overrightarrow{P_2Q_2}$ and $\overrightarrow{P_2Q_2} = \overrightarrow{P_3Q_3}$, the requirement is $\overrightarrow{P_1Q_1} = \overrightarrow{P_3Q_3}$. Suppose first that P_1, Q_1, and P_3 are not collinear (Fig. 4–13). Then we assert $P_1Q_1 \parallel P_3Q_3$ and $P_1P_3 \parallel Q_1Q_3$. The first claim follows from parallelism being an equivalence relation. Moreover, $P_1P_3 \parallel Q_1Q_3$, by Lemma 4–2.1, is equivalent to δ_a. Thus the transitivity of vector equality for all vectors follows from the validity of δ_a in the plane.

Now consider the case in which P_1, Q_1, and P_3 are collinear (Fig. 4–14). Then it is required that $\overrightarrow{P_1Q_1} = \overrightarrow{P_3Q_3}$ through the mediation of a vector $\overrightarrow{P_4Q_4}$

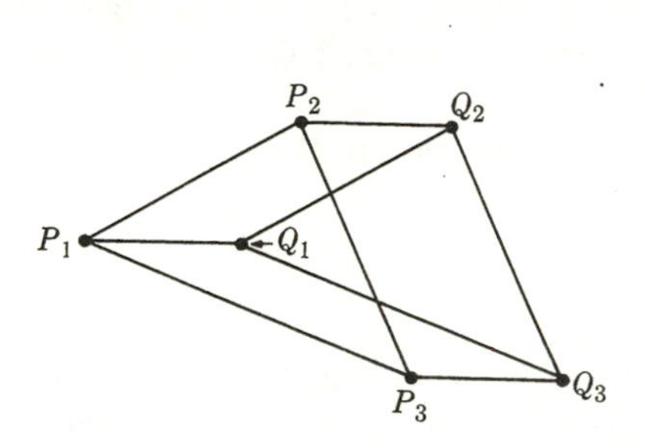

Figure 4-13

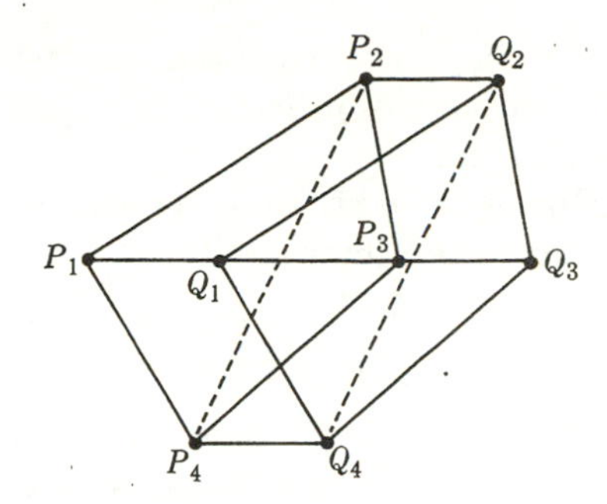

Figure 4-14

for which $\overrightarrow{P_1Q_1} = \overrightarrow{P_4Q_4} = \overrightarrow{P_3Q_3}$, and independently of its choice. This means that $P_1P_4 \parallel Q_1Q_4$ has to imply $P_3P_4 \parallel Q_3Q_4$. This again follows from δ_a, since δ_a for the triangles $P_1P_2P_4$ and $Q_1Q_2Q_4$ states $P_2P_4 \parallel Q_2Q_4$, and then $P_3P_4 \parallel Q_3Q_4$ follows from δ_a for the triangles $P_2P_3P_4$ and $Q_2Q_3Q_4$. This procedure is impossible when P_2, Q_2 and P_4 are collinear, a case which will be left for the exercises.

We have proved the following theorem.

Theorem 4-2.2. In an affine plane in which PP 1 through 3 and δ_a are valid, vector equality is an equivalence relation.

The addition of two vectors $\overrightarrow{P_1P_2}$ and $\overrightarrow{P_2P_3}$ will be defined by $\overrightarrow{P_1P_2} + \overrightarrow{P_2P_3} = \overrightarrow{P_1P_3}$. If the addition of two vectors $\overrightarrow{P_1P_2}$ and $\overrightarrow{Q_2Q_3}$ is required and if $P_2 \neq Q_2$, then there is exactly one point P_3 such that $\overrightarrow{P_2P_3} = \overrightarrow{Q_2Q_3}$, and then again $\overrightarrow{P_1P_2} + \overrightarrow{Q_2Q_3} = \overrightarrow{P_1P_3}$ (Fig. 4-15). The uniqueness of $\overrightarrow{P_2P_3}$ follows from the incidence axioms. An exceptional case would be that in which P_2, Q_2, and Q_3 were collinear, and in this case the procedure of Fig. 4-14 would have to be used, and uniqueness of $\overrightarrow{P_2P_3}$ secured by virtue of δ_a. Whatever vectors $\overrightarrow{P_1'P_2'}$ and $\overrightarrow{P_2'P_3'}$ we choose among all those which are, respectively, equal to $\overrightarrow{P_1P_2}$ and $\overrightarrow{P_2P_3}$, their sum will always be $\overrightarrow{P_1'P_3'}$, in view of δ_a.

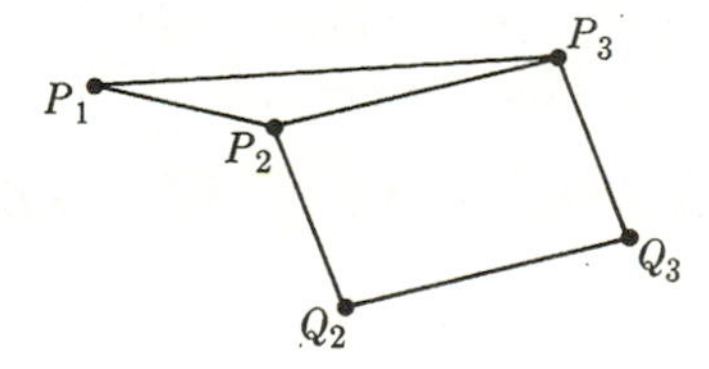

Figure 4-15

The equivalence obtained in Theorem 4-2.2 makes it possible to introduce equivalence classes of equal vectors. From now on, we discard the temporary definition of a vector as an ordered couple of points and define a vector as one of these equivalence classes. This brings us closer to the customary definition of vectors.

We may now compare our results about vectors with the requirements Af 1 through 4 of Section 2–4, and it can easily be verified that these requirements hold.

How is the addition of vectors related to the addition in $(\Sigma, +)$? In Fig. 4–5, we have the vector addition

$$\overrightarrow{[0, 0][0, a]} + \overrightarrow{[0, 0][0, c]} = \overrightarrow{[0, 0][0, a + c]}, \tag{1}$$

for all a and c in Σ. Since every vector on a line through Q_2 is equal to exactly one vector $\overrightarrow{[0, 0][0, a]}$ for some $a \in \Sigma$, Eq. (1) yields an isomorphism between $(\Sigma, +)$ and all the vectors on lines through Q_2.

Theorem 4–2.3. If δ_a holds in an affine plane, the vectors form an abelian group under addition.

Proof. Closure and existence of the null vector are obvious. For a verification of associativity we consider three vectors $\overrightarrow{AB}$, $\overrightarrow{BC}$, and $\overrightarrow{CD}$. The definition of vector addition immediately implies

$$(\overrightarrow{AB} + \overrightarrow{BC}) + \overrightarrow{CD} = \overrightarrow{AD} = \overrightarrow{AB} + (\overrightarrow{BC} + \overrightarrow{CD}).$$

For vectors represented by point couples in more general position, δ_a yields the same result. The additive inverse of $\overrightarrow{AB}$ is $\overrightarrow{BA}$.

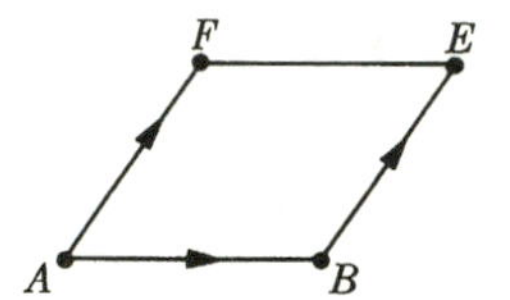

Figure 4–16

For the proof of commutativity,

$$\overrightarrow{AB} + \overrightarrow{CD} = \overrightarrow{CD} + \overrightarrow{AB},$$

we have to distinguish between two cases. First (Fig. 4–16), assume that the well-defined vector $\overrightarrow{BE} = \overrightarrow{CD}$ is not collinear with $\overrightarrow{AB}$. Then $\overrightarrow{AB} + \overrightarrow{BE} = \overrightarrow{AE}$. Now define F by $\overrightarrow{AF} = \overrightarrow{BE}$. This implies $\overrightarrow{AB} + \overrightarrow{CD} = \overrightarrow{AB} + \overrightarrow{BE} = \overrightarrow{AE}$ and

$$\overrightarrow{CD} + \overrightarrow{AB} = \overrightarrow{BE} + \overrightarrow{AB} = \overrightarrow{AF} + \overrightarrow{FE} = \overrightarrow{AE},$$

in view of δ_a, which completes the proof in this case. In the second case $\overrightarrow{AB}$ and $\overrightarrow{BE}$ are collinear. Let G be a point not on AB. Then

$$\begin{aligned}\overrightarrow{CD} + \overrightarrow{AB} &= \overrightarrow{BE} + \overrightarrow{AB} = (\overrightarrow{BG} + \overrightarrow{GE}) + \overrightarrow{AB} = \overrightarrow{BG} + (\overrightarrow{GE} + \overrightarrow{AB}) \\ &= \overrightarrow{BG} + (\overrightarrow{AB} + \overrightarrow{GE}) = (\overrightarrow{BG} + \overrightarrow{AB}) + \overrightarrow{GE} \\ &= (\overrightarrow{AB} + \overrightarrow{BG}) + \overrightarrow{GE} = \overrightarrow{AG} + \overrightarrow{GE} \\ &= \overrightarrow{AE} = \overrightarrow{AB} + \overrightarrow{BE} = \overrightarrow{AB} + \overrightarrow{CD}.\end{aligned}$$

Corollary. If δ_a holds, then in every ternary ring $(\Sigma, *)$ defined over the affine plane, $(\Sigma, +)$ is an abelian group.

Linearity and right distributivity of ternary rings. The property δ_a has further consequences for the ternary rings.

Theorem 4–2.4. If δ_a holds, then every ternary ring coordinatizing the affine plane is linear, $a * b * c = ab + c$.

Proof (Fig. 4–17). We construct the points $[a, a * b * c]$ and $[ab, ab + c]$. Then, by δ_a, their join passes through $[0]$ and hence $a * b * c = ab + c$.

Theorem 4–2.5. If every ternary ring defined in an affine plane is linear, then δ_a is valid throughout this plane.

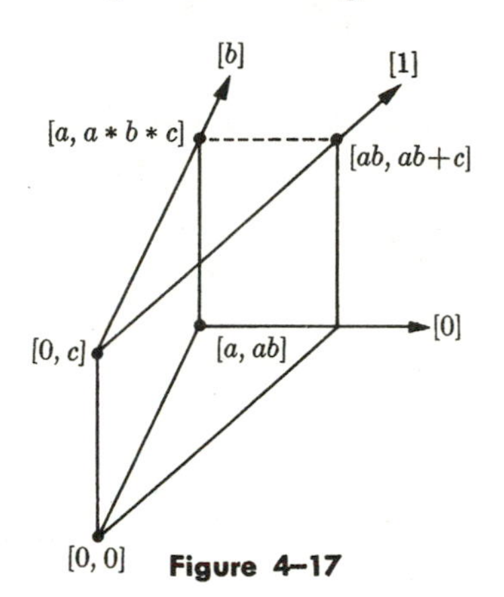

Figure 4–17

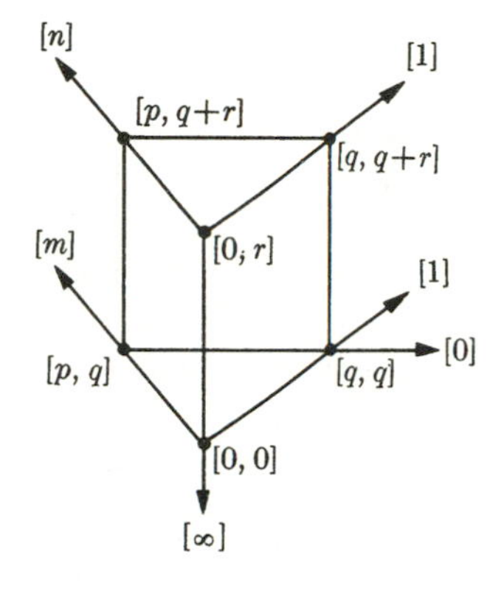

Figure 4–18

Proof. In order to prove the validity of δ_a for two triangles, perspective with improper center C, choose a coordinate system such that $C = [\infty]$ and one of the triangle vertices is at $[0, 0]$. The improper points $[0]$ and $[1]$ are chosen so as to be the intersections of two pairs of corresponding triangle sides. Let the triangle vertices, in terms of the resulting linear ternary ring, be (Fig. 4–18) $[0, 0]$, $[p, q]$, $[q, q]$, $[0, r]$, $[p, q + r]$, and $[q, q + r]$. The remaining sides $[p, q + r][0, r]$ and $[p, q][0, 0]$ have, respectively, the equations $y = xn + r$ and $y = xm$. By substituting the coordinates of the points $[p, q + r]$ and $[p, q]$, respectively, in these equations, we get $q = pn$ and $q = pm$, and consequently $m = n$ because $p \neq 0$. Hence these lines are parallel and δ_a is valid.

Theorem 4–2.6. If δ_a is valid throughout a plane, then every ternary ring defined in this plane is right distributive, $(a + b)c = ac + bc$.

Proof (Fig. 4–19). We know from Theorem 4–2.4 that $ac + bc = a * c * bc$. We have to prove that $a * c * bc = (a + b)c$. We construct these two values. Then δ_a for the triangles $[0, b][a, a + b][0, bc]$ and $[b, b][a + b, a + b][b, bc]$ yields $[a, a + b][0, bc] \parallel [a + b, a + b][b, bc]$. Now we use Lemma 4–2.1 on the

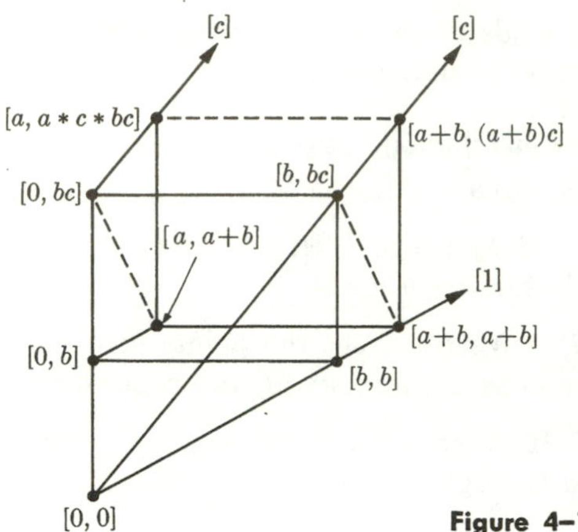

Figure 4–19

triangles $[a, a + b][0, bc][a, a * c * bc]$ and $[a + b, a + b][b, bc][a + b, (a + b)c]$, and obtain that $[a, a * c * bc]$, $[a + b, (a + b)c]$ and $[0]$ are collinear, on the line

$$y = (a + b)c = a * c * bc.$$

This completes the proof.

Veblen-Wedderburn systems

Definition. A *Veblen-Wedderburn system* $(\Sigma, +, \cdot)$ consists of the set Σ (containing at least two elements 0 and 1) closed under the binary operations $(+)$ and $(\cdot)$, subject to the following postulates, for all a, b, c in Σ.

VW 1. $(\Sigma, +)$ is an abelian group with identity element 0.
VW 2. $(\Sigma', \cdot)$ is a loop with identity element 1.
VW 3. $(a + b)c = ac + bc$.
VW 4. $a \cdot 0 = 0$.
VW 5. If $a \neq b$, the equation $xa = xb + c$ has a unique solution x in Σ.

These systems are named after the American mathematicians O. Veblen (1880–1960) and J. H. M. Wedderburn (1884–1948), who introduced them in 1907.

Theorem 4–2.7. If δ_a is valid in an affine plane, every ternary ring of this plane is linear and is a Veblen-Wedderburn system.

Proof. The ternary ring is linear in view of Theorem 4–2.4. VW 1 follows from the corollary of Theorem 4–2.3, VW 2 follows from the definition of the multiplication, and VW 3 is a consequence of Theorem 4–2.6. VW 4 is formula (6) of Section 4–1. As for VW 5, $xa = b + c$ can be written as $x * a * 0 = x * b * c$, and this equation has a unique solution by virtue of TR 5.

In accordance with the last theorem, an affine plane throughout which δ_a holds will be called a *Veblen-Wedderburn plane.*

Central collineations. Our next concern will be with collineations in Veblen-Wedderburn planes. As we saw, the rudimentary affine plane is particularly lacking in collineations. The introduction of δ_a will turn out to have increased the number of collineations in the plane, and the more restrictions we impose on the plane the richer it will become in collineations.

Collineations having a center, like the perspectivities of Section 3–4, will be called *central collineations.* Theorems 3–4.4 and 3–4.5 and their corollaries are valid for central collineations, because their proofs used no properties of perspectivities other than those implied by their being central collineations.

We start with two lemmas.

Lemma 4–2.8. In a projective plane the validity of $\triangle(S, l)$ (that is, the Desargues property with center S and axis l) is necessary and sufficient for the existence of a collineation with axis l and center S mapping an arbitrary given point A onto another given point A' on AS. The points A and A' must be $\neq S$ and not on l.

Proof (Fig. 4–20). Assume that for given axis l and center S all the collineations exist. Let ABC and $A'B'C'$ be two triangles with none of their vertices on l, with S on AA', BB', and CC', and such that $AB \times A'B'$ and $AC \times A'C'$ lie on l. Then there is a collineation α with center S and axis l such that $A\alpha = A'$, implying $B\alpha = B'$, $C\alpha = C'$. Also

$$(BC)\alpha = B\alpha C\alpha = B'C', \qquad \text{and} \qquad (BC \times l)\alpha = B'C' \times l.$$

Therefore $BC \times B'C'$ lies on l, and $\triangle(S, l)$ is valid.

Conversely (Fig. 4–21), assume the validity of $\triangle(S, l)$. Let A and A' be points not on l and $\neq S$ and such that A, A', and S are collinear. Let B be another point, neither on l nor on AA'. We are looking for a collineation α

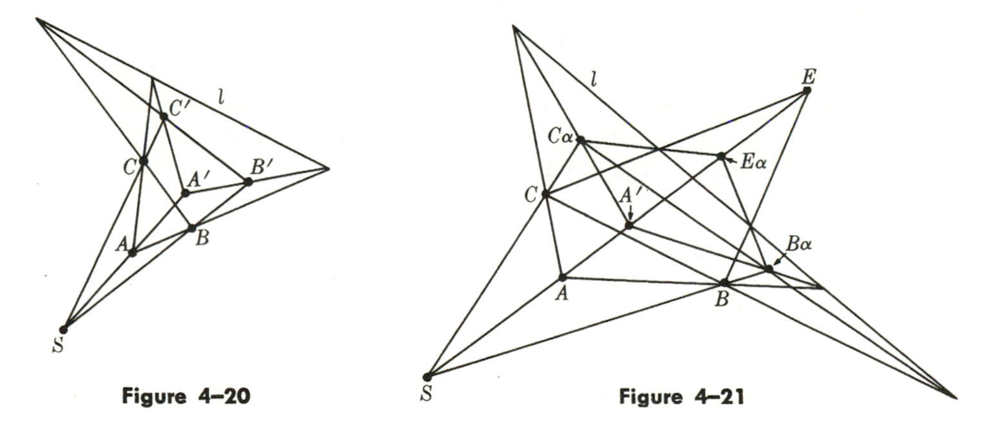

Figure 4–20 **Figure 4–21**

with $A\alpha = A'$, center S and axis l. $B\alpha$ must lie on BS as well as on $(AB \times l)A'$ and is therefore

$$B\alpha = BS \times [(AB \times l)A'].$$

This procedure provides an image for every point not on AA', assuming that the points of l are invariant under α. Now let E be a point on AA', distinct from A and S and not on l. We construct $E\alpha = [(BE \times l)B\alpha] \times AE$. This construction is meaningful only if we can prove that it is independent of the choice of B. Let C replace B and again

$$C\alpha = CS \times [(AC \times l)A'].$$

Is $E\alpha$ on $(CE \times l)C\alpha$? By $\Delta(S, l)$ for the triangles ABC and $A'B\alpha C\alpha$, the point $BC \times B\alpha C\alpha$ lies on l. On applying $\Delta(S, l)$ to the triangles CBE and $C\alpha B\alpha E\alpha$, we find that $C\alpha E\alpha \times CE$ lies on l, which proves that indeed $E\alpha$ on $(CE \times l)C\alpha$. This completes the construction of images for all points of the plane.

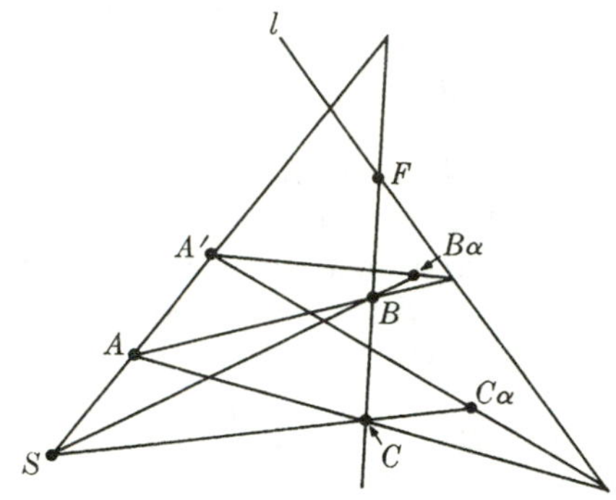

Figure 4–22

We still have to show that α is a collineation, that is, that α takes three arbitrary collinear points into collinear points. Let A, A', l, and S (Fig. 4–22) be defined as above. If B and C are two points not on AA', let $BC \times l = F$. We construct $B\alpha$ and $C\alpha$. $\Delta(S, l)$ for the triangles ABC and $A'B\alpha C\alpha$ yields that $B\alpha$, $C\alpha$, and F are collinear; that is, the collinearity of B, C, and F implies collinearity of their images. Every point on $BF = BC$ goes into a point on $B\alpha F = B\alpha C\alpha$, as asserted. The only case not settled by this proof is that where B or C lies on AA'. In this case we replace A and A' by some other point, not on AA', and its image and proceed in the same way.

The following lemma is important for later applications.

Lemma 4–2.9. If in a projective plane all minor central collineations with axes k and l exist, so do all minor central collineations with every line through $k \times l$ as axis.

Proof (Fig. 4–23). Let j be any line distinct from k and l through $k \times l = S$, and A a point on k, distinct from S. If B is a point of l distinct from S, and

$C = AB \times j$, then, by hypothesis, there exists a minor central collineation α with axis k and center A such that $C = B\alpha$. This implies $l\alpha = j$. By Lemma 4–2.8, the minor Desargues property holds for l as axis and any point of l as center. The collineation α takes these Desargues figures into minor Desargues figures with axis j, and hence the minor Desargues property holds for the axis j. Then, by Lemma 4–2.8, all minor central collineations with j as axis exist, except perhaps those whose center is S. Now let B' be a point on l, distinct from B and S, and $B'A \times j = C'$. Then C' is distinct from C and S. In view of our results so far, there are two minor central collineations with axis j: β with center C such that $B\beta = A$ and β' with center C' such that $A\beta' = B'$. By Theorem 3–4.5, $\beta\beta'$ is a minor central collineation with axis j. Since $B\beta\beta' = B'$, the center of $\beta\beta'$ must be $BB' \times j = S$. Thus $\beta\beta'$ provides the only lacking minor central collineation, and the proof is complete.

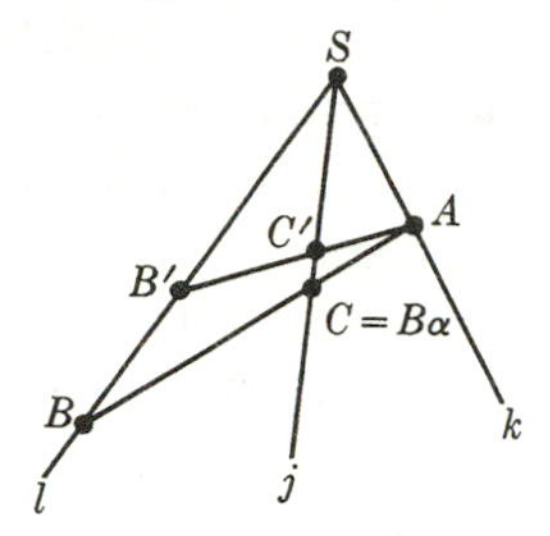

Figure 4–23

Theorem 4–2.10. A plane is a Veblen-Wedderburn plane if and only if for every two points A and A' there exists a minor central collineation α with the improper line l as axis and such that $A\alpha = A'$. The center of α is $AA' \times l$.

Proof. By Lemma 4–2.8 the existence of α is equivalent to the validity of $\Delta(S, l)$ for all points S of the improper line l. This is precisely δ_a throughout the plane, which means that the plane is a Veblen-Wedderburn plane.

The collineations α are exactly the translations of the affine plane. Hence, in view of Theorem 4–2.10, Veblen-Wedderburn planes are also called *translation planes*.

Exercises

1. Prove that δ_a is a consequence of its converse.
2. Prove that δ_a implies the transitivity of vector equality in the case where P_2, Q_2, and P_4 in Fig. 4–14 are collinear.
3. Prove that the postulates Af 1 through 4 of Section 2–4 are satisfied by the vectors of the rudimentary affine plane.
4. Show that Pappus' theorem (p. 154) is not valid throughout Moulton's plane.
5. In Moulton's plane find two points A and A', an axis and a center on AA' but coinciding with neither A nor A', so that no minor central collineation with this center and this axis maps A on A'. Prove the impossibility of finding such a collineation.

6. Show that every Veblen-Wedderburn system can be considered as a ternary ring.
7. Are the following transformations collineations in a Veblen-Wedderburn plane if a and b are distinct from 0 and 1?

   (a) $x \to y - a, y \to x + a$ (b) $x \to ax, y \to yb$
   (c) $x \to x + a, y \to y + b$ (d) $x \to y, y \to x$.

8. Show directly that the "hexagon figure" representing the uniqueness of the additive inverse for every element of Σ follows from δ_a.
9. Draw the "Reidemeister figure" (after the German mathematician K. Reidemeister) which represents the associativity of $(\Sigma, +)$. Show directly that it is a consequence of δ_a.
10. Draw the "Thomsen figure" (after the German mathematician G. Thomsen) representing the commutativity of $(\Sigma, +)$.
11. Prove: If in a parallel three-net the Thomsen figure of Exercise 10 holds, so does the Reidemeister figure of Exercise 9.
12. Prove: If in a parallel three-net the Reidemeister figure of Exercise 9 holds, then all the loops isotopic to $(\Sigma, +)$ are isomorphic.
13. (a) Find the figure representing the fact that $ab = 1$ implies $ba = 1$.
    (b) Prove: If this figure holds throughout the three-net for $(\Sigma', \cdot)$, then $(\Sigma', \cdot)$ is *power associative*, that is, every element of $(\Sigma', \cdot)$ generates an abelian group.
14. (a) Draw the figure representing the commutativity of $(\Sigma', \cdot)$.
    (b) Is its validity a consequence of δ_a?
    (c) Is this "major Thomsen figure" equivalent to the Thomsen figure of Exercise 10?

4–3 MOUFANG PLANES

The minor Desargues property and left distributivity. We shall impose a new restriction on the projective plane. In the Veblen-Wedderburn plane we required the validity of the minor Desargues property δ only for the case δ_a in which the axis, that is, the line joining the intersections of the corresponding pairs of triangle sides, is the improper line of the plane. In this section we will assume the general validity of δ in the plane, including also the case δ_a. The result will be a specialization of the Veblen-Wedderburn systems which coordinatize the plane and a considerable increase in the number of collineations.

A projective plane in which δ holds is called a *Moufang plane*, after the German mathematician Ruth Moufang, who first discussed these planes in 1933.

At first we will show that δ implies other special forms of the Desargues property.

Lemma 4–3.1. The converse of δ is equivalent to δ.

Proof. This is a generalization of Lemma 4–2.1, with the improper line replaced by an arbitrary line.

Lemma 4–3.2. If the minor Desargues property holds, so does the special Desargues property, in which one pair of corresponding triangle sides intersect on the join of two corresponding triangle vertices.

Proof (Fig. 4–24). Let ABC and $A'B'C'$ be the triangles of the Desargues figure and let $AB \times A'B' = F$ lie on CC'. The assertion is the collinearity of D, E, and F. By Lemma 4–3.1, we may apply the converse of δ to the triangles $BB'D$ and $AA'E$. For the corresponding sides we have $BB' \times AA' = S$, $BD \times AE = C$, $B'D \times A'E = C'$. Indeed S, C, and C' are collinear, and $AB \times A'B' = F$ lies on their join. The converse of δ states that AB, $A'B'$, and DE meet in one point, or, in other words, that DE is collinear with F, as required.

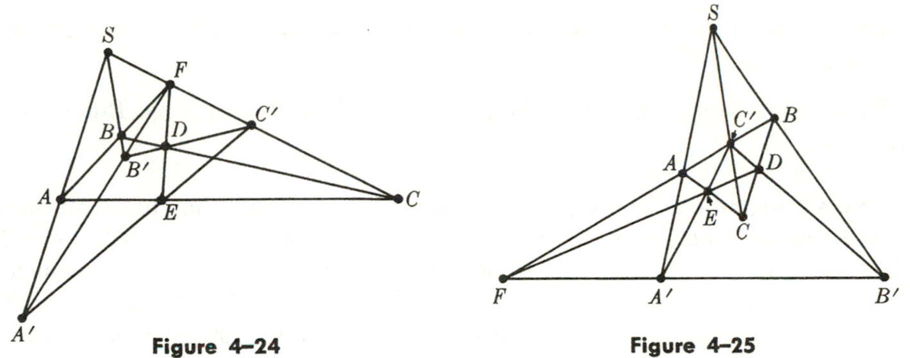

Figure 4–24

Figure 4–25

Lemma 4–3.3. The property δ implies the special form of the Desargues property in which one vertex of one triangle lies on a side of the other triangle.

Proof (Fig. 4–25). The vertex C' lies on the side AB. The collinearity of D, E, and $F = AB \times A'B'$ is required. We use δ for the triangles ECD and $A'SB'$. The lines $A'E, CS$, and DB' indeed intersect in C', and C' lies on the join of $EC \times A'S = A$ and $CD \times SB' = B$. Hence A, B, and $ED \times A'B' = F'$ are collinear. But this means $F = F'$, and the collinearity of D, E, and F follows.

Theorem 4–3.4. In a Moufang plane every ternary ring is left distributive, that is, $a(b + c) = ab + ac$.

Proof (Fig. 4–26). Since in the plane δ_a is valid as a special case of δ, the ternary ring is linear, and the assertion may be written

$$a(1 * b * c) = a * b * ac. \tag{1}$$

We construct

$$[a, a(1 * b * c)] = P$$

and

$$[a, a * b * ac] = Q.$$

We have to prove that

$$P = Q.$$

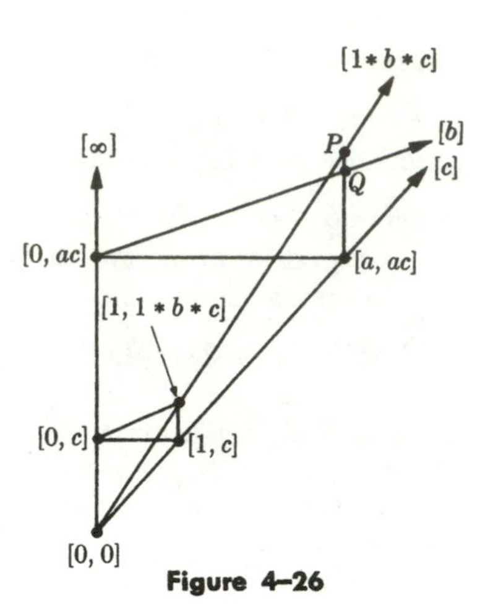

Figure 4-26

But this is precisely the statement of the Desargues property for the triangles $[0, c][1, c][1, 1 * b * c]$ and $[0, ac][a, ac]Q$, the center $[0, 0]$ and axis $[0][\infty]$. By Lemma 4-3.2, this Desargues property is valid in view of the triangle sides $[1, c][1 * b * c]$ and $[a, ac]Q$ intersecting on the join of vertices $[0, c]$ and $[0, ac]$.

Definition. A Veblen-Wedderburn system satisfying the left distributive law is called a *division ring*.

Thus every ternary ring of a Moufang plane is a division ring. But we can say more.

Inverse properties. Every nonzero element b of a ternary ring has a multiplicative *right inverse* b^R, such that $bb^R = 1$, and a *left inverse* b^L, such that $b^Lb = 1$. The *right inverse property* is said to hold if, for all nonzero b, b^R and b^L coincide (now called the *inverse* b^{-1}) and if $(ab)b^{-1} = a$ for all a. Analogously, the *left inverse property* requires the existence of an inverse for all nonzero b and moreover $b^{-1}(ba) = a$.

Theorem 4-3.5. Every ternary ring coordinatizing a Moufang plane has the left inverse property and the right inverse property.

A division ring in which the two inverse properties hold is called an *alternative division ring*. The reason for this name will become clear later.

Proof. (i) The right inverse property (Fig. 4-27). As a temporary notation we write b^{-1} instead of b^L. It will appear later that this is justified. After constructing the points $[a, a]$ and $[ab, (ab)b^{-1}]$, we have to prove that their join passes through $[0]$. Let $[ab, (ab)b^{-1}][b^{-1}, b^{-1}] \times [0, 0][0] = S$. Then Lemma 4-3.3 applies to the triangles

$$[b^{-1}, b^{-1}][1, b^{-1}][ab, (ab)b^{-1}] \qquad \text{and} \qquad [b^{-1}, 1][1, 1][ab, ab],$$

because the vertex $[b^{-1}, b^{-1}]$ lies on the side $[1, 1][ab, ab]$. Consequently $[b^{-1}, 1][ab, ab]$ passes through S. Now we use the same lemma on the triangles

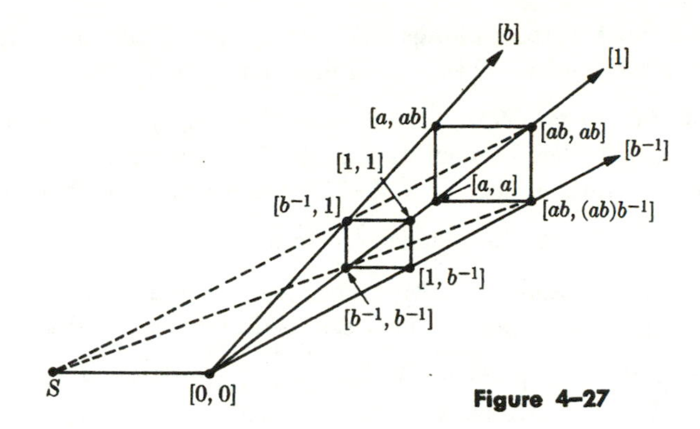

Figure 4–27

$[b^{-1}, 1][a, ab][ab, ab]$ and $[b^{-1}, b^{-1}][a, a][ab, (ab)b^{-1}]$, in which the vertex $[ab, ab]$ lies on the side $[b^{-1}, b^{-1}][a, a]$. We obtain the collinearity of

$$[ab, (ab)b^{-1}], \quad [a, a], \quad \text{and} \quad [0],$$

implying $(ab)b^{-1} = a$.

In order to justify the use of b^{-1} for the left inverse of b, we put $a = 1$ in $(ab)b^L = a$, and get $bb^L = 1$. This makes b^L also the right inverse.

(ii) The left inverse property (Fig. 4–28). We construct $[a^{-1}, a^{-1}(ab)]$ and $[1, b]$ and have to prove that their join passes through $[0]$. In view of Lemma 4–3.3, in the triangles $[a, 1][1, 1][a, ab]$ and $[1, a^{-1}][a^{-1}, a^{-1}][1, b]$, the sides $[1, 1][a, ab]$ and $[a^{-1}, a^{-1}][1, b]$ intersect at a point R on the improper line. Then, by the same lemma, applied to the triangles

$$[1, 1][1, ab][a, ab] \qquad \text{and} \qquad [a^{-1}, a^{-1}][a^{-1}, a^{-1}(ab)][1, b],$$

the side $[a^{-1}, a^{-1}(ab)][1, b]$ must pass through $[1, ab][a, ab] \times [\infty]R = [0]$. This completes the proof.

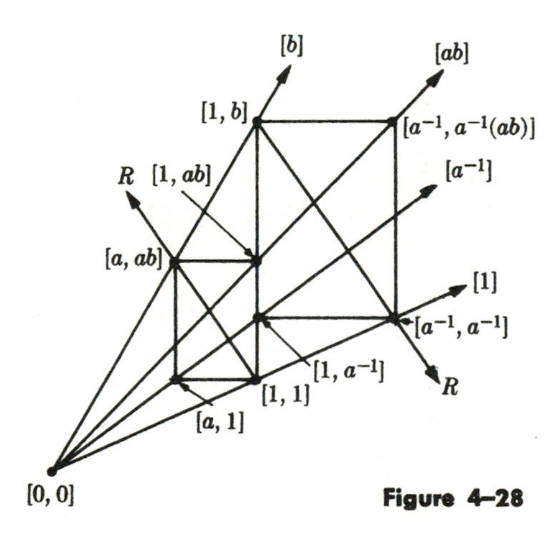

Figure 4–28

Collineations of Moufang planes. Again we ask about the collineations of the Moufang plane, and we obtain the following theorem.

Theorem 4-3.6. A projective plane is a Moufang plane if and only if, for every given line l and two points A and A' not on l, there is a minor central collineation with l as axis and $AA' \times l$ as center, which maps A on A'.

Proof. This is a direct consequence of Lemma 4-2.8.

Theorem 4-3.7. A plane is a Moufang plane if and only if each of its coordinatizing ternary rings is linear and is an alternative division ring.

Proof. Every Moufang plane is a Veblen-Wedderburn plane, and the theorems about Veblen-Wedderburn planes together with Theorems 4-3.4 and 4-3.5 yield the required result that in a Moufang plane every ternary ring is linear and is an alternative division ring.

In order to prove the converse we will show instead that, in a plane coordinatized by an alternative division ring, all minor central collineations exist. By Theorem 4-3.6 this amounts to the same as the converse assertion.

The alternative division ring is a Veblen-Wedderburn system, and by Theorem 4-2.10 all the minor central collineations with the improper line as axis exist. We will exhibit a collineation β which maps the improper line on the y-axis (a motivation for the choice of β will appear in Exercise 3). By doing so, we prove the existence of all minor central collineations with the y-axis as axis. We define the points

$$[a, b]\beta = [a^{-1}, a^{-1}b], \qquad \text{for } a \neq 0,$$
$$[0, b]\beta = [b], \qquad [b]\beta = [0, b], \qquad [\infty]\beta = [\infty].$$

The lines $y = xm + n$ are mapped on $y = xn + m$, $x = c$ on $x = c^{-1}$, when $c \neq 0$, the y-axis on the improper line, and the improper line on the y-axis. In order to verify that β is a collineation, we have to show that incidences are preserved. We will do this only for the lines $y = xm + n$; the other cases can be checked easily. A point $[a, b]$, with $a \neq 0$, lies on such a line if $b = am + n$. It is claimed that the image point $[a, b]\beta = [a^{-1}, a^{-1}b]$ lies on the image line $y = xn + m$, or that $a^{-1}(am + n) = a^{-1}n + m$. But this follows readily from the left distributivity and the left inverse property. For a point $[0, b]$ we have to prove that its image $[b]$ lies on the image line $y = xn + m$, if $b = 0m + n$. This means $b = n$, and $[n]$ is the improper point of the image line. Finally, if a point $[b]$ lies on $y = xm + n$, then $m = b$, and the image point $[0, b] = [0, m]$ indeed lies on $y = xn + m$.

We repeat the procedure, using another collineation α which maps the y-axis on the x-axis. We define α as follows (cf. Exercise 3). The points:

$$[a, b]\alpha = [b, a],$$
$$[b]\alpha = [b^{-1}] \qquad \text{for } b \neq 0,$$
$$[0]\alpha = [\infty], \qquad [\infty]\alpha = [0].$$

The line $x = c$ becomes $y = c$, $y = c$ becomes $x = c$, and $y = xm + n$ goes into $y = xm^{-1} - nm^{-1}$, for $m \neq 0$. The improper line is invariant. Then α preserves incidences as a consequence of the right distributivity and the right inverse property. Thus we have established the existence of all minor central collineations having the x-axis, the y-axis, or the improper line as axis. By Lemma 4-2.9 this implies the existence of all minor central collineations whose axes are lines through $[\infty]$, $[0, 0]$ or $[0]$. Furthermore, every line of the plane passes through the intersection of two of these lines, and thus all minor central collineations exist for every line as axis. This completes the proof.

Alternative laws. A few properties of alternative division rings will be established.

Theorem 4-3.8. The right alternative law, $(ba)a = ba^2$, and the left alternative law, $a(ab) = a^2b$, hold in every alternative division ring.

Proof. We will prove the right alternative law only. The proof for the left alternative law is analogous. For $a = 0$, 1, or -1, the theorem is trivial, and we exclude these cases. We have for any x, by left distributivity,

$$(xa)[a^{-1} - (a + 1)^{-1}] = x - (xa)(a + 1)^{-1}.$$

Then the right inverse property yields

$$\{(xa)[a^{-1} - (a + 1)^{-1}]\}(a + 1) = x(a + 1) - xa = x.$$

By repeated use of the right inverse property, this yields

$$xa = [x(a + 1)^{-1}][a^{-1} - (a + 1)^{-1}]^{-1}. \tag{2}$$

In particular, for $x = a + 1$, Eq. (2) becomes

$$(a + 1)a = [a^{-1} - (a + 1)^{-1}]^{-1}. \tag{3}$$

We substitute Eq. (3) into Eq. (2) and get

$$xa = [x(a + 1)^{-1}][(a + 1)a].$$

Now we set $x = b(a + 1)$. Then $[b(a + 1)]a = b[(a + 1)a]$, $(ba + b)a = b(a^2 + a)$, $(ba)a + ba = ba^2 + ba$, and finally $(ba)a = ba^2$.

It can be shown that in the axioms of an alternative division ring whose characteristic is not 2 the two inverse properties may be replaced by the two alternative laws, and that then the inverse properties follow from the axioms. The proof is rather lengthy and will not be given here. It can be found in the work of G. Pickert, quoted in the references at the end of this chapter.

Are the alternative division rings indeed special Veblen-Wedderburn systems, or is every Veblen-Wedderburn system an alternative division ring? In other words, does the validity of δ_a imply the general validity of δ? Do all minor central collineations exist for every line as axis if they exist for the improper line?

The answer is provided by the following example of a Veblen-Wedderburn system that does not satisfy the alternative laws and therefore, by Theorem 4–3.8, cannot be an alternative division ring.

We construct an algebra of rank 4 over the reals, and use the multiplication table of the quaternion algebra in Section 3–8, with the modifications that $ij = 2k$ (instead of $ij = k$) and $ji = -2k$ (instead of $ji = -k$). The axioms VW 1, 3, and 4 follow immediately from the axioms Al. The verification of VW 2 and 5 requires some calculation and will be left for the exercises. Now $(ij)j = 2kj = -2i$. On the other hand, $ij^2 = i(-1) = -i \neq -2i$.

The alternative laws are special cases of associativity. This raises the question whether associativity is already implied by them or, in other words, whether every alternative division ring is a skew field. A counterexample is provided by the Cayley algebra of Section 3–9, in which the alternative laws hold, but not the full associative law. We should, therefore, expect a further restriction of the projective plane by imposing stronger Desargues properties and thereby requiring the existence of more collineations, so that the coordinates form a skew field. This, indeed, will be done in the next section.

A good deal more is known about the subject of this section. For instance, if every ternary ring of a projective plane is a division ring, then the right inverse property holds in each of them. Every division ring with only one inverse property is already an alternative division ring. These results will not be proved here, and the reader is referred to the list of books at the end of the chapter. The most interesting result among those works is that by R. H. Bruck, E. Kleinfeld, and the Russian mathematician L. A. Skornyakov, which states that an alternative division ring is either a skew field or "a Cayley-Dickson division algebra over its center." Without going into detail concerning the quotation, it suffices to say that the significance of this result lies in the fact that it determines which are the nonassociative alternative division rings, and shows that only relatively few alternative division rings are nonassociative. In other words, a Moufang plane has almost as many collineations as the desarguesian plane, which will be treated in the next section.

Exercises

1. Carry out the proof of Lemma 4–3.1.
2. Complete the proof of Theorem 4–3.7 by proving that the collineation α preserves incidences.
3. If the plane in the proof of Theorem 4–3.7 were P^2, how would the projectivities α and β be expressed in homogeneous coordinates?

4. (a) Draw the figures for the left inverse property and for the left alternative law.
   (b) What conclusions can you draw from their mutual resemblance?
   (c) Is the same true for the right inverse property and the right alternative law?
5. In order to show the mutual independence of the left and right inverse properties, find a finite loop having either property without satisfying the other.
6. Show that the left and right inverse properties imply

$$(ab)^{-1} = b^{-1}a^{-1}.$$

7. The identity $(ab)a^R = b$ is called the "crossed-inverse property." Does this property imply
   (a) the uniqueness of the inverses?
   (b) $a^L(ba) = b$?
   (c) $(ab)^R = a^R b^R$?
   (d) the right inverse property?
8. (a) Draw the figure for the crossed-inverse property (Exercise 7).
   (b) Compare it with the figure for commutativity. Does either the commutativity or the crossed-inverse property imply the other?
9. In a group, a subset H is said to be generated by a set S of elements if each of the elements of H can be written as a product of elements of S and their inverses.
   (a) Give a meaningful definition applying to a loop instead of a group.
   (b) Show that the same definition may be used for loops as for groups if the loop has the left or the right inverse property.
10. In the example of a Veblen-Wedderburn system not satisfying the alternative laws, verify the validity of VW 2 and 5.
11. Why does the algebra of Exercise 10 not violate the Hurwitz-Albert theorem?
12. In a Moufang plane a collineation with axis $y = x$ and center $[1, 1]$ maps the point $[\infty]$ on $[1, 0]$. Find the images of
    (a) $[0, 1]$ (b) $[0]$ (c) the line $x = 0$.

4-4 DESARGUESIAN PLANES

Desargues and associativity. We require the validity of the full Desargues property, Δ, as an axiom for projective planes, and call them *desarguesian.* Since δ_a and δ are but special cases of Δ, all the ternary rings coordinatizing desarguesian planes are linear and are also alternative division rings. Moreover, the following theorem can be stated.

Theorem 4–4.1. The multiplication of the coordinates in a desarguesian plane is associative.

Corollary. All ternary rings of a desarguesian plane are linear and skew fields.

Proof (Fig. 4–29). We construct $[a, a(bc)]$ and $[ab, (ab)c]$, and have to prove that their join passes through $[0]$. We apply Δ to the triangles $[1, b][b, b]$ $[b, bc]$ and $[a, ab][ab, ab][ab, (ab)c]$ and obtain

$$[1, b][b, bc] \parallel [a, ab][ab, (ab)c].$$

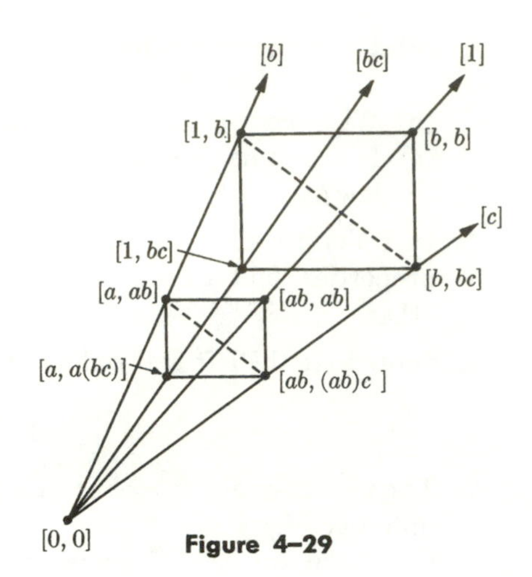

Figure 4–29

Then Δ for the triangles $[1, b][1, bc][b, bc]$ and $[a, ab][a, a(bc)][ab, (ab)c]$ yields

$$[a, a(bc)][ab, (ab)c] \parallel [1, bc][b, bc],$$

which is equivalent to the assertion.

Collineations in desarguesian planes.

Theorem 4–4.2. In a desarguesian plane three distinct points A, A', and S and a line l are given such that A and A' are collinear with S, but neither of them lies on l. Then there exists a collineation α with axis l and center S such that $A\alpha = A'$.

Proof. Lemma 4–2.8.

Theorem 4–4.3. If one ternary ring of a projective plane is linear and a skew field, then the plane is desarguesian.

Proof. The theorem could be proved by a straightforward calculation of the type performed in elementary analytic geometry, the only restriction being that steps involving multiplicative commutativity have to be avoided. However, we shall give a different proof, which sheds light on the transformation-theory background of the theorem. We will show the existence of all the central collineations, as mentioned in Theorem 4–4.2, by describing one special type of major central collineation and by then reducing to this case the proof for the existence of every arbitrary central collineation.

The following is a major central collineation, say ϕ_a, for every nonzero value of a (for the motivation of this choice cf. Exercise 2). For points:

$$[c, d]\phi_a = [ac, ad], \qquad [d]\phi_a = [d], \qquad [\infty]\phi_a = [\infty],$$

and for lines:

$$(x = c) \to (x = ac), \qquad (y = xm + n) \to (y = xm + an).$$

The improper line is invariant under ϕ_a. We check the invariance of incidence only for the lines $y = xm + n$ and the points $[c, d]$. The remaining cases are easy. Suppose that $[c, d]$ lies on $y = xm + n$. Then

$$\phi_a\colon [c, d] = [c, cm + n] \to [ac, a(cm + n)].$$

We have to show that $a(cm + n) = (ac)m + an$. But this is true in a skew field. The improper line is preserved pointwise, and thus it is the axis. The point $[0, 0]$ is the center because every line through it, with an equation $y = xm$ or $x = 0$, is preserved. There are no other central collineations with this axis and this center, because every such collineation would necessarily have $[d] \to [d]$ and $[\infty] \to [\infty]$, and if $[c, d] \to [e, f]$, then $[c, d]$, $[e, f]$, and $[0, 0]$ have to be collinear. This implies that for some m, $d = cm$ and $f = em$, and hence $fd^{-1} = emm^{-1}c^{-1} = ec^{-1} = a$, say, and $f = ad$, $e = ac$. Thus $[c, d] \to [ac, ad]$ for some nonzero a. The rules for the lines follow easily.

Now we have to show the existence of all major collineations for arbitrarily chosen axis and center. Let A and B be two points on the axis, let C be the center, and let A, B, C not all be collinear. Not all three points $[0]$, $[\infty]$, and $[0, 0]$ can be on any one of the lines AB, BC, and CA. Suppose $[0, 0]$ is not on AB. As a consequence of the arguments of Theorem 4-3.7, there is a minor central collineation α such that $C\alpha = [0, 0]$, with AB as axis, and therefore $A\alpha = A$, $B\alpha = B$. For the same reason there is a minor central collineation β with axis $[0, 0]A$, mapping B on $[\infty]$, and preserving $[0, 0]$ and A. Finally there exists a minor central collineation γ with axis $[0, 0]\,[\infty]$ taking A into $[0]$ and preserving $[0, 0]$ and $[\infty]$. If P and Q are arbitrary distinct points, collinear with C, but not coinciding with C and not on AB, then $P\alpha\beta\gamma$ and $Q\alpha\beta\gamma$ are proper and collinear with $[0, 0]$, and there is a ϕ_a which maps $P\alpha\beta\gamma$ on $Q\alpha\beta\gamma$. Thus $\alpha\beta\gamma\phi_a(\alpha\beta\gamma)^{-1}$ is the collineation with center C and axis AB, mapping P on Q. If $[0, 0]$ is on AB, the proof must be slightly modified.

Thus all major central collineations exist for any given axis and center. All the minor central collineations exist for every center and axis in the plane, in view of Theorem 4-3.6. Thus, by Lemma 4-2.8, the Desargues property holds, and the plane is desarguesian. This completes the proof.

Theorem 4-4.4. In a desarguesian plane there exists a collineation mapping any given quadrangle on any given image quadrangle.

Proof. Using $\alpha\beta\gamma$ of the last proof, we get a collineation θ for three of the vertices and three image vertices. Choose the three image vertices as the reference points Q_0, Q_1, Q_2 of a ternary ring. Let the fourth desired image point have the coordinates $[a, b]$ in the ternary ring, while the image of the fourth vertex under θ is $[a', b']$. Both points are proper and a, b, a', and b' are nonzero, because in a quadrangle no three vertices can be collinear. There is a collineation $\pi_{a,b}$ taking the point $[1, 1]$ into $[a, b]$. This collineation operates on the points:

$$[c, d]\pi_{a,b} = [ca, db], \qquad [d]\pi_{a,b} = [a^{-1}db], \qquad [\infty]\pi_{a,b} = [\infty],$$

and on the lines:

$$(y = xm + n) \to (y = xa^{-1}mb + nb), \qquad (x = c) \to (x = ca),$$

and the improper line is preserved. (For motivation cf. Exercise 4.) Indeed $[1, 1]\pi_{a,b} = [a, b]$, and the remaining reference points are preserved. Again, we check only the incidence invariance for points $[c, d]$ and lines $y = xm + n$. We have $[c, cm + n] \to [ca, (cm + n)b]$ and

$$(ca)(a^{-1}mb) + nb = (cm + n)b,$$

which establishes the incidence. Now, $\pi_{a',b'}^{-1}\pi_{a,b}$ is a collineation which maps $[a', b']$ on $[a, b]$ and preserves the points Q_0, Q_1, and Q_2. Thus $\theta\pi_{a',b'}^{-1}\pi_{a,b}$ is the required collineation.

Projective three-space. The last theorem conforms with Theorem 3-2.5. This shows that one of the fundamental properties of the projective plane over the reals is satisfied already in projective planes over an arbitrary skew field.

For a further investigation of the role of the Desargues property, we digress from the treatment of projective planes and briefly discuss (rudimentary) projective three-spaces. We define them by the following axioms, again using points and lines as undefined concepts.

PS 1. For every two points there is exactly one line joining them.

PS 2. There is a line and a point not on it.

PS 3. Each line has at least three points.

PS 4. (Fig. 4-30). If A, B, and C are noncollinear points and D a point on BC, with $B \neq D \neq C$, and if E is a point on AC with $A \neq E \neq C$, then there is a point F on AB such that D, E, and F are collinear.

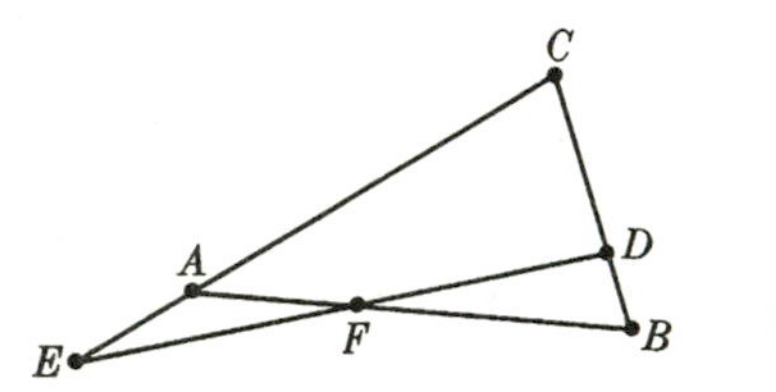

Figure 4-30

Definition. If P is a point not on the line QR, then the set of all points on all the lines PS, with S on QR, is called a *plane*. Points in the same plane are said to be *coplanar*.

PS 5. For every three noncollinear points there is a fourth point such that not all four are coplanar.

PS 6. Every two planes have a common line.

A few theorems follow which can be derived directly from these axioms.

Theorem 4–4.5. In every plane of a projective three-space, the axioms PP 1, 2, 3 hold.

Proof. Axioms PP 1 and PP 3 follow from PS 1, PS 2, PS 3 and PS 4. For the proof of PP 2, suppose the plane π to be generated by the point R and the line l (Fig. 4–31). Let $j = AB$ and $k = CD$ be the two lines whose intersection is to be proved, with A, B, C, and D arbitrary points of π. Furthermore let $RA \times l = P_1$, $RB \times l = P_2$, $RC \times l = P_3$, and $RD \times l = P_4$. We use PS 4 repeatedly on different triangles. The triangle RP_3P_4 yields the existence of $k \times l = K$, from the triangle RP_1P_2 we obtain $j \times l$, from RP_1P_3 we have $j \times RP_3$, and finally the triangle P_3CK yields the existence of the point $k \times j$. This proves the theorem.

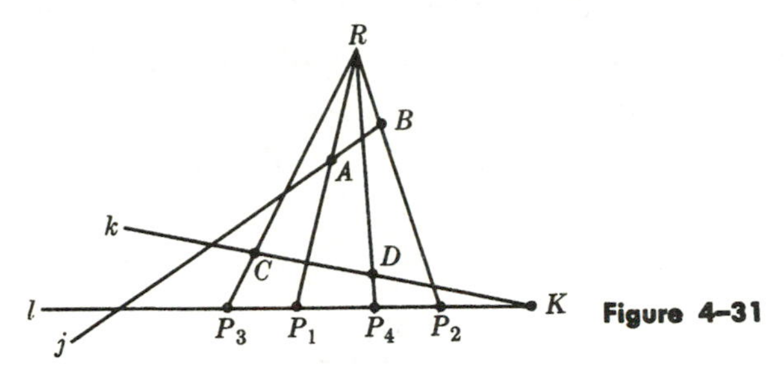

Figure 4–31

Theorem 4–4.6. If a plane π is generated by a point A and a line BC, then B and AC generate π, and so do C and AB.

Proof. Let P be any point of π. Then, by Theorem 4–4.5, $BP \times AC$ exists, and by the definition of the plane, P belongs to the plane defined by B and AC. The remaining case is treated analogously.

Corollary 1. Three noncollinear points determine a unique plane which contains them. Any three noncollinear points of a plane π determine π.

Corollary 2. Two intersecting lines determine a unique plane which contains them.

Corollary 3. If two points belong to a plane π, then all points of their join belong to π.

Theorem 4–4.7. A plane π and a line l, not in π, have a common point.

Proof. Let π' be a plane containing l. Then, by PS 6, π and π' have a common line, say k. Now, by Theorem 4–4.5, k and l have a common point, which belongs to π as well as to π'. This proves the theorem.

Now it is obvious that a projective three-space consists of all points on all lines joining points of a plane to a fixed point outside the plane (whose existence is assured by PS 5).

Desargues in projective three-space. Now we are ready to state *and to prove* the Desargues property for projective three-space.

Theorem 4–4.8 (Desargues). In projective three-space, let ABC and $A'B'C'$ be two triangles such that AA', BB', and CC' are distinct and meet at a point S. Then $AB \times A'B'$, $BC \times B'C'$, and $AC \times A'C'$ exist and are collinear.

Proof. Let the plane ABC be called ϕ, and let the plane $A'B'C'$ be ϕ'. We distinguish between two cases. First case: $\phi \neq \phi'$. By Corollary 2 of Theorem 4–4.6, the lines SC and SB determine a plane π_1, containing S, C, B, C', and B'. Also, S, A, C, A', and C' lie in a plane π_2, and S, B, A, B', and A' in a plane π_3. By Theorem 4–4.5, in π_1, the point $BC \times B'C'$ exists, in π_2, the point $AC \times A'C'$ exists, and in π_3, the point $AB \times A'B'$ exists. Now, by PS 6, ϕ and ϕ' intersect in a line, say l. Since $BC \in \phi$ and $B'C' \in \phi'$, we have $BC \times B'C' \in l$, and likewise, $AC \times A'C' \in l$ and $AB \times A'B' \in l$. This completes the first case.

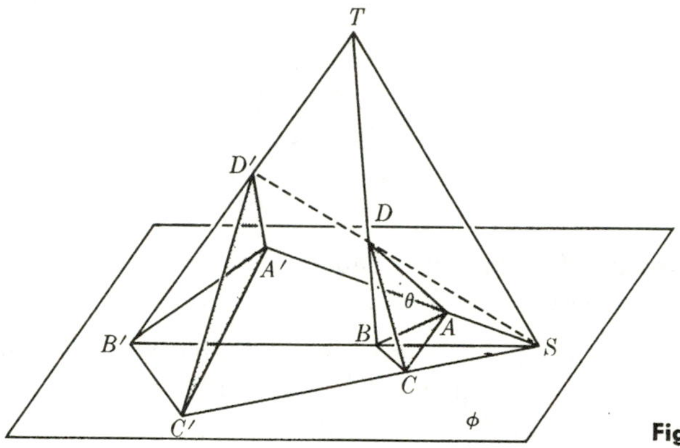

Figure 4–32

Now assume $\phi = \phi'$ (Fig. 4–32). Then our statement is precisely the plane Desargues property Δ. Let T be a point outside $\phi = \phi'$, and let $\theta \neq \phi$ be a plane through AC, but not through T. Then BT pierces θ in a point, say D. The line SD is coplanar with $B'T$, because both lie in the plane STB'. Let $SD \times B'T = D'$. Now the triangles ACD and $A'C'D'$ give rise to a figure of the type treated in the first case, because AA', CC' and DD' meet in S. Hence $AC \times A'C'$, $AD \times A'D'$, and $CD \times C'D'$ lie on a line, say k, which cannot contain T. Let ψ be the plane generated by k and T. The planes ABD and $A'B'D'$ meet in a line containing $AD \times A'D'$ and T, and this line therefore belongs to ψ. Also the planes BCD and $B'C'D'$ meet in a line containing $CD \times C'D'$ and T, and this line, therefore, lies in ψ. Thus the points $AB \times A'B'$, $BC \times B'C'$, and $AC \times A'C'$ all belong to ψ as well as to ϕ. Hence they are all on the common line of ψ and ϕ. This completes the proof. (To summarize the last steps, we could say that $\psi \cap \phi$ is the image of k under a projection from T on ϕ.)

The resulting situation is most peculiar. As we saw, it was impossible to prove Δ by means of PP 1 through 3. On the contrary, in Section 4–2 we were able to set up a counter-example in which PP 1 through 3 were valid but not Δ. Now it turns out that PS 1 through 6 do imply Δ. Thus the Desargues property in the plane can be proved if the plane is regarded as part of a projective three-space, but not in the plane by itself. We will express this result in a slightly different form.

Theorem 4–4.9. A projective plane can be imbedded into a projective three-space only if it is desarguesian.

Desargues properties in affine planes. We will now attack a different problem, namely that of coordinatizing the *affine* plane by means of *affine* specializations of the Desargues property. To a certain extent this was done in Section 4–2. There we were able to show that the validity of δ_a makes the coordinatizing ternary ring linear and a Veblen-Wedderburn system (Theorem 4–2.7). What affine Desargues property has to be postulated in addition to δ_a for obtaining left distributivity and multiplicative associativity? When we look at Fig. 4–26, which served for proving the left distributivity, and at Fig. 4–29, for multiplicative associativity, we see that in both cases an affinely specialized $\Delta(S, l)$ was used, namely one in which l is improper whereas S is proper. In Section 3–4 we called this property the affine Desargues property. We will designate it by Δ_a. We have, therefore, the following.

Theorem 4–4.10. If in an affine plane δ_a and Δ_a hold, the coordinatizing ternary ring is linear and a skew field.

For later application we state the affine equivalence of Δ_a and its converse, in analogy with Lemmas 4–2.1 and 4–3.1.

Lemma 4–4.11. The property Δ_a holds if and only if the following is correct: If in two distinct triangles ABC and $A'B'C'$ corresponding sides are parallel, and if $AA' \times BB' = S$, then S, C, and C' are collinear.

The proof will be left as an exercise.

EXERCISES

1. Complete the proof of Theorem 4–4.3.
2. If the proof of Theorem 4–4.3 would have to be performed in P^2, how would ϕ_a be expressed in homogeneous coordinates?
3. Complete the proof of Theorem 4–4.4.

4. If the proof of Theorem 4-4.4 were performed in P^2, what would $\pi_{a,b}$ be in homogeneous coordinates?
5. Complete the proof of Theorem 4-4.6.
6. Prove: In a desarguesian plane all the coordinatizing skew fields are isomorphic.
7. In a desarguesian plane which is not coordinatized by a field, prove that there is at least one proper line whose equation cannot be written in the form $y = mx + n$, nor in the form $x = a$.
8. In a desarguesian plane find the equation of the join of
   (a) $[a, b]$ and $[c, d]$, with $a \neq c$,
   (b) $[a, b]$ and $[m]$.
9. Find the intersection of the lines $y = xm + n$ and $y = xp + q$, with $m \neq p$,
   (a) in a Moufang plane,
   (b) in a desarguesian plane,
   (c) in P^2. Simplify the expressions as far as possible.
   (d) Can you determine the intersection in a Veblen-Wedderburn plane?
10. In a desarguesian plane find the vertices of the triangle whose sides have the equations $y = xa + b$, $y = xc + d$ and $y = xc + e$. No two of a, b, c, d, and e are equal.
11. If in a desarguesian plane a collineation with axis $x = 0$ and center $[1, 1]$ maps the point $[1, 0]$ on $[1, p]$ with $0 \neq p \neq 1$, find the images of

    (a) $[0]$ (b) $[1, p]$ (c) $[0, p]$ (d) the line $y = x$.
12. In a desarguesian plane a collineation maps $[1, 1]$ on $[-1, 1]$, and $[0]$, $[\infty]$, and $[0, 0]$ are invariant. Find the images of all the points and all the lines of the plane and verify that collinearity is preserved.
13. Prove Lemma 4-4.11.

4-5 PAPPIAN PLANES

Pappus and commutativity. In this section we add another axiom, the validity of the Pappus property, Π (p. 154). A projective plane satisfying PP 1 through 3 and this new axiom will be called a *pappian plane*.

Theorem 4-5.1. A projective plane is pappian if and only if the multiplication in each of its ternary rings is commutative.

Proof (Fig. 4-33). In an arbitrary ternary ring construct $[a, ab]$ and $[b, ba]$. The assertion is that $ba = ab$; in other words, that $[0]$ lies on $[a, ab][b, ba]$. This follows from Π. For easier reference we draw Fig. 4-34. Let $[0, 0] = A$, $[a, a] = B$, $[b, b] = C$, $[\infty] = A'$, $[1, b] = B'$, $[1, a] = C'$, $[a, ab] = D$, $[b, ba] = E$, and $[0] = F$. Then the statement of Π is that D, E, and F are collinear. This is exactly our assertion.

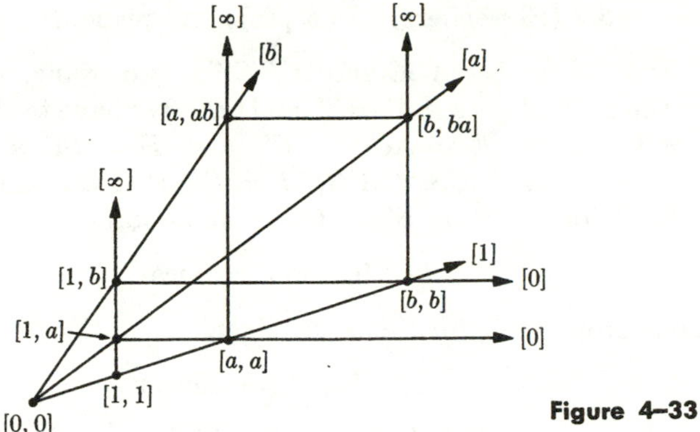

Figure 4-33

Conversely, suppose that every ternary ring in the plane has commutative multiplication. A Pappus figure (Fig. 4-34) is given, and the collinearity of D, E, and F has to be proved. Choose A, F, A' and $AB \times A'B'$, respectively, as the reference points $[0, 0]$, $[0]$, $[\infty]$, and $[1, 1]$ of a coordinate system.* Then B and C will be the points $[a, a]$ and $[b, b]$ for some elements a and b of one of the ternary rings. The commutativity of the multiplication in this ring yields $ab = ba$, which, in turn, implies the collinearity of D, E, and F.

Pappus and Desargues. The commutativity of all the ternary rings or, equivalently, the validity of Π, has an interesting consequence. In Theorem 4-1.4 we saw that a loop which, together with all its isotopes, is commutative, has to be associative. Now the isotopes of the multiplicative loop of a ternary ring are multiplicative loops of other ternary rings of the plane (cf. Exercise 15 of Section 4-1), and in a pappian plane all these loops are commutative. Hence we obtain the result that multiplication in a ternary ring of a pappian plane is always associative. This raises the question: Is every pappian plane desarguesian? The answer was given in 1905 by the German mathematician G. Hessenberg.

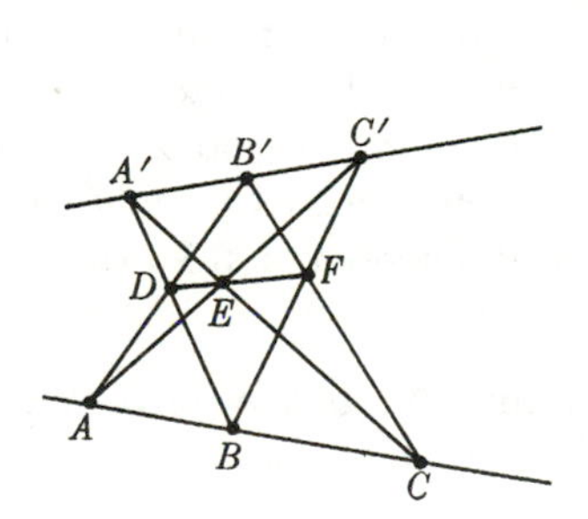

Figure 4-34

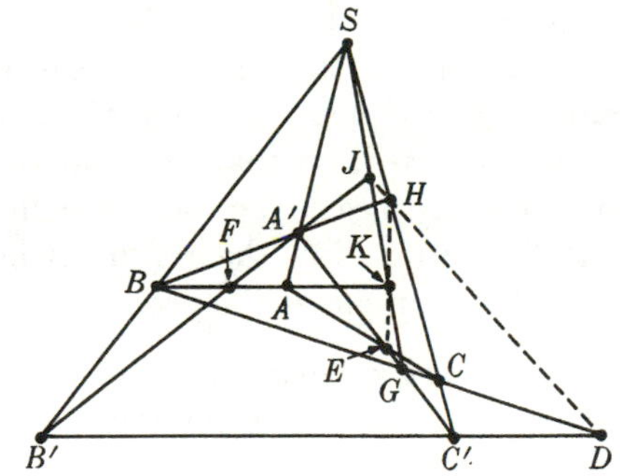

Figure 4-35

* If A, F, and A' are collinear, the points of the figure have to be renamed. The case where D is on CC', E on BB', and F on AA', requires a special proof.

Theorem 4–5.2 (Hessenberg). In a projective plane, Π implies Δ.

Proof (Fig. 4–35). Let ABC and $A'B'C'$ be two triangles, with S, the intersection of the distinct lines AA', BB', and CC'. We have to show the collinearity of $F = AB \times A'B'$, $E = AC \times A'C'$, and $D = BC \times B'C'$. We define $G = BC \times A'C'$, $H = CS \times A'B$, $J = SG \times A'B'$, and $K = SG \times AB$. Now we use Π for the figure $B'BS$, $GC'A'$ and obtain

$$D, H, \text{ and } J \text{ collinear.} \tag{1}$$

The Pappus property Π, for $SA'A$, BCG makes

$$H, E, \text{ and } K \text{ collinear.} \tag{2}$$

Finally we apply Π to BHA', JGK. We obtain the collinearity of $BG \times HJ$, $HK \times A'G$, and $BK \times A'J = F$. But, by Eq. (1), $BG \times HJ = D$, and by Eq. (2), $HK \times A'G = E$. Hence D, E, and F are collinear, which completes the proof. Some complications might arise for special cases when points of Fig. 4–35 coincide. All these cases can be settled satisfactorily by relatively easy, but tedious, arguments.*

An alternative proof of this theorem will be given later.

We have immediately the following theorem.

Theorem 4–5.3. In a pappian plane every ternary ring is a field.

Proof. By Theorem 4–5.2 the plane is desarguesian. The corollary to Theorem 4–4.1 makes every ternary ring a skew field. But in view of Theorem 4–5.1 this skew field is commutative and therefore a field.

Conversely, we have the following.

Theorem 4–5.4. If, in a projective plane, one ternary ring is linear and a field, the plane is pappian.

Proof. By Theorem 4–4.3, the plane is desarguesian. Let a Pappus figure be given, as in Fig. 4–34, and then it is asserted that D, E, and F are collinear. By Theorem 4–4.4, there is a collineation that maps four given points, no three of which are collinear, on four given image points with the same property. Thus we may assume that $A = [0, 0]$, $D = [1, 1]$, $A' = [\infty]$, and $F = [0]$, because the collineation preserves collinearity and hence also the validity or nonvalidity of Π. Now the argument in the proof of Theorem 4–5.1 can be repeated. This completes the proof.

In Hessenberg's Theorem we learned that Π implies Δ. Is the converse also true? In other words, are the two equivalent? The answer is no.

---

* How careful one has to be in checking these special cases becomes evident from the fact that one of them was overlooked in all textbooks and articles for 48 years! See A. Cronheim's paper in *Proc. Amer. Math. Soc.*, **4**, 219 (1953).

Theorem 4–5.5. The property Δ does not imply Π.

Proof. By Theorem 4–1.5, there is a projective plane coordinatized by any given ternary ring. The quaternion algebra is a linear ternary ring and a skew field. In view of Theorem 4–4.3, a plane constructed over this skew field is desarguesian. However, the quaternion algebra is noncommutative, and according to Theorem 4–5.1 the plane over it cannot be pappian. Thus we obtain an example of a desarguesian, nonpappian, projective plane, which proves the theorem.

A special situation exists in finite planes. An algebraic theorem due to J. H. M. Wedderburn (1905) states that every finite skew field is a field. Therefore, every finite desarguesian plane is pappian. A later result, connected with the names of E. Artin, F. W. Levi, and M. Zorn, goes even further in proving that every finite alternative division ring is a field. Thus Π holds in every finite Moufang plane. A detailed proof of these theorems can be found in M. Hall's *Theory of Groups* (see references).

Pappus in affine planes. As we did in Section 4–4, we will develop affine specializations of Π for the purpose of coordinatizing an affine plane. The *Thomsen property*, Θ, is the special Pappus property in which the two points A' and F are improper. It reads, therefore (Fig. 4–36):

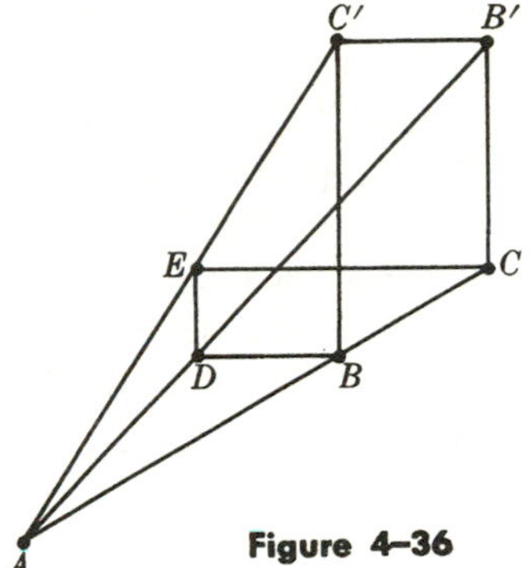

Figure 4–36

Θ. If A, B, and C are points on a line l, and B' and C' points on a line $\neq l$ such that $BC' \parallel B'C$, let E be on AC' and D on AB' such that $CE \parallel BD \parallel B'C'$. Then $DE \parallel BC'$.

Looking at Fig. 4–33, we see that the form of Π as used in the proof of Theorem 4–5.1 was, indeed, Θ.* Thus the commutativity of the multiplication in the affine plane is assured. In an affine plane in which δ_a, Δ_a, and Θ hold, every ternary ring is a field. However, in analogy with Theorem 4–5.2, we will show that δ_a and Δ_a are not needed, because they are consequences of another affine specialization of Π. In this *affine Pappus property*, Π_a, the points D, E, and F of Fig. 4–34 are improper. We have, then (Fig. 4–37):

Π_a. If A, B, and C are points on a line l, and A', B', and C' points on a line $\neq l$, and if $AB' \parallel A'B$ and $AC' \parallel A'C$, then $BC' \parallel B'C$.

Theorem 4–5.6. Π_a affinely implies Δ_a and δ_a.

Proof. We start with Δ_a (Fig. 4–38). Let ABC and $A'B'C'$ be the Desargues triangles, let S be on AA', BB', and CC', and moreover let $AB \parallel A'B'$, $BC \parallel B'C'$. We have to show $AC \parallel A'C'$. Let AB intersect the parallel to BC through S

---

* See also Exercise 14 of Section 4–2.

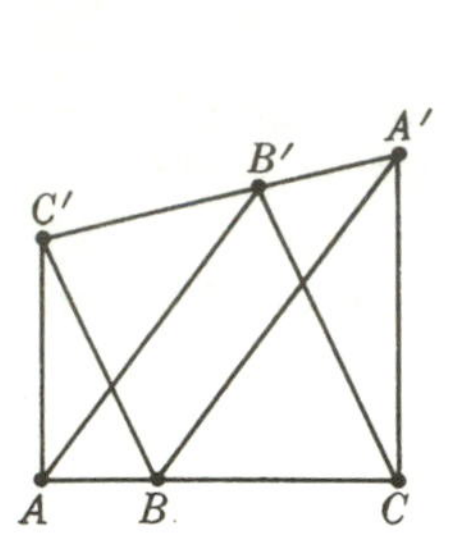

Figure 4-37

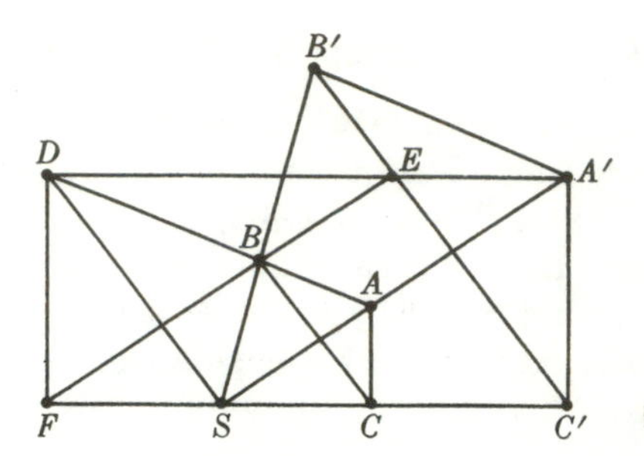

Figure 4-38

at D, and let $A'D \times B'C' = E$. It is easy to show that the nonexistence of intersections at D or E would lead to contradictions. We apply Π_a to the two collinear triples A', D, E and B, B', S. Since $A'B' \parallel DB$ and $DS \parallel EB'$, we obtain $A'S \parallel EB$. Then BE and SC must intersect, otherwise $SA \parallel SC$, a contradiction. Let $BE \times SC = F$. Now since $AS \parallel BF$ and $BC \parallel SD$, we apply Π_a to the collinear point triples F, S, C and A, B, D and get $AC \parallel DF$. Finally, since $SD \parallel C'E$ and $EF \parallel SA'$, Π_a for D, E, A' and C', S, F yields $DF \parallel A'C'$. But also $AC \parallel DF$, and hence $AC \parallel A'C'$, and Δ_a holds.

Now suppose δ_a does not hold. Then, by Lemma 4-2.1, its converse is not valid either, and for two distinct triangles ABC and $A'B'C'$ with corresponding parallel sides and $AA' \parallel BB'$, $CC' \times AA' = S$ exists. But this conflicts with Lemma 4-4.11 by which BB' also has to pass through S, so that $AA' \parallel BB'$ is impossible. Thus δ_a has to be valid, and the proof is complete.

We saw that every ternary ring in an affine plane is a field if Θ and Π_a are valid. However, even these requirements can be cut down, namely by elimination of Θ.

Theorem 4-5.7. In an affine plane Π_a implies Θ.

Proof (Fig. 4-39). For Θ we construct the figure $SDB'CEA'B$ such that $B'D \parallel A'B$, $BD \parallel B'C \parallel A'E$ and such that $A'B'$, DE, and BC intersect at S. We claim $CE \parallel A'B$. Let $BD \times A'B' = C'$, $B'D \times BC = A$. Then, by Π_a for ABC, $A'B'C'$, we get $AC' \parallel A'C$. In view of Theorem 4-5.6, we may use Δ_a for the triangles $AC'D$ and $CA'E$ in which $AC' \parallel A'C$ and $C'D \parallel A'E$. We obtain $AD \parallel CE$, which was asserted.

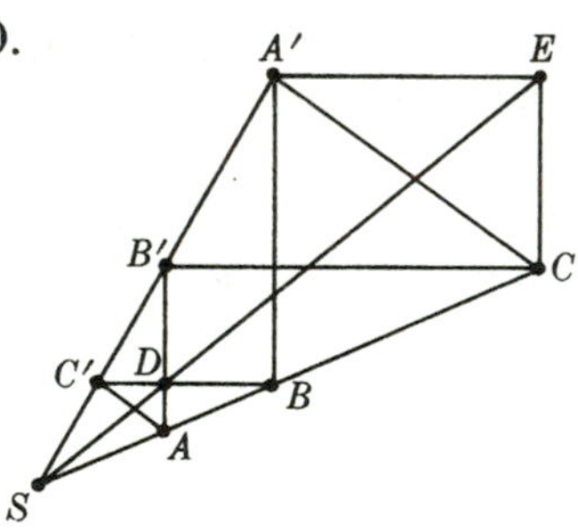

Figure 4-39

Corollary. In an affine plane in which Π_a holds, every ternary ring is linear and a field.

The last theorem provides us with an alternative simple proof of Hessenberg's theorem, 4-5.2. For, by Theorem 4-4.3, if a single ternary ring of a projective plane is linear and a skew field, $\triangle$ holds throughout the plane. Now, Π implies Π_a, and by Theorem 4–5.6, Π_a implies $\triangle_a$ and δ_a which, in turn, assure the linearity and skew-field properties of one ternary ring. Thus $\triangle$ is a consequence of Π.

4–6 AXIOMS OF ORDER AND CONTINUITY. PLANES OVER THE REALS

The real field. We saw that a projective plane satisfying PP 1 through 3 and the Pappus axiom can be coordinatized by a field. In Chapters 2 and 3 we gave special attention to planes over the field of reals. Thus it is reasonable to ask about additional axioms which make the coordinatizing field isomorphic to the field of real numbers. In order to be able to do this, we first have to study properties that characterize the real field. We start with a few definitions.

A skew field F is said to be *ordered* if it contains a subset P of so-called *positive elements* which satisfies the order axioms.

Or 1. P is closed under addition and multiplication.

Or 2. For every element $x \in F$, one and only one of three alternatives holds: either $x \in P$, or $x = 0$, or $-x \in P$ (the law of trichotomy).

In view of Or 2, for any two elements a and b of F, one and only one of three alternatives holds: either $a - b \in P$, or $a - b = 0$, or $b - a \in P$. In these cases we say $a > b$, $a = b$, or $a < b$, respectively. (The statements $a > b$ and $b < a$ are equivalent.) Thus $x \in P$ means $x > 0$, and $-x \in P$ means $x < 0$.

Not every skew field is ordered. For instance, finite fields are not ordered (see Exercise 1).

A *lower bound* of a subset S of an ordered field F is an element $b \in F$ for which $b \leqq s$ for all $s \in S$. The element b is the *greatest lower bound* if there is no lower bound greater than b. An ordered field F is called *complete* if every nonempty set of positive elements of F has a greatest lower bound in F. The field of reals is easily seen to be a complete ordered field.

Theorem 4–6.1. A field is complete and ordered if and only if it is isomorphic to the field of reals.

The proof will not be given here. It can be found in Birkhoff and MacLane, *A Survey of Modern Algebra*, New York: Macmillan, 1953.

Separation. Geometric axioms leading to the ordering property of a field are provided in projective planes by axioms of *separation*. It may seem more natural to use the concept of betweenness instead. However, as we saw in Chapter 3, betweenness is not a projective concept. It belongs instead to affine geometry.

If A, B, C, and D are distinct collinear points, we define (cf. Exercise 2 of Section 3–2) a relation of separation, σ, between them, written $\sigma(AB|CD)$ (in words "AB separate CD") such that the following separation axioms hold.

Sp 1. Every line contains at least four distinct points.

Sp 2. $\sigma(AB|CD)$ implies $\sigma(AB|DC)$ and $\sigma(CD|AB)$.

Sp 3. One and only one of the relations $\sigma(AB|CD)$, $\sigma(BC|DA)$, $\sigma(CA|BD)$ holds.

Sp 4. If $\sigma(AB|CD)$ and $\sigma(BC|DE)$, then $\sigma(CD|EA)$.

Sp 5. If $\sigma(AB|CD)$ and if the four collinear points A', B', C', and D' are images of A, B, C, and D, respectively, under a point perspectivity, then $\sigma(A'B'|C'D')$.

A projective plane in which Sp 1 through 5 are valid is called an *ordered projective plane.*

Remark. The axioms Sp 1 through 4 are not *independent,* that is, part of them may be implied by the remaining axioms. For instance, in Sp 3, "one and only one" could be replaced by "at least one," without weakening the axiom system. Such a procedure may be considered as an aesthetic blemish, but sometimes a system of axioms with redundancies is more workable and avoids many lengthy and tedious discussions. On the other hand, if this indulgence is carried too far, one may reach an extreme in which all the theorems of the subject appear as axioms.

An immediate consequence of Sp 2 is the equivalence of $\sigma(AB|CD)$, $\sigma(BA|CD)$, $\sigma(AB|DC)$, $\sigma(BA|DC)$, $\sigma(CD|AB)$, $\sigma(CD|BA)$, $\sigma(DC|AB)$, and $\sigma(DC|BA)$. We will make frequent use of this fact.

A *segment* is defined in the following way. If A, B, and C are distinct collinear points, AB/C ("AB without C") is the set of all points X such that $\sigma(AB|XC)$. The mention of C in AB/C is necessary because AB alone could have meant two different segments, one of which contains C whereas the other does not.

Separation and order. Our next aim is to show that in an ordered pappian plane the field $(\Sigma, +, \cdot)$ is ordered, that is, satisfies Or 1 and 2. We start with some lemmas. In order to simplify the notation we will use b for $[0, b]$ and ∞ for $[\infty]$ when no ambiguity is to be feared.

Lemma 4–6.2. If a_i, with $i = 1, 2, 3, 4$, are distinct elements of Σ, and if $0 \neq p \in \Sigma$, then $\sigma(a_1a_2|a_3a_4)$ implies $\sigma(a_1 + p, a_2 + p|a_3 + p, a_4 + p)$ and $\sigma(a_1p, a_2p|a_3p, a_4p)$. One of the points $[0, a_i]$ may be replaced by $[\infty]$, and then $a_i + p$ and a_ip have to be replaced by ∞.

Proof (Fig. 4–40). A point perspectivity with center [1] maps each point $[0, a_i]$ on $[p, p + a_i]$, and then a point perspectivity with center [0] maps this point on $[0, p + a_i]$. The additive statement then follows from Sp 5. For the multiplicative assertion (Fig. 4–41), a point perspectivity with center [0] maps every $[0, a_i]$ on $[a_i, a_i]$. This point, in turn, is taken by a point perspectivity with center $[\infty]$ into $[a_i, a_ip]$. Finally, a third point perspectivity with center [0] maps $[a_i, a_ip]$ on $[0, a_ip]$. Then Sp 5 yields the multiplicative statement. The case $a_i = \infty$ is trivial.

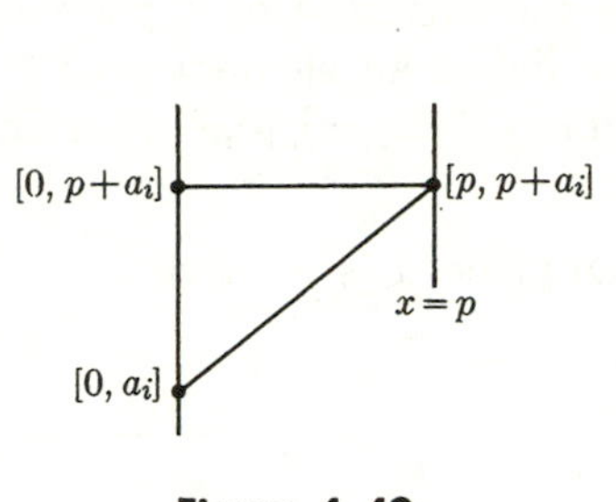

Figure 4-40

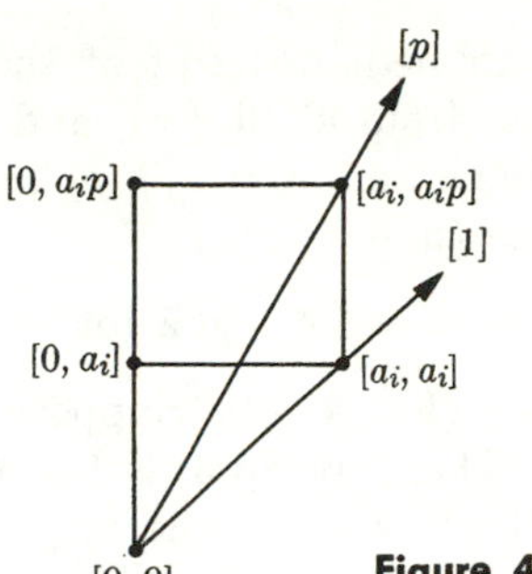

Figure 4-41

Lemma 4-6.3. In $(\Sigma, +)$ of an ordered pappian plane the characteristic is not 2.

Proof. Suppose the characteristic were 2. By Sp 1, there exists an element a of Σ which is distinct from 0 and 1. There are three possibilities, in view of Sp 3:

Case (i)

$$\sigma(0a|1\infty). \tag{1}$$

Case (ii)

$$\sigma(01|a\infty).$$

Case (iii)

$$\sigma(1a|\infty 0). \tag{2}$$

Case (i). We have, by Lemma 4-6.2,

$$\sigma(1, a + 1|0, \infty), \tag{3}$$

and

$$\sigma(a + 1, 1|a, \infty), \tag{4}$$

in view of $1 + 1 = a + a = 0$. From Eqs. (1) and (3), by Sp 4, $\sigma(a0|\infty 1)$ and $\sigma(0, \infty|1, a + 1)$ yield

$$\sigma(\infty, 1|a + 1, a). \tag{5}$$

In view of Sp 3, Eqs. (4) and (5) contradict each other.

Case (ii). Interchange 1 and a in the proof of case (i).

Case (iii). By Lemma 4-6.2, Eq. (2) yields

$$\sigma(a + 1, 0|\infty, a), \tag{6}$$

and

$$\sigma(0, a + 1|\infty, 1). \tag{7}$$

By Sp 4, Eqs. (2) and (6), in the form $\sigma(1a|\infty 0)$ and $\sigma(a, \infty|0, a + 1)$, yield $\sigma(\infty, 0|a + 1, 1)$ which, in view of Sp 3, contradicts Eq. (7).

An immediate result of the last lemma is the existence of a point $[0, -1]$ distinct from $[0, 0]$, $[\infty]$, and $[0, 1]$. Now we define an element p of Σ to be positive if and only if $[0, p]$ belongs to the segment $[0, 0][\infty]/[0, -1]$ or, in other terms, $\sigma(p, -1|0, \infty)$.

Lemma 4-6.4. In Σ for an ordered pappian plane, 1 is positive.

Proof (Fig. 4-42). Suppose 1 were not positive. Then, by Sp 3, either $\sigma(-1, 0|\infty, 1)$ or $\sigma(0, 1|-1, \infty)$. In the first case Sp 5 yields successively $\sigma([1, 0], 0|[0], [-1, 0])$, $\sigma([1, 1], 0|[1], [-1, -1])$, and $\sigma(1, 0|\infty, -1)$, a contradiction by Sp 3. The second case, in the same manner, leads to

$$\sigma(0, [-1, 0]|[1, 0], [0]),$$
$$\sigma(0, [-1, -1]|[1, 1], [1]),$$

and $\sigma(0, -1|1, \infty)$, again a contradiction.

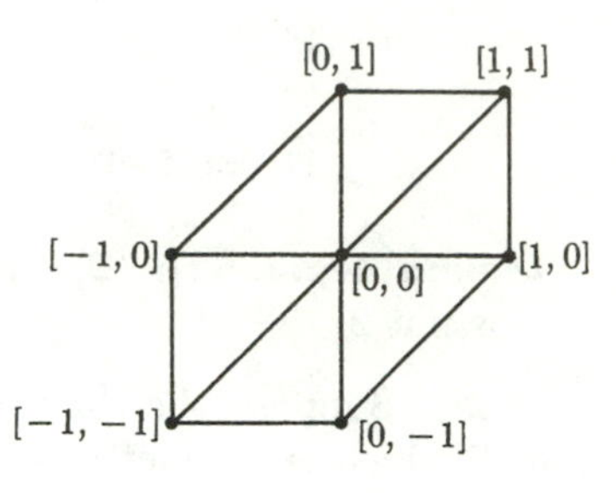

Figure 4-42

We come now to the proof of the validity of the additive part of Or 1.

Theorem 4-6.5. If a and b are positive elements of Σ in an ordered pappian plane, then $a + b$ is positive.

Proof (Fig. 4-43). The figure reflects the fact that in a field $a - 1 = a + (-1) = (-1) + a$. Now, by Sp 5, the positiveness of a, that is,

$$\sigma(-1, a|0, \infty), \tag{8}$$

implies $\sigma(-1, C|A, [1])$ and successively $\sigma(a, C|D, [0])$, and

$$\sigma(a, -1|a - 1, \infty). \tag{9}$$

The positiveness of b implies $\sigma(-1, b|0, \infty)$, which, by Lemma 4-6.2, brings about

$$\sigma(a - 1, a + b|a, \infty). \tag{10}$$

In view of Sp 4, Eqs. (10) and (9), written as $\sigma(a + b, a - 1|\infty, a)$ and $\sigma(a - 1, \infty|a, -1)$, imply

$$\sigma(\infty, a|-1, a + b). \tag{11}$$

Now Eqs. (11) and (8), in the form $\sigma(a + b, -1|a, \infty)$ and $\sigma(-1, a|\infty, 0)$, imply

$$\sigma(a, \infty|0, a + b) \tag{12}$$

by Sp 4. Finally, application of Sp 4 to $\sigma(-1, a|\infty, 0)$, from Eq. (8), and $\sigma(a, \infty|0, a + b)$, from Eq. (12), results in $\sigma(\infty, 0|a + b, -1)$. Hence $a + b$ is positive.

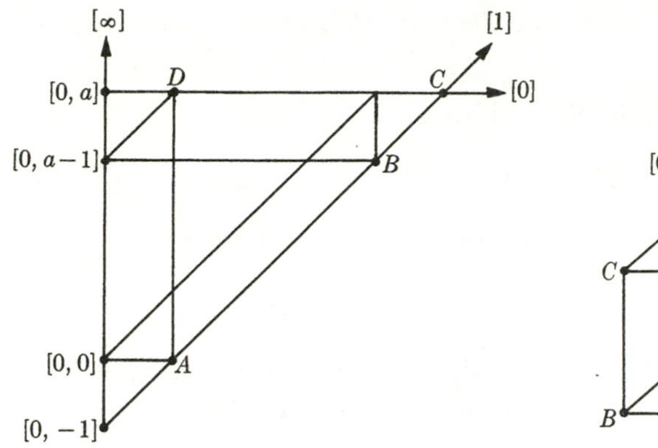

Figure 4-43

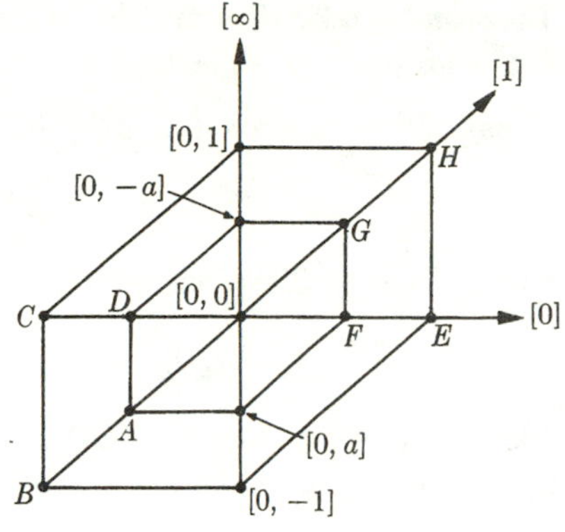

Figure 4-44

The following theorem will deal with a verification of Or 2.

Theorem 4-6.6. If a is an element of Σ in an ordered pappian plane, then exactly one of three alternatives holds: either $a = 0$, or a is positive, or $-a$ is positive.

Proof (Fig. 4-44). If $a = 0$, a is not positive, because $\sigma(0, -1|0, \infty)$ is impossible, and for the same reason $-a$ is not positive. If $a = -1$, it is not positive since $\sigma(-1, -1|0, \infty)$ is false. But then $-a = 1$ is positive, in view of Lemma 4-6.4. Now let $a \neq 0$ and $a \neq -1$. Then there are three cases: (i) $\sigma(a, -1|0, \infty)$, (ii) $\sigma(a, \infty|-1, 0)$, and (iii) $\sigma(a, 0|\infty, -1)$. By Sp 3, one and only one of these three relations holds.

In Case (i) a is positive, by the definition of positiveness, and $-a$ cannot be positive too, because then, by Theorem 4-6.5, $a + (-a) = 0$ would have to be positive.

Case (ii) (Fig. 4-44). By repeated use of Sp 5, we then obtain $\sigma(a, \infty|-1, 0) \Rightarrow \sigma(A, [1]|B, 0) \Rightarrow \sigma(D, [0]|C, 0) \Rightarrow$

$$\sigma(-a, \infty|1, 0). \tag{13}$$

Lemma 4-6.4 means $\sigma(-1, 1|0, \infty)$, and this, in conjunction with Eq. (13) in the form $\sigma(1, 0|\infty, -a)$, yields $\sigma(0, \infty|-a, -1)$, in view of Sp 4. Thus $-a$ is positive. This precludes a being positive.

Case (iii). As a consequence of Lemma 4-6.2, $\sigma(a, 0|\infty, -1)$ implies $\sigma(-a, 0|\infty, 1)$. By Lemma 4-6.4, $\sigma(1, -1|0, \infty)$. Sp 4, applied to the two last relations in the form $\sigma(-1, 1|\infty, 0)$ and $\sigma(1, \infty|0, -a)$, results in

$$\sigma(\infty, 0|-a, -1),$$

and $-a$ is positive.

The remaining part of Or 1 will be settled in the following.

Theorem 4–6.7. If a and b are positive elements of Σ in an ordered pappian plane, then ab is positive.

Proof. By Lemma 4–6.2, $\sigma(b, -1|0, \infty)$ implies

$$\sigma(ab, -a|0, \infty). \tag{14}$$

Since, in view of Theorem 4–6.6, $-a$ is not positive, by Sp 3 either

$$\sigma(-a, 0|-1, \infty) \qquad \text{or} \qquad \sigma(-a, \infty|-1, 0)$$

holds. In the first case, Eq. (14) together with $\sigma(-a, 0|\infty, -1)$, by Sp 4 yields $\sigma(0, \infty|-1, ab)$, as claimed. In the second case, Eq. (14), in the form

$$\sigma(ab, -a|\infty, 0),$$

combined with $\sigma(-a, \infty|0, -1)$, by Sp 4 results in $\sigma(\infty, 0|-1, ab)$. This is the desired result.

Continuity in pappian projective planes. In an ordered pappian plane the coordinatizing field is ordered. We will add another geometric axiom, a continuity postulate, which will make the field complete and, therefore, isomorphic to the field of reals.

The following is a projective continuity axiom.

PC. Let S be a nonempty set of points of a segment AC/B, with the property that AC/B contains points L and U such that $\sigma(AP|LU)$ holds for all points P of S. Then in AC/B there are points G and H such that $\sigma(LP|GU)$ and $\sigma(UP|HL)$ for all permitted choices of P, L, and U.

Theorem 4–6.8. A projective plane satisfying the postulates PP 1 through 3, Π, Sp 1 through 5, and PC, is coordinatized by a field which is isomorphic to the field of reals.

Proof. We know that the field is ordered and have to prove only its completeness. But this follows immediately from PC when we choose $A = [0, 0]$, $B = [0, -1]$, and $C = [\infty]$. Then AC/B becomes the set of all $[0, p]$ with positive p, and PC implies the existence of a greatest lower bound G (or H) and of a least upper bound H (or G) for every subset S of AB/C.

The converse of Theorem 4–6.8 states that P^2 over the reals satisfies PP 1 through 3, Π, Sp 1 through 5, and PC. This is obvious in view of the developments of Chapter 3.

The real affine plane. How would the various axioms have to be modified for the affine plane? The incidence axioms PP 1 through 3 will have to be changed, because in an affine plane improper points can no longer be considered as points. Lines that formerly met in an improper point will now be called *parallel*, a term which will mean here merely "not intersecting."

PP 1 through 3 in their affinely modified form become, respectively, the three following axioms of a rudimentary affine plane (cf. Exercises 3 and 4 of Section 4-1 and Exercise 8 of this section).

AP 1. For every two points there is exactly one line incident with both.
AP 2. (Euclid's parallel axiom). Given a point P and a line l not through P, there exists a unique line through P parallel to l.
AP 3. There exist three points which are not incident with a line.

In the affine plane, Π will be replaced by Π_a. Concerning the order axioms it has to be noted that in the affine plane the straight lines are no longer closed. The removal of the improper point makes the affine line into an open line. The separation relation of four collinear points has to be replaced by a betweenness relation of three distinct collinear points. $\sigma(AB|CD)$, with D the improper point of the line, will be written $\beta(ACB)$; in words, "C between A and B" (Fig. 4-45). The axioms Sp 1 through 5 then change respectively into the five betweenness axioms of an *ordered affine plane* (cf. Exercise 9):

Bt 1. Every line contains at least three distinct points.
Bt 2. $\beta(ABC)$ implies $\beta(CBA)$.
Bt 3. One and only one of the relations $\beta(ABC)$, $\beta(BCA)$, and $\beta(CAB)$ holds if A, B, and C are distinct and collinear.
Bt 4. If $\beta(ACB)$ and $\beta(BEC)$, then $\beta(ACE)$.

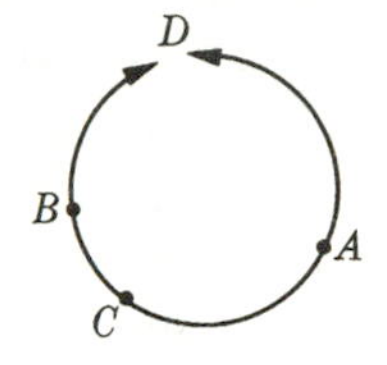

Figure 4-45

In the affine plane, the segment AB/D will, naturally, always exclude the improper point D and can, therefore, be called simply $|AB|$. Thus in the affine plane the segment $|AB|$ consists of all the points C satisfying $\beta(ACB)$. A *half-line* A/B consists of all the points C satisfying $\beta(BAC)$. In words, A/B is "the half-line originating from A, without B." With these modified terms, Bt 5 and PC for the affine plane will be as follows.

Bt 5. Let A, B, and C, lie on a line j and A', B', and C' on another line k, and let S be a point not on j or k, such that AA', BB', and CC' either are parallel or lie on the half-lines SAA', SBB', and SCC' originating from S. Then $\beta(ABC)$ implies $\beta(A'B'C')$.

We add an affine continuity axiom.

AC. Let S be a nonempty set of points of a half-line A/B, with the property that A/B contains at least one point L such that $\beta(ALP)$ for all points $P \in S$. Then in A/B there is a point G such that $\beta(LGP)$ for all permitted choices of P and L.

We have constructed the affine plane over the reals. The axioms used were (a) the incidence and parallel postulates AP, (b) Pappus' affine postulate Π_a, (c) the betweenness postulates Bt, (d) the affine continuity postulate AC. Based on these axioms, plane affine geometry can be developed along the lines of Chapter 2. Thus we have come to the end of the axiomatic development of real affine plane geometry. Certainly, more convenient axiom systems could be devised for the real affine plane, had we not insisted on considering affine geometry as a special case of projective geometry and, therefore, taken the projective plane and its axioms as our starting point. Indeed, ample treatment of such a direct axiomatic approach to the euclidean plane can be found in the literature to which we will refer at the end of this chapter.

EXERCISES

1. Show that:
   (a) a field with a finite characteristic cannot be ordered,
   (b) finite fields cannot be ordered.
2. Prove that between two elements of an ordered skew field there is always
   (a) another element,
   (b) an infinity of other elements.
3. A skew field is said to be *archimedean* (after Archimedes of Syracuse, 287–212 B.C.) if it is ordered, and if for two arbitrarily given positive elements p and q it is possible to add $p + p + \cdots + p$ a sufficient number of times such that the result is $>q$. Show that every complete ordered field is archimedean.
4. Define geometrically an archimedean plane, that is, a plane that is coordinatized by an archimedean skew field.
5. To show that every archimedean skew field F is a field, complete the details of the following sketch of a proof. Let $1 + \cdots + 1$ (n times) $= n$ be called a non-negative integer. If a and b are noncommuting positive elements of F, and $ab - ba > 0$, let

   $$c = (a + b + 1)^{-1}(ab - ba).$$

   Then there exists $d \in F$ such that $d > 0$, $d < 1$, $d < c$, and there are non-negative integers p and q such that $pd < a \leqq (p + 1)d$ and $qd < b \leqq (q + 1)d$. Combine these inequalities and get

   $$ab - ba < (p + q + 1)d^2.$$

   But $(p + q + 1)d < a + b + 1$, and consequently $ab - ba < (a + b + 1)c$, a contradiction.
6. Show that Sp 1 can be replaced by the weaker axiom, "There is a line which contains at least four distinct points," without weakening the axiom set Sp.

7. Show that in Sp 3, "one and only one" may be replaced by "at least one," without weakening the axiom set Sp.
8. Let an affine plane, as defined by AP 1 through 3, be augmented by exactly one improper point on each line such that:
   (a) parallels have an improper point in common,
   (b) there exists an improper line containing all improper points and no other points. Prove that this augmented plane satisfies PP 1 through 3.
9. Prove that:
   (a) Bt 1 through 5 hold in an ordered projective plane in which one line has been singled out as improper line,
   (b) Sp 1 through 5 hold in an augmented affine plane such as defined in Exercise 8.
10. Prove: If ABC is a triangle and $\beta(BCD)$ and $\beta(AEC)$, then there is a point F on DE such that $\beta(AFB)$. (In some textbooks this statement substitutes Bt 5.) Show also that $\beta(DEF)$.
11. Show that PC in a projective plane is equivalent to AC in the affine plane in the same sense as in Exercises 8 and 9.
12. Which of the axioms of the real affine plane hold in Moulton's plane (Section 4-2) and which are violated?
13. Which of the axioms of the real affine plane are valid in $A^2(p)$, where p is an odd prime, and which are violated?

4-7 EUCLIDEAN AND HYPERBOLIC PLANES

Congruence and euclidean planes. In order to develop the affine plane into a euclidean plane in an axiomatic way, we need the concept of congruence. In the affine plane we encountered a special case of congruence, namely that of equality of vectors. This special type of congruence enables us to take care of segments on parallel lines only. Thus we have to introduce axioms of congruence and then verify that they are valid for vector equality. We define a relation "$\equiv$" (congruent) between affine segments satisfying the following congruence axioms.

Cg 1. If A and B are distinct points and C/P a half-line, then there is precisely one point D on C/P such that $|AB| \equiv |CD|$.

Cg 2. If $|AB| \equiv |CD|$ and $|CD| \equiv |EF|$, then $|AB| \equiv |EF|$.

Cg 3. $|AB| \equiv |BA|$.

Cg 4. If $\beta(ABC)$ and $\beta(A'B'C')$ and $|AB| \equiv |A'B'|$ and $|BC| \equiv |B'C'|$, then $|AC| \equiv |A'C'|$.

Cg 5. (Fig. 4-46). Let ABC and $A'B'C'$ be triangles whose corresponding sides are congruent. If D and D' are points such that $\beta(BCD)$ and $\beta(B'C'D')$ and if $|BD| \equiv |B'D'|$, then $|AD| \equiv |A'D'|$.

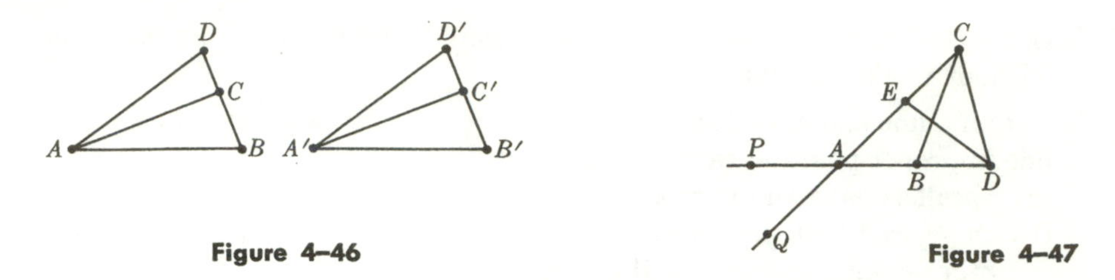

Figure 4-46

Figure 4-47

In the following we will make a few observations which can easily be proved. The proofs will not be worked out explicitly, but some of them will be given as exercises.

(1) Congruence is an equivalence relation; that is, it satisfies the postulates Ev.

(2) Distances $|AB|$ and $|CD|$, as defined in Section 2-7, are equal if and only if the segments $|AB| \equiv |CD|$, provided the axioms of the real affine plane hold. Thus no confusion will be caused if in the following we use $|AB|$ for the segment as well as for the distance, and if we write $|AB| = |CD|$ instead of $|AB| \equiv |CD|$.

(3) Every two half-lines A/P and A/Q determine an *angle*. We say that two angles with vertices A and A' are congruent or equal if there are points B, C, B', and C' such that $B \in A/P, C \in A/Q, B' \in A'/P', C' \in A'/Q', |AB| = |A'B'|$, $|AC| = |A'C'|$, and $|BC| = |B'C'|$ (Fig. 4-47). Now the equality of the angles is independent of the choice of B, C, B', and C'. For suppose that, respectively, D, E, D', and E' had been chosen instead, such that $|AD| = |A'D'|$ and $|AE| = |A'E'|$. Then, by Cg 5, first $|EB| = |E'B'|$, and then $|DE| = |D'E'|$.

(4) The "congruence theorems" for triangles of high-school geometry follow easily from the axioms of the euclidean plane, that is, AP, Bt, AC, and Cg. The proof will be left for the exercises.

(5) For vectors over the reals, $\overrightarrow{AB} = \overrightarrow{CD}$ implies $|AB| = |CD|$.

In view of these remarks it is reasonable to assume that high-school plane geometry can be developed from the axioms of the euclidean plane. This applies to the well-known theorems about angles, parallels, congruence of triangles and polygons, and similarity of triangles and polygons. We will not go into the details of this development, which is lengthy and tedious, but refer the interested reader to the literature.

The axiom Π_a turns out to be a consequence of the remaining axioms. We will show only that it follows from the theorems of high-school plane geometry.

Theorem 4-7.1. The property Π_a is implied by the axioms AP, Bt, AC, and Cg.

Proof. We distinguish between the case (i) in which $ABC \parallel A'B'C'$ (Fig. 4-48) and the case (ii) where $AB \times A'B' = S$ (Fig. 4-49). We assume $AB' \parallel A'B$ and $BC' \parallel B'C$, and show $AC' \parallel A'C$.

Case (i). $ABA'B'$ and $BCB'C'$ are parallelograms and thus $|A'B'| + |B'C'| = |A'C'|$ and $|AB| + |BC| = |AC|$ are parallel and equal segments. Hence $ACA'C'$ is a parallelogram and $AC' \parallel A'C$.

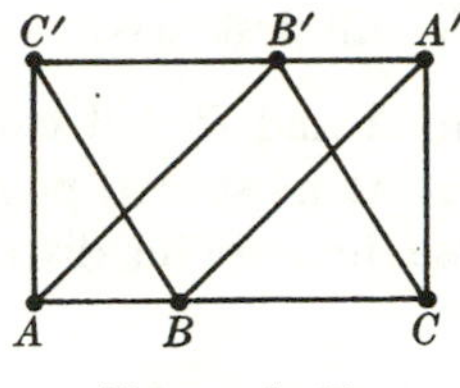

Figure 4-48

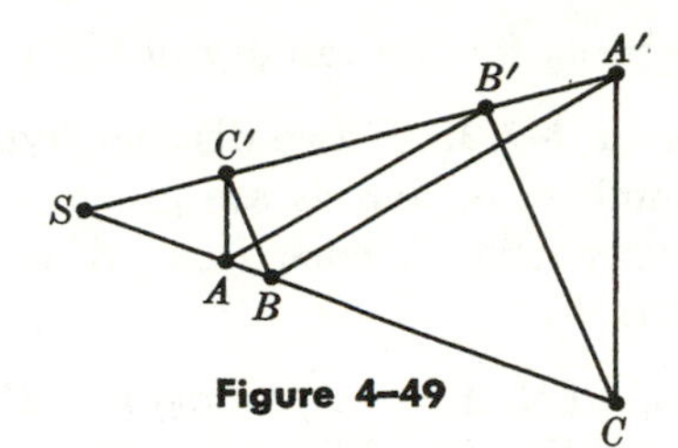

Figure 4-49

Case (ii). We have $|SA|/|SB| = |SB'|/|SA'|$ and $|SB|/|SC| = |SC'|/|SB'|$, and consequently by multiplication, $|SA|/|SC| = |SC'|/|SA'|$, which implies $AC' \parallel A'C$.

To summarize, the euclidean plane is based on the axioms AP 1 through 3, Bt 1 through 5, AC, and Cg 1 through 5.

Axioms for the real hyperbolic plane. Our next concern will be the real hyperbolic plane. We met it in two distinct models: The euclidean Poincaré model of Section 1-9 and the projective Cayley-Klein model of Section 3-7. Which of the axioms of the euclidean plane are satisfied in the hyperbolic plane? We will answer this question by studying the Cayley-Klein model and then show the equivalence of the two models.

Theorem 4-7.2. The axioms AP 1, AP 3, Bt 1 through 5, Cg 1 through 5, and AC are satisfied in the hyperbolic plane, whereas AP 2 does not hold.

Proof. We deal with the Cayley-Klein model. Its points are all the real points inside the unit circle, and thus AP 1, AP 3, and Bt 1 are obviously satisfied. Hyperbolic points are collinear if and only if they lie on the same affine line, because the improper line of the projective plane certainly does not have any hyperbolic points. Hence the betweenness relation β on the hyperbolic lines can be defined as coinciding with β on the affine lines. Then Bt 1 through 5 are satisfied. Also AC can be verified without difficulty.

As to the congruence axioms, we define $|AB|$ as the hyperbolic distance $d(A, B)$, in accordance with Eqs. (1) or (3) of Section 3-7. Then Cg 2 and Cg 3 are obviously satisfied. If C/P is a half-line intersecting the absolute conic at M, and if the second intersection of CM with the conic is N, then $(CD|MN)$ ranges continuously through all real values between 1 and ∞ when D runs along C/P from C to M. Thus $d(C, D)$ ranges continuously from $\frac{1}{2}\log 1 = 0$ to ∞ and assumes the given value $d(A, B)$ in exactly one point D of C/P, in accordance with the corollary to Theorem 3-3.2. This proves the validity of Cg 1.

In Exercise 3 of Section 3-3 it was proved that

$$(AB|MN)(BC|MN) = (AC|MN).$$

Consequently, if $\beta(ABC)$ holds, then $d(A, B) + d(B, C) = d(A, C)$, which takes care of Cg 4.

The proof for the validity of Cg 5 is harder. We will first need two lemmas.

Lemma 4-7.3. If two distinct hyperbolic points A and B and two positive numbers d_1 and d_2 are given, then there exist at most two points whose hyperbolic distance from A is d_1 and whose hyperbolic distance from B is d_2.

Proof. Let $A = [a_0, a_1, a_2]$ and $B = [b_0, b_1, b_2]$ in the Cayley-Klein model. Then, by Eq. (3) of Section 3-7, every point $X = [x_0, x_1, x_2]$ at a hyperbolic distance d_1 from A satisfies

$$\frac{(a_0x_0 - a_1x_1 - a_2x_2)^2}{(a_0^2 - a_1^2 - a_2^2)(x_0^2 - x_1^2 - x_2^2)} = \cosh^2 d_1.$$

The denominator cannot vanish, because A and X are hyperbolic points and do not lie on the absolute conic. Thus

$$(a_0x_0 - a_1x_1 - a_2x_2)^2 = k(x_0^2 - x_1^2 - x_2^2), \tag{1}$$

where $k \neq 0$ and depends only on A and d_1. The locus which could be called "the hyperbolic circle about A with radius d_1" is, therefore, represented by a quadratic function.

Now let $\cosh d_1 = c \cosh d_2$, which defines c uniquely, because the real cosh function does not assume the value zero. Thus

$$\frac{a_0x_0 - a_1x_1 - a_2x_2}{\sqrt{(a_0^2 - a_1^2 - a_2^2)(x_0^2 - x_1^2 - x_2^2)}} = c\,\frac{b_0x_0 - b_1x_1 - b_2x_2}{\sqrt{(b_0^2 - b_1^2 - b_2^2)(x_0^2 - x_1^2 - x_2^2)}},$$

or

$$a_0x_0 - a_1x_1 - a_2x_2 = c'(b_0x_0 - b_1x_1 - b_2x_2), \tag{2}$$

where c' depends only on A, B, and c. This is the equation of a projective straight line. The required points $[x]$ have to satisfy Eq. (1) as well as Eq. (2). A simultaneous system of a linear and a quadratic equation cannot have more than two solutions. This proves Lemma 4-7.3.

Lemma 4-7.4. If ABC and $A'B'C'$ are two triangles in the hyperbolic plane, such that $d(A, B) = d(A', B')$, $d(B, C) = d(B', C')$, and $d(C, A) = d(C', A')$, then there is a hyperbolic isometry mapping A on A', B on B', and C on C'.

Proof (Fig. 4-50). Let AB and $A'B'$ intersect the absolute conic in M, N, and M', N', respectively, where M and M' are chosen so that $\beta(MAB)$ and $\beta(M'A'B')$. If the absolute polarity takes AB into P and $A'B'$ into P', let PA and $P'A'$ intersect the absolute conic at E, F, and E', F', respectively. By Lemma 3-5.7, there is exactly one hyperbolic isometry α with $M\alpha = M'$, $N\alpha = N'$, and $E\alpha = E'$ and consequently $F\alpha = F'$, $P\alpha = P'$, and $A\alpha = A'$;

and exactly one isometry γ with $M\gamma = M'$, $N\gamma = N'$, and $E\gamma = F'$ and consequently $F\gamma = E'$, $P\gamma = P'$, and $A\gamma = A'$. Since

$$d(A, B) = \tfrac{1}{2}|\log_e (AB|MN)|$$
$$= d(A', B') = \tfrac{1}{2}|\log_e (A'B'|M'N')|,$$

and since $\beta(MAB)$ and $\beta(M'A'B')$ hold, we have $(AB|MN) = (A'B'|M'N')$, and therefore $B\alpha = B\gamma = B'$. Now, α and γ are isometries, and we get

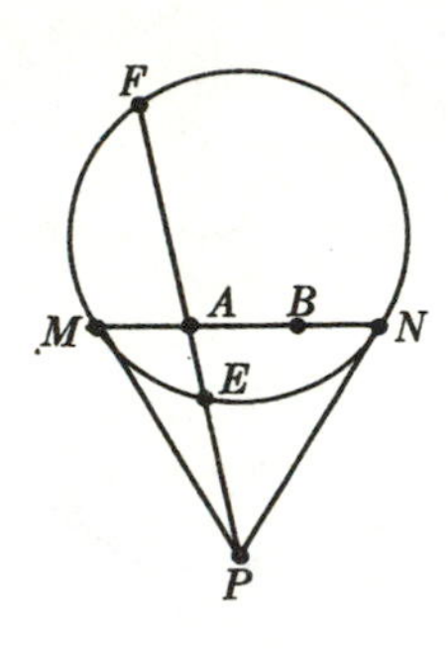

Figure 4-50

$$d(C\gamma, A') = d(C\alpha, A') = d(C, A) = d(C', A')$$

and $d(C\gamma, B') = d(C\alpha, B') = d(C, B) = d(C', B')$. But, by Lemma 4-7.3, there are at most two points at a distance $d(C, A)$ from A' and at a distance $d(B, C)$ from B'. Hence they must be $C\alpha$ and $C\gamma$, and C' is either $C\alpha$ or $C\gamma$, and either α or γ is the required hyperbolic isometry.

We use Lemma 4-7.4 for proving that Cg 5 is satisfied. There is a hyperbolic isometry mapping the triangle ABC on $A'B'C'$. Since $d(B, D) = d(B', D')$ and since $\beta(BCD)$ and $\beta(B'C'D')$ hold, D is carried by this isometry into D', and $d(A, D) = d(A', D')$, as claimed.

Euclid's parallel axiom AP 2 does not hold in the hyperbolic plane as there are two parallels to a given line through a given point not on this line. This justifies the attribute "noneuclidean" for hyperbolic geometry. The proof of Theorem 4-7.2 is now complete.

So far we have dealt only with the Cayley-Klein model of the hyperbolic plane. Let us now compare this model with the Poincaré model, in which the hyperbolic points are represented by the points $x + iy$ with $y > 0$ in the complex number plane. The hyperbolic lines of this model were the upper half-lines of the euclidean lines parallel to the y-axis and the upper half-circles with centers on the x-axis. In Theorem 3-5.6 and its Corollary 1, it was shown that the isometry groups of the two models are isomorphic. This is sufficient for considering the models as representing the same geometry. However, our present purpose calls rather for the exhibition of an explicit bijective mapping of one model on the other.

Equivalence of the two models of the hyperbolic plane. As illustrated in Fig. 4-51, a hemisphere of radius 1 is placed so that it touches the euclidean plane at the origin of an orthonormal coordinate system. By a perpendicular parallel projection the points inside the unit circle are mapped onto the points of the hemisphere. The following equations for the mapping of a point $P = [x, y]$ on a point $P' = [x', y', z']$ may be read easily from the figure, if the three-dimensional orthonormal coordinate system of the x', y', z', with its origin at the

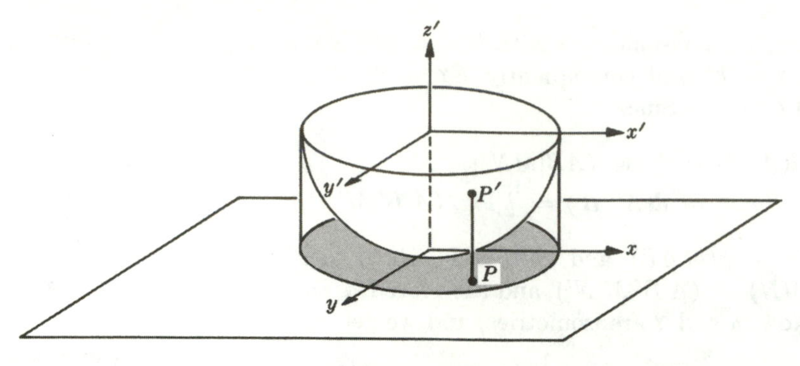

Figure 4-51

hemisphere center is introduced, as shown in the figure. We obtain

$$x' = x, y' = y, z' = -\sqrt{1 - x^2 - y^2}.$$

As a second mapping (Fig. 4-52) we consider a stereographic projection from the point $[1, 0, 0]$ onto the tangent plane touching the sphere at $[-1, 0, 0]$. On this tangent plane an orthonormal coordinate system is introduced with its origin at the contact point. The relation between P' and the coordinates x'', y'' of the image point P'' is

$$x'' = \frac{2y'}{1 - x'}, \qquad y'' = \frac{-2z'}{1 - x'},$$

(see Exercise 15). Combination of the two mappings yields

$$x'' = \frac{2y}{1 - x}, \qquad y'' = \frac{2\sqrt{1 - x^2 - y^2}}{1 - x}.$$

It is geometrically obvious that this mapping $P \to P''$ is bijective and takes the interior of the unit circle onto the upper half-plane. It is easy to verify this algebraically by finding the inverse mapping

$$x = \frac{x''^2 + y''^2 - 4}{x''^2 + y''^2 + 4}, \qquad y = \frac{4x''}{x''^2 + y''^2 + 4}. \tag{3}$$

What are the images of the hyperbolic lines in the Cayley-Klein model, that is, the secant segments of the lines which intersect the unit circle? Let the equation of the line be $ax + by + c = 0$. The euclidean distance between the origin and the line is $|c|/\sqrt{a^2 + b^2}$ and has to be <1. Thus

$$c^2 < a^2 + b^2. \tag{4}$$

By Eq. (3) we have

$$a(x''^2 + y''^2 - 4) + 4bx'' + c(x''^2 + y''^2 + 4) = 0,$$

$$x''^2(a + c) + y''^2(a + c) + 4bx'' + 4(c - a) = 0. \tag{5}$$

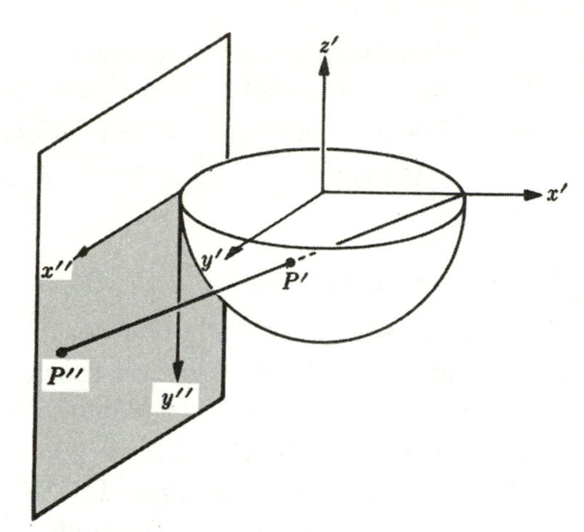

Case (1). $a + c = 0$. By Eq. (4), $b \neq 0$, and Eq. (5) becomes $x'' = 2a/b$, which is a line parallel to the y-axis, a hyperbolic line of the Poincaré model. On the other hand, every line $x'' = 2d$ corresponds, by means of $a = -c = d$ and $b = 1$ to just one line $dx + y - d = 0$.

Case (2). $a + c \neq 0$. Then Eq. (5) assumes the form

$$\left(x'' + \frac{2b}{a + c}\right)^2 + y''^2 = 4\,\frac{a - c}{a + c} + \frac{4b^2}{(a + c)^2} = 4\,\frac{a^2 + b^2 - c^2}{(a + c)^2}\cdot$$

The right-hand side is positive, in view of Eq. (4), and Eq. (5) represents a circle whose center lies on the x''-axis. Its upper half is a hyperbolic line of the Poincaré model. Conversely, if a circle

$$(x'' - s)^2 + y''^2 = r^2 > 0$$

is given, then there is a line,

$$(r^2 - s^2 + 4)x - 4sy + 4 - r^2 + s^2 = 0,$$

which corresponds to it and contains a segment inside the unit circle, because

$$a^2 + b^2 = (r^2 - s^2 + 4)^2 + 16s^2 > (4 - r^2 + s^2)^2 = c^2.$$

Exercises

1. Prove that congruence of segments is an equivalence relation.
2. Prove the congruence theorems for triangles in the euclidean plane by using the axioms AP, Bt, AC, and Cg. The theorems are: Two triangles ABC and $A'B'C'$ have equal corresponding sides and angles if
   (a) $|AB| = |A'B'|$, $|BC| = |B'C'|$, $\measuredangle ABC = \measuredangle A'B'C'$,
   (b) $|AB| = |A'B'|$, $\measuredangle ABC = \measuredangle A'B'C'$, $\measuredangle CAB = \measuredangle C'A'B'$,
   (c) $|AB| = |A'B'|$, $|BC| = |B'C'|$, $|CA| = |C'A'|$.

3. Formulate theorems dealing with angles in analogy with the statements Cg 1 through 4 which deal with segments. Prove these theorems.
4. Prove that congruence of angles is an equivalence relation.
5. Let AB and CD be two distinct lines, $AB \times CD = S$, and $\beta(ASB)$ and $\beta(CSD)$. Prove:
   (a) $\sphericalangle ASB = \sphericalangle CSD$,
   (b) $\sphericalangle ASC = \sphericalangle BSD$, using only the axioms AP, Bt, AC, and Cg.

   Note that a sum of angles has to be properly defined.
6. Prove that an external angle of a triangle is greater than each of the internal angles not adjacent to it. Do you need Euclid's parallel axiom to prove this?
7. Let A, B, C, be a collinear, $\beta(ABC)$. If D and E are points on the same side of AB such that $\sphericalangle CBE = \sphericalangle CAD$, prove that $AD \parallel BE$. Do you need Euclid's parallel axiom for this proof? How do you define "on the same side of AB?"
8. Let A, B, and C be collinear, $\beta(ABC)$. If D and E are points on the same side of AB such that $AD \parallel BE$, prove that $\sphericalangle CBE = \sphericalangle CAD$. Do you need Euclid's parallel axiom for this proof?
9. Prove that opposite sides of a parallelogram are equal. Is this theorem correct in the hyperbolic plane?
10. If $\overrightarrow{AB} = \overrightarrow{CD}$, prove that $|AB| = |CD|$. Is this true when AP 2 does not hold?
11. Discuss the following statement: Since Π_a can be replaced by the axioms Cg, and since Cg hold in the real hyperbolic plane, so does also Π_a.
12. (a) Find the points in the Cayley-Klein model which correspond to the points i and $2i$ in the Poincaré model.
    (b) Find their hyperbolic distance in both models and compare.
13. (a) Find the equations of the sides and the diagonals of the quadrangle $[2, 1, 0][2, 0, 1][2, -1, 0][2, 0, -1]$ in the Cayley-Klein model.
    (b) Find the midpoint of the quadrangle.
    (c) Find the corresponding quadrangle in the Poincaré model and verify that the midpoint corresponds to the midpoint determined in (b).
14. Which point of the Cayley-Klein model corresponds to the point $z = \infty$ of the Poincaré model?
15. Derive the relation between the coordinates of P' and P'' on p. 252.
16. Show that an additional model of hyperbolic geometry (the *conformal unit circle model*) may be obtained from Poincaré's model by the transformation

$$z \to \frac{z(1+i) + (1-i)}{-z(1+i) + (1-i)}.$$

   What are the hyperbolic lines in this model? Which lines are parallel?

4-8 FINAL REMARKS

At the end of the axiomatic treatment an overall review is appropriate. We started out from a rudimentary plane which satisfied only incidence axioms. The coordinatizing algebraic structure was a ternary ring, and the plane was devoid of collineations. Gradually we added more and more restrictions in the form of incidence properties which we postulated, the algebraic structure became less unwieldy, and the plane became richer in collineations. Eventually the Pappus postulate was added and turned out to imply all the previously adjoined, weaker, incidence properties. The ternary ring became a field. Order axioms and a continuity axiom made the coordinate field isomorphic to the field of reals. This brought us back to the starting point of Chapter 2, and consequently the developments of Chapters 2 and 3 for the real field may be based on the axioms of incidence, order and continuity, and the Pappus postulate.

For the foundations of the euclidean plane, congruence axioms were adjoined to those used for the real projective and affine planes. These congruence axioms, together with the axioms of incidence, order, and continuity made the Pappus axiom redundant.*

Then we checked the real hyperbolic plane as represented by the Poincaré and Cayley-Klein models, and found that in it all the axioms of the euclidean plane, except one, were satisfied. This axiom was Euclid's parallel axiom which required the existence of a unique parallel to a given line through any given point not on the line.

The nonvalidity of Euclid's axiom is a remarkable property of the hyperbolic plane. For many centuries it was an unsolved problem whether Euclid's parallel axiom was a consequence of the remaining axioms of the euclidean plane or not. "Proofs" that the axiom was actually a theorem were developed and later found to contain flaws. Only the introduction of hyperbolic geometry by Lobachevsky and Bolyai (the German mathematician C. F. Gauss, 1777–1855, had also discovered it without publishing his results) provided the final answer, namely that since a logically consistent geometry could be based on the remaining axioms without satisfying the parallel axiom, this axiom could not depend on them.

Elliptic geometry also is a "noneuclidean" geometry in the sense that there are no parallels at all in the elliptic plane, because every two lines have a point of the plane in common. Hence the axiom cannot hold.

At first the noneuclidean geometries were considered merely as counterexamples for proving the independence of axioms. This role was most important in view of the fact that until that time Euclid's axioms were regarded as absolute

---

*In J. Algebra **3** (1966), 304–359, the Israeli mathematician S. A. Amitsur has succeeded in determining all incidence properties in desarguesian, non-pappian planes. He obtained this result by an investigation of identities in division algebras. One important conclusion of his work is, that in an *ordered* desarguesian plane any noncontradictory incidence property, which is not a consequence of Δ, implies II.

truths. Thus the appearance of a new geometry in which part of these "truths" did not hold signified a shattering of beliefs that were firmly established in mathematics and philosophy. Later the new geometries attracted the interest of the mathematical world in their own right. In our treatment we did not exhibit any axiom system for the hyperbolic plane. It can be shown that such a system is provided by the axioms of the euclidean plane (AP 1 and 3, Bt, AC, Cg) in which the parallel axiom AP 2 is replaced by the axiom, "Given a point P and a line l not through P, there exist at least two lines through P which do not intersect with l."

The first complete treatment of the axiomatics of euclidean geometry was published in 1899 by the German mathematician David Hilbert (1862–1943). In his book *Grundlagen der Geometrie*, Hilbert stated his axioms and proved that they were consistent, independent, and categorical. In order to prove the independence of his axioms he constructed various new geometries, which in each case did not satisfy one of the axioms or axiom groups, while the rest of the axioms held. Again history repeated itself. While Hilbert used these geometries only as counter-examples, later mathematicians were attracted to such geometries as new research objects, and a whole new field of mathematics emerged: the study of non-desarguesian geometries and of general projective planes. This tendency had an analog in algebra, where nonassociative generalizations of groups and fields were widely studied in accordance with the strong trend toward generality throughout mathematics.

Our treatment of the axiomatic foundations was by no means complete. We did not strive for independence of our axioms, and in fact sometimes used axioms which were much stronger than necessary. Moreover, by separating the different geometries entirely, far more convenient axiom systems can be devised. Thus a separate development starting with affine geometry, independent of projective geometry, leads conveniently, by adjunction of the congruence axioms, to "absolute geometry," which is the common part of euclidean and hyperbolic geometry. Then the branching into euclidean geometry or hyperbolic geometry is obtained by the adjunction of a single axiom, the parallel axiom or its hyperbolic counterpart, respectively.

The interested reader can find these subjects in the following list of references for further reading.

Rudimentary planes

L. M. Blumenthal, *A Modern View of Geometry*, San Francisco: Freeman, 1961. Elementary and rather comprehensive.

R. H. Bruck, *Recent Advances in the Foundations of Euclidean Plane Geometry*, Slaught Memorial Paper No. 4, *Amer. Math. Monthly* **62**, 2 (1955). An expository article, elementary.

M. Hall, *The Theory of Groups*, New York: Macmillan, 1959. Chapter 20 gives a moderately advanced treatment of projective planes.

G. Pickert, *Projektive Ebenen*, Berlin: Springer, 1955. An advanced most comprehensive work.

B. Segre, *Lectures on Modern Geometry*, Rome: Edizioni Cremonese, 1961. A moderately advanced treatment.

Axiomatic foundations

K. Borsuk and W. Szmielew, *Foundations of Geometry*, Amsterdam: North-Holland, 1960. Comprehensive and detailed.

H. S. M. Coxeter, *Introduction to Geometry*, New York: John Wiley, 1961. A moderately easy-to-read work covering most aspects of geometry.

H. G. Forder, *The Foundations of Euclidean Geometry*, Cambridge: Cambridge University Press, 1927. Reprint, New York: Dover, 1958. This work is most detailed and rigid, and no steps are omitted.

D. Hilbert, *The Foundations of Geometry*, trans. E. J. Townsend, Chicago: Open Court, 1902. Reprint, La Salle, Ill.: Open Court, 1959. A classic.

G. deB. Robinson, see p. 197.

Noneuclidean geometry

H. S. M. Coxeter, see p. 198.

H. Schwerdtfeger, see p. 198.

F. Klein, see p. 198.

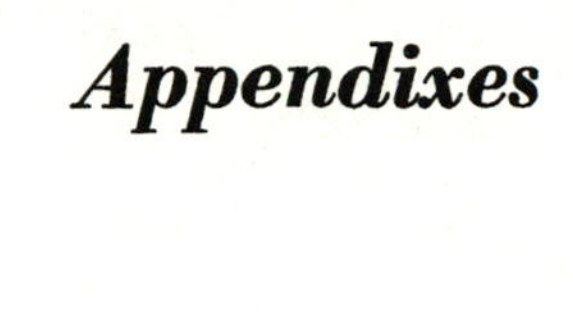

Appendixes

List of Notations

A^n, affine n-space over a field

$A^2(p)$, two-dimensional affine geometry over $GF(p)$

arg z, argument of z

$\beta(ABC)$, B is between A and C

C, the complex field

C_e, the conic $x_0^2 + x_1^2 + x_2^2 = 0$

C_E, the degenerate conic $x_0^2 = 0$

C_h, the conic $x_0^2 - x_1^2 - x_2^2 = 0$

$d(P, Q)$, the distance between P and Q

det, determinant

dim, dimension

δ_{ik}, the Kronecker delta

E^n, euclidean n-space

GF, Galois field

I, identity transformation

$\mathrm{Im}(z)$, imaginary part of z

I_n, $n \times n$ identity matrix

$N(\mathbf{q})$, norm of $\mathbf{q}$

P^n, projective n-space over a field

R, the real field

$\mathrm{Re}(z)$, the real part of z

$\sigma(\mathrm{AB} \mid \mathrm{CD})$, A and B separate C and D

$T_{\mathbf{v}}$, translation through $\mathbf{v}$

V^n, n-dimensional vector space

$(V^n)^*$, dual of V^n

$V^2(p)$, V^2 over $GF(p)$

$a \times b$, intersection of the lines a and b

$\triangleleft$, normal subgroup of

$\measuredangle$, angle

$[\mathbf{x}]$, the set $\{r\mathbf{x}\}$ for all scalars $r \neq 0$; a projective point

$\langle \mathbf{u} \rangle$, a projective hyperplane

$a{*}b{*}c$, ternary operation

a/b, $a\backslash b$, $a \vdash c$, $a \dashv c$, operations in loops

AB/C, the projective segment AB without C

$|AB|$, affine segment

A/B, the affine half-line starting from A, not containing B

$\equiv$, congruent

$\cong$, isomorphic

$(AB \mid CD)$, cross ratio

$\parallel$, parallel

$\perp$, perpendicular

$\cup$, set union

$\cap$, set intersection

$\subset$, subset of

$\in$, element of

List of Axioms

List of Groups

Index

Index

Mathematics

FUNCTIONAL ANALYSIS (Second Corrected Edition), George Bachman and Lawrence Narici. Excellent treatment of subject geared toward students with background in linear algebra, advanced calculus, physics and engineering. Text covers introduction to inner-product spaces, normed, metric spaces, and topological spaces; complete orthonormal sets, the Hahn-Banach Theorem and its consequences, and many other related subjects. 1966 ed. 544pp. 6⅛ x 9¼. 0-486-40251-7

ASYMPTOTIC EXPANSIONS OF INTEGRALS, Norman Bleistein & Richard A. Handelsman. Best introduction to important field with applications in a variety of scientific disciplines. New preface. Problems. Diagrams. Tables. Bibliography. Index. 448pp. 5⅜ x 8½. 0-486-65082-0

VECTOR AND TENSOR ANALYSIS WITH APPLICATIONS, A. I. Borisenko and I. E. Tarapov. Concise introduction. Worked-out problems, solutions, exercises. 257pp. 5⅜ x 8¼. 0-486-63833-2

AN INTRODUCTION TO ORDINARY DIFFERENTIAL EQUATIONS, Earl A. Coddington. A thorough and systematic first course in elementary differential equations for undergraduates in mathematics and science, with many exercises and problems (with answers). Index. 304pp. 5⅜ x 8½. 0-486-65942-9

FOURIER SERIES AND ORTHOGONAL FUNCTIONS, Harry F. Davis. An incisive text combining theory and practical example to introduce Fourier series, orthogonal functions and applications of the Fourier method to boundary-value problems. 570 exercises. Answers and notes. 416pp. 5⅜ x 8½. 0-486-65973-9

COMPUTABILITY AND UNSOLVABILITY, Martin Davis. Classic graduate-level introduction to theory of computability, usually referred to as theory of recurrent functions. New preface and appendix. 288pp. 5⅜ x 8½. 0-486-61471-9

ASYMPTOTIC METHODS IN ANALYSIS, N. G. de Bruijn. An inexpensive, comprehensive guide to asymptotic methods–the pioneering work that teaches by explaining worked examples in detail. Index. 224pp. 5⅜ x 8½ 0-486-64221-6

APPLIED COMPLEX VARIABLES, John W. Dettman. Step-by-step coverage of fundamentals of analytic function theory–plus lucid exposition of five important applications: Potential Theory; Ordinary Differential Equations; Fourier Transforms; Laplace Transforms; Asymptotic Expansions. 66 figures. Exercises at chapter ends. 512pp. 5⅜ x 8½. 0-486-64670-X

INTRODUCTION TO LINEAR ALGEBRA AND DIFFERENTIAL EQUATIONS, John W. Dettman. Excellent text covers complex numbers, determinants, orthonormal bases, Laplace transforms, much more. Exercises with solutions. Undergraduate level. 416pp. 5⅜ x 8½. 0-486-65191-6

RIEMANN'S ZETA FUNCTION, H. M. Edwards. Superb, high-level study of landmark 1859 publication entitled "On the Number of Primes Less Than a Given Magnitude" traces developments in mathematical theory that it inspired. xiv+315pp. 5⅜ x 8½. 0-486-41740-9